# UREMIC TOXINS

# ADVANCES IN EXPERIMENTAL MEDICINE AND BIOLOGY

## Recent Volumes in this Series

# UREMIC TOXINS

Edited by

## Severin Ringoir and
## Raymond Vanholder

University of Ghent
Ghent, Belgium

and

## Shaul G. Massry

University of Southern California
Los Angeles, California

PLENUM PRESS • NEW YORK AND LONDON

Library of Congress Cataloging in Publication Data

Ghent Symposium on Uremic Toxins (1986)
   Uremic toxins.

   "Proceedings of the Ghent Symposium on Uremic Toxins, held October 3–4, 1986, in Ghent, Belgium"—T.p. verso.
   Includes bibliographies and index.
   1. Uremia—Congresses. 2. Renal insufficiency—Complications and sequelae—Congresses. I. Ringoir, S. II. Vanholder, Raymond. III. Massry, Shaul G. IV. Title. [DNLM: 1. Toxins—congresses. 2. Uremia—congresses. WJ 348 G412u 1986]
   RC915.G48   1986                                616.6′35                                87-25936
   ISBN-13: 978-1-4684-5447-5        e-ISBN-13: 978-1-4684-5445-1
   DOI:   10.1007/978-1-4684-5445-1

Proceedings of the Ghent Symposium on Uremic Toxins, held October 3–4, 1986, in Ghent, Belgium

© 1987 Plenum Press, New York
Softcover reprint of the hardcover 1st edition 1987
A Division of Plenum Publishing Corporation
233 Spring Street, New York, N.Y. 10013

To our patients who taught us the art of the
practice of medicine
and
to our students and associates who stimulated
our academic endeavors

## PREFACE

The present book contains the Proceedings of a two day Symposium on Uremic Toxins organized at the University of Ghent in Belgium.

A series of guest lectures, free communications and posters have been presented. An international audience of 163 scientists from 16 nationalities listened to and discussed extensively a spectrum of topics brought forward by colleagues and researchers who worked for many years in the field of Uremic Toxins.

There is a striking contrast between all the new dialysis strategies available in the work to "clean" the uremic patients and the almost non-progression of our knowledge on uremic toxins in the past decade.

In this sense the symposium was felt by all participants as a new start for the research in the biochemical field of the definition of uremia. If the present volume would stimulate new work in this field in order to define uremia, or identify the uremic toxins, the purpose of the organizers would be maximally fulfilled.

Severin Ringoir<br>
Raymond Vanholder<br>
Shaul G. Massry

# CONTENTS

# PARATHYROID HORMONE: A UREMIC TOXIN

Shaul G. Massry

University of Southern California, School
of Medicine, Los Angeles, California

The uremic syndrome is mirky
With investigation very quirky
But let no technique be untried
Let no method be denied
Let no idea be untold
Eventually the truth will unfold

George E. Schreiner
modified by Michael Kaye

## Uremic Toxins: What are they?

A Sundry of clinical, biochemical and radiographic abnormalities occur in patients with advanced renal failure. These are listed in Table 1, and they constitute the uremic syndrome. Their pathogenesis have been attributed to uremic toxins.

Support for the "uremic toxin theory" was provided by the observations that uremic sera may adversely affect a wide variety of biologie systems in vitro. The list of the compounds that have been incriminated as uremic toxins is large, and some of them are shown in Table 2. However, a cause and effect relationship between these various compounds and the mainfestations of the uremic syndrome has not been established in most instances.

Several reasons may be responsible for the failure to identify a uremic toxin that could clearly be considered the culprit in the genesis of the clinical manifestations of the uremic syndrome. First, the tendency to extrapolate an adverse effect of uremic sera in an vitro system to the clinical level has resulted in erroneous conclusions. Second, rigorous criteria were not used before considering a substance to be a toxin. A compound should satisfy at least five criteria before being considered responsible for the genesis of one or more of the uremic signs and symptom (1). These criteria include that: a) the nature and structure of the compound should be known, b) its blood level should be elevated, c) a

1

relation between it and the uremic manifestations must be demonstrated, d) an improvement in the signs and symptoms of uremia should follow a reduction in its blood levels, and e) its administration to experimental animals with normal renal function should produce derangements similar to those seen in uremic patients. Third, the emphasis was placed on products of protein breakdown as the possible uremic toxins while the role of the hormonal disturbances in the genesis of the uremic syndrome is not fully explored.

It is evident, therefore, that the exploration of uremic toxicity must employ a more critical approach which must encompass an integrated research endeavor utilizing at least three levels of investigations. First, one must examine the effect of a potentially toxic compound on in vitro system to understand the mechanisms of its action on various organs and to explore whether such effects have relevance to any of the uremic manifestations. Second, one must investigate the effect of excess amount of the compound in question on the genesis of the various components of the uremic syndrome in experimental animals with chronic renal failure, and examine, whenever feasible, its effect on the function of various organs in animals with normal renal function. Third, one must evaluate in patients with renal failure the consequences of the reduction in the blood levels of the toxic compound.

In order to update the clinician and the investigator dealing with the uremic syndrome and its pathogenesis, it is timely to review the relevant information on the various compounds currently considered potential uremic toxins. When it was appreciated that kidney failure is associ- ated with accumulation of urea in body fluids, the medi- cal community implicated the elevated blood levels of urea in the genesis of the uremic syndrome. The advance of the theory of the toxicity of middle molecules (2) has stimulated a tremendous amount of research, but their nature and role in uremic toxicity is as yet not elucidated. Other information has suggested that ele- vated aluminum burden in patients with renal failure can cause dialysis encephalopathy (3) and can produce osteo- malacia (4) and aluminum is now considered the uremic toxin.

A great interest has developed in the understanding of the relentless progression of renal insuffciency in patients with renal diseases. The exact mechanism of this phenomenon is not fully understood and its basis may be multifactoral. Any factor that may participate in, or induce, the progression of renal failure in patients with renal disease should be considered a uremic toxin.

In an editorial published in 1977, Massry suggested that the elevated blood levels of parathyroid hormone (PTH) may exert widespread adverse effects on many organs (5). The Massry hypothesis implicated PTH as a major uremic toxin participating in the genesis of many of the manifestations of the uremic syndrome. Since that time, a large number of studies have provided evidence for the multi-effects of PTH and on its role in the overall pathogenesis of uremia.

**Table 1. Components of the Uremic Syndrome**

---

Encephalopathy
Neuropathy
Dialysis dementia
Bone disease
Soft tissue calcification
Soft tissue necrosis
Pruritus
Hyperlipidemia
Glucose intolerance
Anemia
Bleeding tendencies
Immunological disturbances
Myocardiopathy
Myopathy
Sexual dysfunction

---

**Table 2. A Partial List of Compounds That Have Been Incriminated as Uremic Toxins**

---

Urea
Guanidinium compounds
  Methylguanidine
  Dimethylguanidine
  Guanidinosuccemic acid
  Guanidinoacetic acid
  Creatinine
Aromatic compounds
  Pheolic and hydroxyphenobic acid
  Aromatic amines
  Indoles
Aliphatic amines
Conjugated aminoacids
Peptides
Middle molecules

---

## Role of Parathyroid Hormone in Uremic Toxicity

One of the major endocrine abnormalities in uremia is hyperplasia of the parathyroid glands, (6-8) which results in a state of secondary hyperparathyroidism with marked elevations in the blood levels of parathyroid hormone (PTH) (9,10). It has been proposed that PTH could be deleterious to many organ systems (5).

There are at least eight theoretical pathways through which PTH may mediate its adverse effects (11). These include: a) increasing intracellular calcium concentration, b) altering the intracellular-extracellular calcium ratio, c) affecting cellular membrane permeability and/or integrity, d) affecting phospholipid turnover e) exaggerated stimulation of cyclic AMP production, f) causing soft tissue calcification either through a direct effect or by raising calcium-phosphorus product in blood, g) entering the cell and affecting function of subcellular structures, and h) augmenting protein catabolism and, therefore, participating in the genesis of wasting syndrome of uremia and contributing modestly to the accumulation of nitrogenous compounds in blood.

PTH is known to augment entry of calcium into a variety of mammalian cells, (12-15) and excess hormone increases the calcium content of various tissues such as skin (16) cornea, (17) blood vessels (18) heart (19) brain (20) and peripheral nerves (21). An increase in intracellular calcium concentration may directly affect cell function, since intracellular calcium is a critical cell messenger (22). In addition, a change in the intracellular calcium concentration and the consequent alteration in intracellular-extracellular calcium ratio may result in alterations in cell membrane permeability and function and, as such, may affect cell function. Also, PTH may affect phospholipids of cell membrane, and alterations in these phospholipids may affect agonist-receptor interaction and cell membrane fluidity. Hirata and Axelrod (23) reported that beta adrenergic agents stimulate the methylation of phosphatidylethanolamine and its transmembrane movement in red cells, and that this affects membrane fluidity. The incorporation of Pi into phosphatidylinositol and phosphatidic acid occurs within a few minutes after the aplication of a stimulus, and persists for as long as the stimulus is applied; this step has been implicated in the interaction between the stimulus and its receptor on the cell surface (24). This effect may be independent of cyclic AMP production and precedes the movement of calcium across cell membrane ((25). Indeed, Green et al (26) showed that phosphatidic acid and phosphatidylinositol are potent calcium inonophores. Studies by Lo et al (27) showed that PTH increased the incorporation of Pi into phosphatidic acid and phosphatidylinositol in renal cortical slices, but cylic AMP did not. Futhermore, Molitoris et al (28) reported that PTH enhanced the release of myoinositol from the kidney. Finally, Brautbar et al (30) reported that PTH affects phospholipid metabolism of human red blood cells. These observations strongly support an action of PTH on phospholipid turnover.

PTH could be a catabolic agent and, as such, play a role in the wasting syndrome of uremia and contribute in a modest manner to the accumulation of nitrogenous compounds in blood. Several observations support a catabolic effect of PTH. Administration of PTH to humans was associated with a negative nitrogen balance (30-32). Furthermore, nitrogen balance in patients with parathyroid adenoma is usally negative (32,33) and becomes significantly more positive after surgery (33). Finally, PTH stimulates the release of alanine and glutamine from muscle (34). The confirmation of a catabolic effect of PTH would provide a link between the toxic theory of PTH and the toxic hypothesis of nitrogenous compounds.

Finally, the blood of uremic patients contains a middle molecule substance which has glycine, glutamic acid, leucine, phenylalanine, tyrosine, and histidine; this compound is called peak 7C and it has been suggested that it is a uremic toxin since it inhibits in vitro lymphoblastic transformation. Others found that this substance impairs glucose utilization and amino acid transport in vitro (35). Frohling et al (35) reported elevated levels of peak 7C in uremic patients and these levels displayed a significant correlation with those of PTH. Also, parathyroidectomy was followed by a decrease in the blood levels of peak 7C.

Thus, it is possible that the elevated blood levels of PTH are not only toxic by directly affecting cell function but may also contribute to uremic toxicity by enhancing catabolism and/or by augmenting the production of peak 7C. It is reasonable, therfore, to suggest that PTH may play a central role in the overall uremic toxicity. It should be emphasized that this in no way means that other hormonal imbalances or end products of protein metabolism do not play a role in the toxicity of uremia.

Available data provide evidence that parathyroid hormone affects the function of a wide variety of organs in uremia and its effects satisfy the criteria for a uremia toxin. The organs affected by parathyroid hormone include brain, peripheral nerves, myocardium, skeletal muscles, white cells, plateletts, red blood cells, erythopocitic system, lipid metabolism, sexual function, skin and bone. For further details of the affect of PTH on these systems the reader is referred to a recent review by Massry (36).

This manuscript will describe the evidence for the role of PTH in the pathogenesis of carbohydrate intolerance in uremia.

## Carbohydrate Intolerance and Hyperparathyroidism in Uremia

Patients with chronic renal failure display abnormalities in carbohydrate metabolism (37-40). They almost always have resistance to the peripheral action of insulin (40,41), while insulin secretion could be normal (39,42), increased (42,44) or decreased (38). Glucose intolerance is, therfore, usually encountered in uremic patients in whom both impaired tissue sensitivity to insulin and impaired secretion of the hormone co-exist (40,45).

Certain data suggest that parathyroid hormone (PTH) may affect carbohydrate metabolism. Patients with primary hyperparathyroidism may have glucose intolerance (46,47). Elevated insulin plasma levels both in the fasting state and in response to glucose (46,47) as well as insulin resistance (47) have been reported in these patients.

It is plausible, therefore, to suggest that the state of secondary hyperparathyroidism which exists in patients with advanced renal failure (6-10) plays an important role in the genesis of the glucose intolerance of uremia.

We examined the role of PTH in the glucose intolerance of uremia utilizing intravenous glucose tolerance test (IVGTT) and euglycemic and hyperglycemic clamp studies as described by De Fronzo et al (48). The investigations were performed in two groups of dogs with comparable degree and duration of chronic renal failure (CRF) produced by 5/6 nephrectomy; one group (6 dogs) with intact parathyroid glands (NPX) and hence secondary hyper-parathyroidism, and the second group (6 dogs) without the parathyroid glands (NPX-PTX) but maintained normocalcemic by high intake of calcium. The details of the experimental procedures and the various techniques and methods utilized in the study were reported elsewhere (49).

The 5/6 nephrectomy resulted in a significant (p<0.01) decrease in creatinine clearance in both the NPX (from 56+2 to 12+4 ml/min) and the NPX-PTX (from 58+3 to 13+3 ml/min) dogs and in a significant increase in serum PTH levels in the NPX animals (from 1.0+0.5 to 37+0.5 pg/ml). There were no significant differences among the plasma concentrations of electrolytes before and after the induction of CRF.

The results of the IVGTT before and 3 months after CRF in NPX and NPX-PTX dogs are shown in figure 1-5. Within 3 min after the injection of the glucose load, the plasma concentrations of glucose reached their peak and decreased thereafter. The NPX animals with intact para-thyroid glands and elevated blood levels of PTH displayed glucose intolerance with the plasma concentrations of glucose being significantly (p<0.01) higher at 20, 30,40,50, and 60 min than those observed before CRF (figure 1). In contrast, there were no significant differences between the results of the IVGTT before and after CRF in the NPX-PTX (figure 2). Thus, these latter animals did not have glucose intolerance.

The K-g rate of glucose (the rate of decline in plasma concentration of glucose) decreased significantly (p<0.01) after CRF in the NPX dogs (from 2.86+0.48 to 1.23+0.18%/min) while K-g rate in the NPX-PTX dogs was not affected by CRF (2.41+0.43 vs. 2.86+0.86%/min, figure 3).

There were significant increments in plasma insuin levels during IVGTT in all animals. In the NPX dogs, plasma insulin concentrations increased from 24+2.3 U/ml to a peak of 105 mU/ml (P<0.01) and remained elevated throughout the study. In the NPX-PTX dogs, the maximum increment in plasma insulin concentration (from 18+1.2 to 229+19.4 mU/ml) was more than twice that observed in the NPX animals (p<0.01); the levels gradually declined but were higher than those in the NPX dogs during the first 30 min (p<0.01) and returned to baseline values by 1 h (figure 4). These differences in plasma insulin were not due to higher plasma glucose concentrations in NPX-PTX dogs, and for any given level of plasma glucose during IVGTT, the plasma insulin was higher in NPX-PTX than in NPX dogs (figure 5).

The results of the studies with hyperglycemic clamp are given in figure 6. The total amount of glucose metabolized during the 20 to 120-min period was significantly (p<0.01) lower by 38% in the NPX compared to the NPX-PTX group (6.64+1.13 vs. 10.74+1.10 mg/kg min). The early, late and total insulin responses were greater in NPX-PTX animals than in NPX dogs. The total response gave values of 147+31 vs. 72+9 U/ml (p<0.025). There was no significant difference between the M/I ratio, a measure of tissue sensitivity to insulin, in the NPX and NPX-PTX dogs (9.9+66 vs. 8.9+1.3 mg/kg min per U/ml).

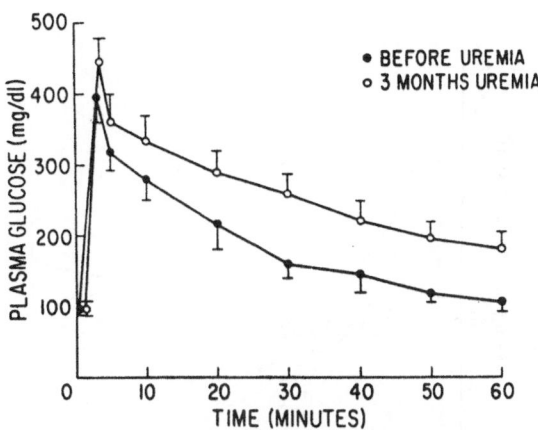

Figure 1. The changes in plasma glucose concentrations during intravenous glucose tolerance tests performed before (●) and after (o) 3
months of CRF in dogs with intact parathyroid glands. Each data
point represents the mean value of 6 dogs and the brackets denote 1
SE. Plasma glucose levels in dogs with CRF were significantly higher
(p<0.01) at 20-60 min. With permission from Akmal et al (49).

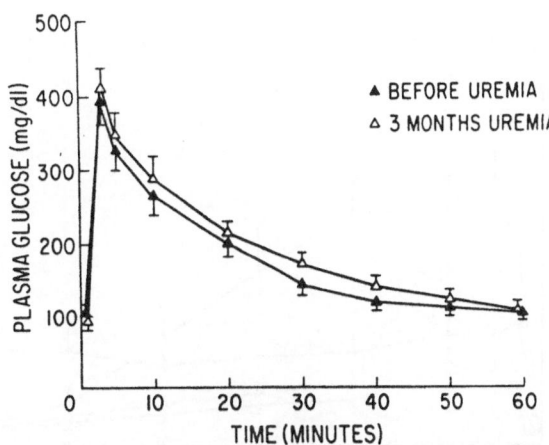

Figure 2. The changes in plasma glucose concentrations during intravenous glucose tolerance tests performed before (▲) and after (Δ) 3
months of CRF in parathyroidectomized dogs. Each data point represents
the mean value of 6 dogs and the brackets denote 1 SE. There was no
significant differences in plasma glucose concentrations before and
after CRF at all times. With permission from Akmal et al (49).

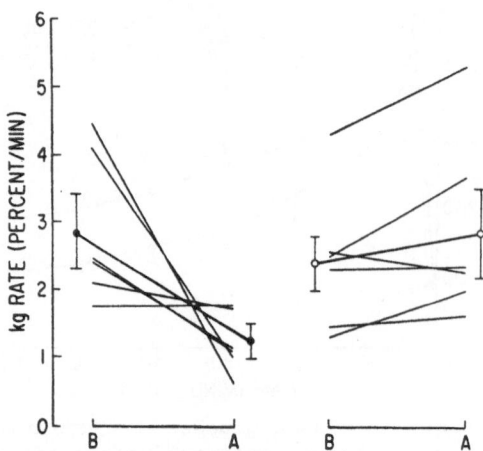

Figure 3. The kg rate before and 3 months after the CRF in NPX (left panel) and NPX-PTX (right panel). B denotes before CRF and A denotes CRF. Each line represents 1 animal and the heavy lines depict the mean values with brackets denoting 1 SE. The kg rate after CRF was significantly (p<0.01) lower than before CRF in NPX dogs. With permission from Akmal et al (49).

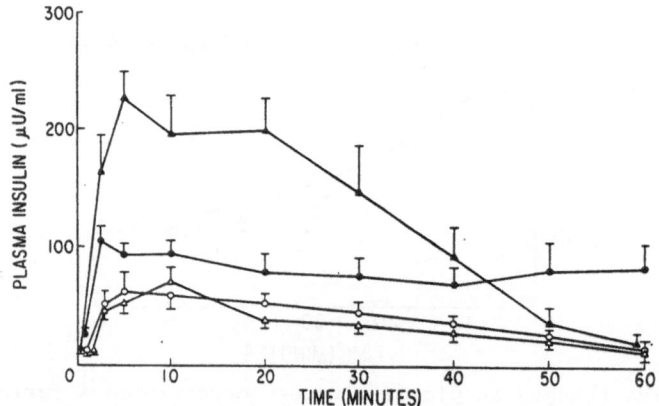

Figure 4. The changes in plasma insulin concentrations during intravenous glucose tolerance tests performed before (open symbols) and after 3 months of CRF (closed symbols) in NPX (o●) and NPX-PTX (ΔΔ). With permission from Akmal et al (49).

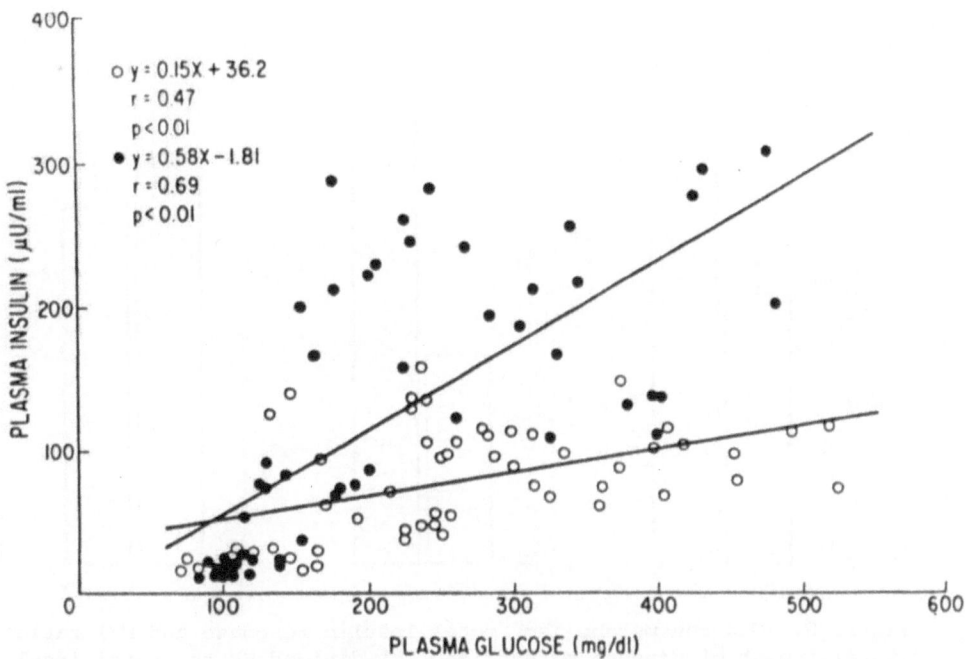

Figure 5. The relationship between plasma insulin and glucose concentrations observed during intravenous glucose tolerance tests
performed in NPX (o) and NPX-PTX (●). With permission from Akmal et
al (49).

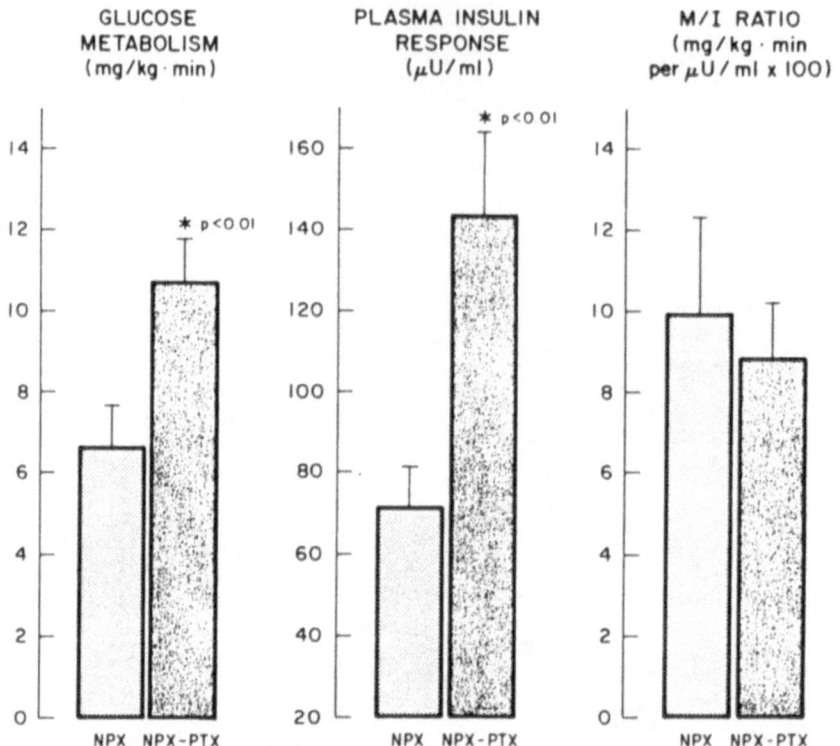

Figure 6. Glucose metabolism, total insulin response and M/I ratio (total amount of glucose metabolized (M) divided by the total insulin response (I) observed during the hyperglycemic clamp in NPX and NPX-PTX dogs. Each column represents the mean of data from 6 NPX and NPX-PTX dogs. The brackets denote 1 SE. Asterisks indicate significant difference from NPX with $p<0.01$. With permission from Akmal et al (49).

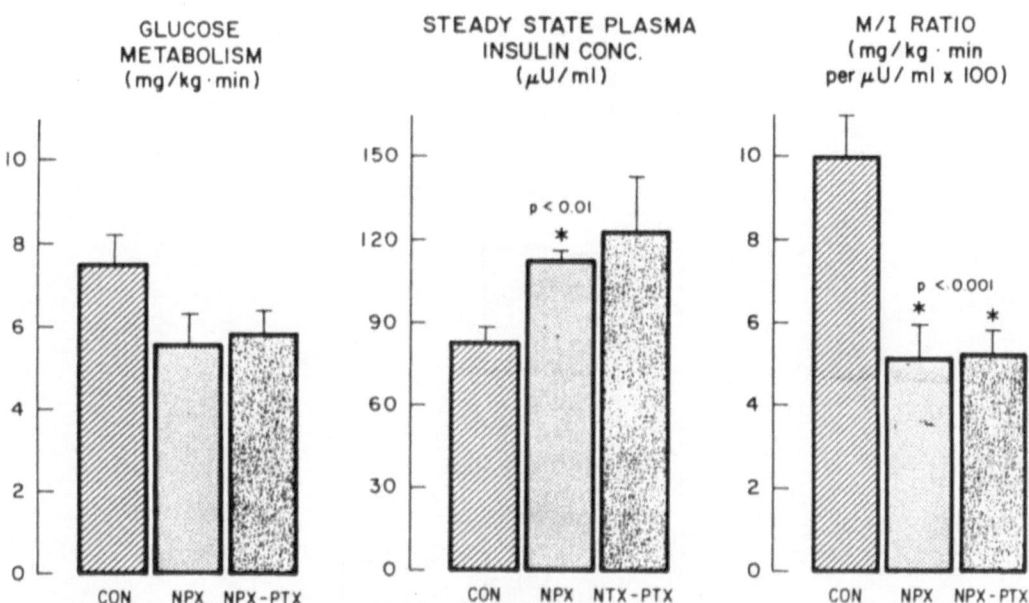

Figure 7.  Glucose metabolism, steady state plasma insulin concentrations and M/I ratio (total amount of glucose metabolized (M) divided
by total insulin response (I) observed during euglycemic insulin clamp
in normal dogs and in NPX and NPX-PTX dogs.  Each column represents
the mean of data from 6 NPX and 7 NPX-PTX dogs.  The brackets denote
1 SE.  CON = Control.  Asterisks denote significant difference (p<0.01)
from control.  With permission from Akmal et al (49).

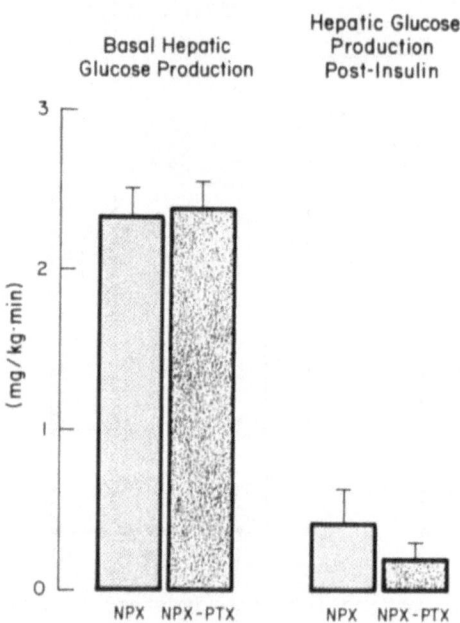

Figure 8. Basal and postinsulin hepatic glucose production observed during euglycemic insulin clamp in 3 NPX and 3 NPX-PTX dogs. The columns represent the mean data and the brackets denote 1 SE. With permission from Akmal et al (49).

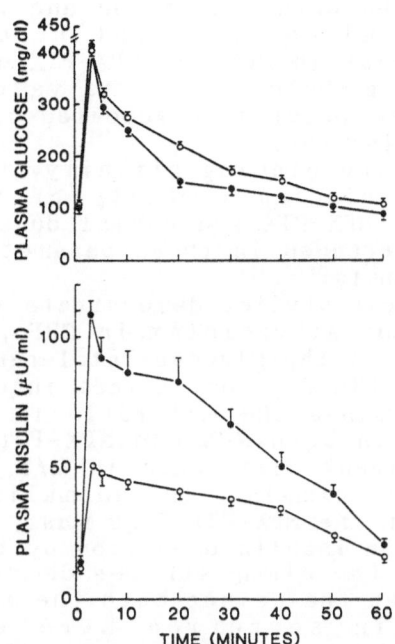

Figure 9.  The changes in plasma glucose and insulin concentration
during intravenous glucose tolerance tests performed in 6 normal dogs
(o) and 6 normocalcemic-normophosphatemic chronic (6 weeks) thyro-
parathyroidectomized dogs with normal renal function (●).  Each
data point is the mean of 6 animals and the brackets denote 1 SE.
Plasma insulin levels were significantly higher (p<0.01) at all
points between 3 and 50 min.  With permission from Akmal et al (49).

The results of the studies with euglycemic clamp are presented in figure 7. The total amount of glucose metabolized during elevated blood levels of insulin in the NPX dogs (5.59+0.71 mg/kg · min) was not different from that in the NPX-PTX animals (5.85+0.47 mg/kg·min). Also the M/I ratio was not different among the two groups of dogs (5.12+0.76 vs. 5.18+0.57 mg/kg · min per U/ml) but both values were significantly (p<0.01) lower than in normal dogs (9.98+1.26 mg/kg · min per U/ml). The metabolic clearance rate of insulin was significantly reduced in both NPX (12.1+0.7 mg/kg · min, p<0.01) and NPX-PTX (12.1+0.9 mg/kg · min, p<0.02) dogs as compared with control animals.

Figure 8 provides the results of the studies evaluating basal hepatic glucose production and the response to insulin utilizing the bolus injection and the continuous infusion of [$^3$H]-glucose. Basal hepatic glucose production was similar in NPX and NPX-PTX dogs (2.33+0.32 vs. 2.38+0.35 mg/kg · min), and the values were not different from those previously reported in normal dogs (2.80+0.20 mg/kg · min) (50).

We also studied the binding affinity, binding sites concentration and binding capacity of monocytes to insulin in the NPX, NPX-PTX and normal dogs. There were no significant differences in these parameters among the three groups of animals.

The results of these studies demonstrate that the state of secondary hyperparathyroidism in CRF plays a major role in the genesis of the glucose intolerance in uremia. However, the excess PTH does not affect insulin action on peripheral tissues since the M/I ratio is significantly lower than normal in both NPX and NPX-PTH animals and there is no significant difference in M/I ratio between these two groups of animals. The normalization of the glucose tolerance in the NPX-PTX dogs must, therefore, be due to improvement in insulin secretion by the β-cells as both the IVGTT and the clamp studies demonstrated. It should also be noted that since both the early and late phases of insulin secretion were enhanced by parathyroidectomy, one must assume that the release of both the stores as well as the newly synthesized insulin is enhanced in the NPX-PTX animals. Finally, it should be mentioned that the effect of parathyroidectomy on plasma insulin concentration during hyperglycemia is independent of the state of CRF inasmuch as plasma insulin levels during IVGTT in chronic normocalcemic-normophosphatemic-thyroparathyroidectomized dogs were twice (p<0.01) the levels in normocalcemic dogs with intact parathyroid glands (figure 9).

In summary, our results indicate that (a) glucose intolerance does not develop with CRF in the absence of PTH; (b) PTH does not affect the metabolic clearance of insulin in CRF, and (c) the normalization of glucose intolerance in CRF in the absence of PTH is due to increased insulin secretion. Thus, the data are consistent with the notion that excess PTH in CRF interferes with the ability of the β-cells to augment insulin secretion appropriately in response to the insulin-resistant state.

The observations of Mak et al. (51) in 8 children

before and after medical suppression of the secondary hyperparathyroidism are consistent with our results. They demonstrated that the glucose intolerance in these children disappeared after the normalization of the blood levels of PTH, and the improvement was due to increased insulin response to hyperglycemia after the treatment of the secondary hyperparathyroidism. They concluded that the higher plasma insulin levels overcame the insulin-resistant state, which was not affected by the suppression of the parathyroid gland activity.

## REFERENCES

1. Massry, SG: Parathyroid hormone and the uremic manifestations. Contrib Nephrol 20:84-91, 1980.
2. Babb, AL, Popovitch RP, Christopher, TG, et al: The genesis of the square meter-hour hypothesis. Trans Amer Soc Artif Internal Organs 17:81-91, 1971.
3. Alfrey, AG, LeGendre, GR, Kaehny, WD: The dialysis encephalopathy. N Engl J Med 294:184-188, 1976.
4. Berlyne, GM: Aluminum toxicity in man. Miner Elect Metab 2:71-73, 1979.
5. Massry, SG: Is parathyroid hormone a uremic toxin? Nephron 19:125-130, 1977.
6. Pappenheimer, AM, Wilens, SL: Enlargement of the parathyroid glands in renal disease. Am J Pathol 11:73-91, 1935.
7. Roth, SI, Marshall, RB: Pathology and ultrastructure of the human parathyroid glands in chronic renal failure. Arch Intern Med 124:390-407, 1969.
8. Katz, AD, Kaplan, L: Parathyroidectomy for hyperplasia in renal disease. Arch Surg 107:51-55, 1973.
9. Berson, SA, Yallow, RS: Parathyroid hormone in plasma in adenomatous hyperparathyroidism, uremia, and bronchogenic carcinoma. Science 154:907-909, 1966.
10. Massry, SG, Coburn, JW, Peacock, M: Turnover of endogenous parathyroid hormone in uremic patients and those undergoing hemodialysis. Trans Am Soc Artif Intern Organs 8:416-422, 1972.
11. Massry, SG, Goldstein, DA: The search for uremic toxin(s) "X":"X"=PTH. Clin Nephrol 11:181-189, 1979.
12. Wallach, S, Ballavia, JV, Shorr J: Tissue distributions of electrolyte, $Ca^{47}$, and $Mg^{28}$ in experimental hyper- and hypoparathyroidism. Endocrinology 78:16-28, 1966.
13. Chausmer, AB, Sherman, BS, Wallach, S: The effect of parathyroid hormone on hepatic cell transport of calcium. Endocrinology 90:663-672, 1972.
14. Borle, AB: Calcium metabolism at the cellular level. Fed Proc 30:1944-1950, 1973.
15. Bogin, E, Massry, SG, Levi, J: Effect of parathyroid hormone on osmotic fragility of human erythrocytes. J Clin Invest 69:1017-1025, 1982.
16. Massry, SG, Coburn, JW, Hartenbower, DL: Mineral content of human skin in uremia: Effect of secondary hyperparathyroidism and hemodialysis. Proc Eur Dial Transplant Assoc 7:146-150, 1970.
17. Berkow, JW, Fine, BS, Zimmerman, LE: Unusual ocular calcification of hyperparathyroidism. Am J Ophthalmol 68:814-824, 1968.

18. Bernstein, DS, Pletka, P, Hartner, RS: Effect of total parathyroidectomy and uremia on the chemical composition of bone, skin, aorta and rats. Israel J Med Sci 7:513-514, 1971.

19. Kraikipanitch, S, Lindeman, RD, Yoenice, AA: Effect of azotemia and myocardial accumulation of calcium. Miner Elect Metab 1:12-20, 1978.

20. Arieff, AI, Massry, SG: Calcium metabolism of brain in acute renal failure. J Clin Invest 53:387-392, 1974.

21. Goldstein, DA, Chui, LA, Massry, SG: Effect of parathyroid hormone and uremia on peripheral nerve calcium and motor nerve conduction velocity. J Clin Invest 62:88-93, 1978.

22. Rasmussen, H: Calcium as intracellular messenger in hormone action, in Massry SG, Letteri, JM, Ritz, E (eds): Regulation of Phosphate and Mineral Metabolism, New York, Plenum, pp 473-491, 1982.

23. Hirata, F, Axelrod, J: Enzymatic methylation of phosphotidylethanolamine increases erthrocyte membrane fluidity. Nature 275:219-220, 1978.

24. Michell, RH: Inositol phospholipids and cell surface receptor function. Biochim Biophys Acta 415:81-147, 1975.

25. Gardner, JD: Regulation of pancreatic exocrine function in vitro: Initial steps in the actions of secretagogues. Ann Rev Physiol 41:55-66, 1979.

26. Green, DE, Fry, M, Blondin, GA: Phospholipids as the molecular instruments of ion and solute transport in biological membrane. Proc Natl Acad Sci 77:257-261, 1980.

27. Lo, H, Lehotay, DC, Katz, D: Parathyroid hormone-mediated incorporation of $^{32}$P orthophosphate into phosphatidic acid and phosphatidylinositol in renal cortical slices. Endocr Res Commun 3:377-385, 1976.

28. Molitoris, BA, Hruska, KA, Fishman, N: Effects of glucose and parathyroid hormone on renal handling of myoinositol by isolated perfused dog kidney. J Clin Invest 63:1110-1118, 1979.

29. Brautbar, N, Chakaborty, J, Coats, J, Massry, SG: Calcium, parathyroid hormone and phospholipid turnover of human red blood cells. Miner Elect Metab 11:111-116, 1985.

30. Albright, F, Bauer, W, Ropes, M: Studies of calcium and phosphorus metabolism. J Clin Invest 7:139-181, 1929.

31. Clarkson, B, Kowlessar, OD, Horwith, M: Clinical and metabolic study of a patient with malabsorption and hypoparathyroidism. Metabolism 9:1093-1106, 1960.

32. Landau, RL, Kappas, A: Anabolic hormones in hyperparathyroidism. Ann Intern Med 62:1223-1233, 1965.

33. King, RG, Stanbury, SW: Magnesium metabolism in primary hyperparathyroidism. Clin Sci 39:281-303, 1970.

34. Garber, AJ; Effect of parathyroid hormone on skeletal muscle protein and amino acid metabolism in the rat. J Clin Invest 71:1806-1821, 1983.

35. Frohling, PT, Kokot, F, Cernacek, P: Relation between middle molecules and parathyroid hormone in patients with chronic renal failure. Miner Elect Metab 7:48-53, 1982.

36. Massry, SG: The toxic effects of parathyroid hormone in uremia. Seminars Nephrol 3:306-327, 1983.

37. Westervelt, FB, Schreiner, GE: The carbohydrate

intolerance of uremic patients. Ann Intern Med 57:266-275, 1962.

38. Hampers, CL, Soeldoner, JS, Doak, PB, Merill, JP: Effect of chronic failure and hemodialysis on carbohydrate metabolism. J Clin Invest 45:1719-1731, 1966.

39. Horton, ES, Johnson, C, Lebovitz, HE: Carbohydrate metabolism in uremia. Ann Intern Med 68:63-74, 1968.

40. De Fronzo, RA, Andres, R, Edgar, P, Waler, WG: Carbohydrate metabolism in uremia. A review Medicine 52:469-481, 1973.

41. De Fronzo, RA, Alverstrand, A, Smith, D, Hendler, R, Hendler, E, Wahren, J: Insulin resistance in uremia. J Clin Invest 67:563-568, 1981.

42. Samaan, NA, Freeman, RM: Growth hormone levels in severe renal failure. Metabolism 19:102-113, 1970.

43. Lowrie, EG, Soeldner, JS, Hampers, CL, Merrill, JP: Glucose metabolism and insulin secretion in uremic, prediabetic, and normal subjects. J Lab Clin Med 76:603-615, 1970.

44. Hutching, RH, Hagstrom, RM, Scribner, BH: Glucose intolerance in patients on long-term intermittent dialysis. An Intern Med 65:275-285, 1966.

45. De Fronzo, RA: Pathogenesis of glucose intolerance in uremia. Metabolism 27:1866-1880, 1978.

46. Ginsberg, H, Olefsky, JM, Reaven, GM: Evaluation of insulin resistance in patients with primary hyperparathyroidism. Proc Exp Biol Med 148:942-945, 1975.

47. Kim, H, Kalkhoff, RK, Costrini, NV, Cerletty, JM, Jacobson, M: Plasma insulin disturbances in primary hyperparathyroidism. J Clin Invest 50:2596-2605, 1971.

48. De Fronzo, RA, Tobin, JD, Andres, R: Glucose clamp technique. A method for quantifying insulin secretion and resistance. Am J Physiol 237:E214-E223, 1979.

49. Akmal, M, Massry, SG, Goldstein, AD, Fanti, P, Weisz, A, DeFronzo, R: Role of parathyroid hormone in the glucose intolerance of chronic renal failure. J Clin Invest 75:1037-1044, 1985.

50. Bevilacqua, S, Barrett, E., Farranini, E, Gusberg, R, Stewart, A, Richardson, L, Smith, D, DeFronzo, R: Lack of effect of parathyroid hormone on hepatic glucose metabolism in the dog. Metabolism 30:469-475, 1981.

51. Mak, RH, Turner, C, Haycock, GB, Chantler, C: Secondary hyperparathyroidism and glucose intolerance in children with uremia. Kidney Int 24:5123-5133.

# RETENTION PATTERNS

A. Schoots[1], R. Vanholder[2], S. Ringoir[2], C. Cramers[1]

[1]Lab. of Instrumental Analysis, Eindhoven University of
Technology, 5600 MB Eindhoven, The Netherlands
[2]Renal Division, University Hospital, B-9000 Ghent,
Belgium

During the last century, there has been a continuous search for
the toxins responsible for the uremic syndrome. Parallel to the
development of analytical chemistry, many substances have been shown to
accumulate in the body fluids of chronic uremic patients [1].
These substances have molecular weights from 23 for sodium to 10,000
for low molecular weight proteins such as $\beta_2$-microglobulin. They
are metal ions, hormones, products of protein, nucleic acid, carbohydrate or lipid metabolism or have dietary origin. Furthermore they
may accumulate both in their native forms and as protein-ligand com-
plexes or hepatic conjugates.
Some of the accumulating substances may be toxic as such, but probably
the cumulative and/or synergistic toxicity of many substances will be
felt by various biochemical processes in uremic patients.

In the past 15 years there has been much attention to the
so-called "middle molecules", as the toxins responsible for neurological disorders in uremic patients. In the "square meter hour" hypothesis
proposed by Babb in 1971 [2], dialysis efficiency of these molecules
was described to be dependent on membrane permeability, membrane
surface area and weekly dialysis time, and independent from dialysate
and blood flow rates. In 1972 the "square meter hour" hypothesis was
changed in the "middle molecule" hypothesis by Babb and co-workers.
Until then, there was clinical evidence for the validity of the hypothesis, based on the application of different dialysis strategies with
low dialysate flow, and large and small membrane surface area.
Since then, the supposed, toxic substances in uremia, called "middle
molecules", were defined as follows: "Middle molecules are substances
that behave in a dialyzer <u>as though</u> their molecular weights were in the
range of 300-2000 daltons". In terms of this definition middle molecules remain elusive substances [3], both conceptually and analyti-
cally, as dialyzability not only depends on solute dimensions and the
pore size of the membrane, but also on electric charge, solvatation,
hydrophobic interaction, boundary layer formation, and protein binding.

Since the proposal of the middle molecule hypothesis there has
been intensive research to separate, isolate, and identify these

substances. These efforts have been directed almost exclusively to molecules with higher molecular weight, which are only a subset of the substances indicated by Babb and Scribner in 1971. Many groups have tried to separate uremic serum or ultrafiltrate components by MOLECULAR WEIGHT, applying the technique of size-exclusion chromatography, often called gel-filtration.

The groups of Stockholm and Paris used Sephadex G-15 gel filtration in conjunction with ion exchange chromatography and initially characterized the separated solutes as peptides with molecular weights of 1100-1300 daltons, based on calibration of the used gel filtration columns [4,5]. The solutes in question later proved to be glucuronides with molecular masses of 371 and 526 respectively. As these relatively low molecular weights indicate, there are some problems with Sephadex gel filtration.

It concerns the anomalous retention behaviour of certain solutes. Aromatic and heterocyclic compounds do adsorb on the gel surface and show elution volumes much larger than expected from the size exclusion calibration curve. On the other hand ionic compounds such as acids may be excluded from the gel particles due to charge interaction or the formation of solvatation layers. The latter solutes therefore elute earlier than would be estimated from the calibration curve, thus behaving as if they had higher molecular weights. When reanalyzing uremic serum gel filtration fractions [6], it was found that in the "middle molecule area" of the gel chromatogram, a number of solutes of relatively low molecular weight appear, that could be analyzed by high performance liquid chromatography, gas chromatography and isotachophoresis (Fig. 1). Among them were acetate, phosphate and sulfate.

Another problem with gel filtration is the low resolving power. In Table 1 the separation efficiencies of some chromatographic techniques are compared. This efficiency can be expressed in terms of a theoretical plate number, as indicated, or the so called peak capacity, being the number of solutes or peaks that can be separated with a given technique, with usual column lengths, within the period of one analysis run.

Table I. COMPARISON OF CHROMATOGRAPHIC TECHNIQUES

| TECHNIQUE | EFFICIENCY (plate number) | PEAK CAPACITY | ANALYSIS TIME(min) |
|---|---|---|---|
| GEL (SEPHADEX) | 700 | 5 | 240 |
| GEL (TSK) | 5,000 | 15 | 25 |
| HPLC | 20,000 | 65 | 25 |
| HPLC(GRADIENT) | 20,000 | 100 | 25 |
| GC(CAPILLARY) | 100,000 | 150 | 25 |

Conditions: usual column lengths; resolution Rs = 1; $0 < K_{av} < 1$ in gel chromatography; $0.5 < k' < 10$ in HPLC and GC.

Gel filtration is an inefficient technique, and within the window of total permeation and total exclusion, only 5 solutes can be totally separated on sephadex gels (within 4 hours) and 15 on the high performance TSK columns(within 25 minutes). On the other hand 100 peaks can be separated with HPLC in the gradient mode, and 150 with isothermal gas chromatography both with analysis times of 25 minutes.

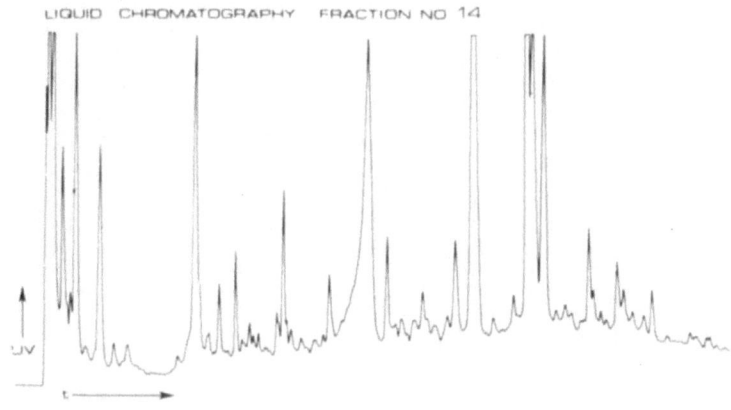

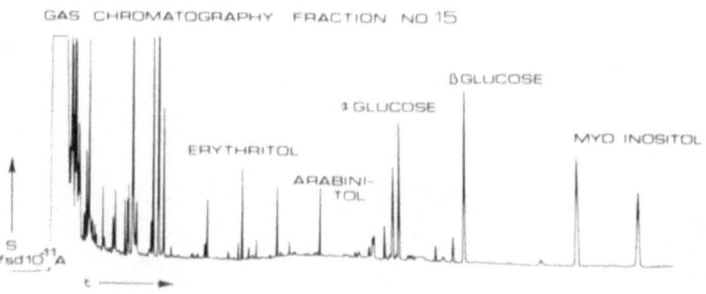

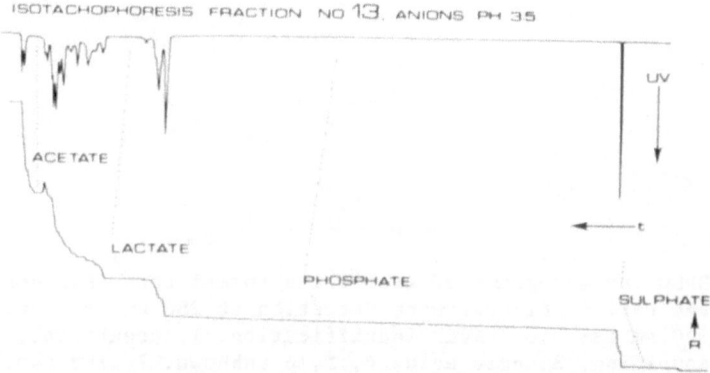

Fig. 1.   Analysis of 'middle molecule' Sephadex G-15 gel filtration
fractions by liquid chromatography, gas chromatography and
isotachophoresis respectively.


Summarizing, the foregoing leads to two conclusions. First, one
should not necessarily look for solutes of a certain molecular weight,
but screen for solutes that have, <u>for some reason</u>, a low dialyzability,
which is a different line of approach. This means there is no particular reason to use gel filtration. Second, this screening should be
performed using separation techniques with high efficiency and peak
capacity.

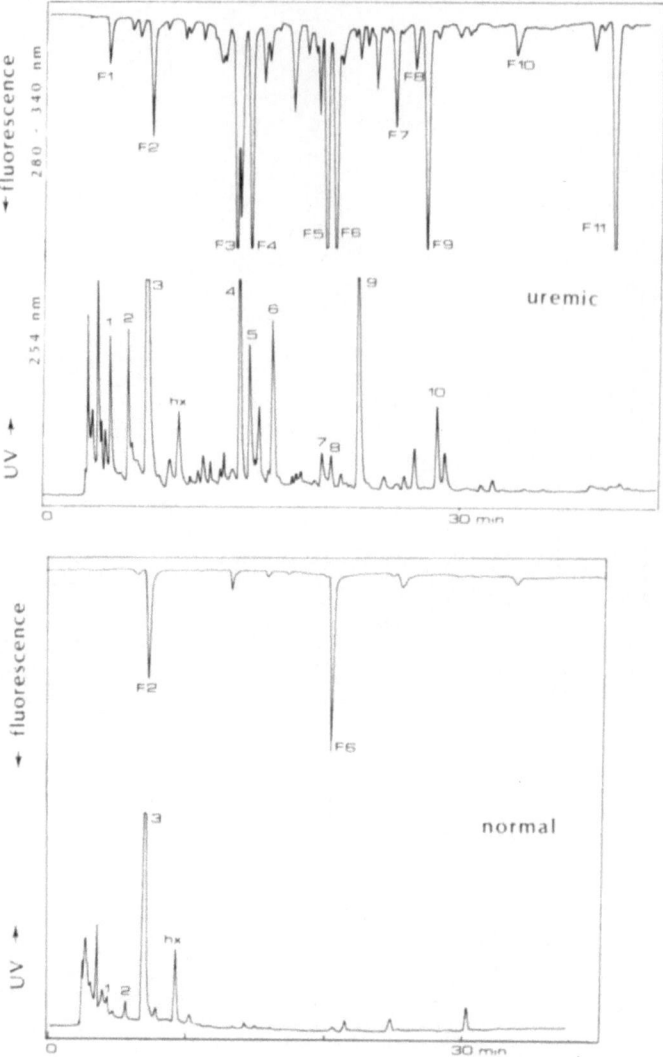

Fig. 2. HPLC chromatograms of uremic and normal sera. For conditions
see ref. [7]. Fluorescence detection at 280 nm excitation and
340 nm emission. Peak identification: 1. creatinine, 2. pseudouridine, 3. uric acid, 4, 5, 6 unknown, 7. indican, 8.
tryptophan, 9. hippuric acid, F1,F3,F4,F7, F8, F9 unknown, F2.
tyrosine, F5. indican, F6. tryptophan, F11. indoleacetic acid.

High performance liquid chromatography (HPLC) in the gradient mode
is a technique that is suitable for screening. It has some advantages:
1)   No chemical reactions with the solutes to be analyzed are neces-
     sary for elution.
2)   There is no thermal decomposition of the labile biochemical
     substances.
3)   The method is non-destructive, so isolation of individual
     substances is possible.
4)   The method has sufficient resolving power.

5)     Solutes with different charge and polarity can be analyzed in the
       same run.
6)     The method is quantitative.
7)     The method can be automated, and both known and unknown solutes
       can be screened.

With the HPLC method some 75 samples can be analyzed weekly,
yielding reproducible quantitative data on more than twenty known and
unknown solutes characteristic for uremia [7]. Fig. 2 shows HPLC
patterns for uremic and normal serum samples. UV and fluorescence
detection are applied simultaneously in a single run. UV-absorbance at
254 nanometer is the best choice as a detection technique as it is the
most universal method which can be used in conjunction with gradient
elution. Fluorescence detection at 280 nm excitation, and 340 nm emission facilitates the quantitation of the numerous fluorescent substances, and especially a number of tryptophan metabolites.

A number of solutes was identified, as given in the figure legend.
The structure of a number of UV-absorbing and fluorescent peaks remains
to be elucidated.
In order to study patient heterogeneity with respect to the profiles
the sera of more than 75 patients on chronic hemodialysis treatment
were analyzed. A relation to residual renal function was obvious,
although to different degrees for the various solutes.
The HPLC method has been applied to studies on the dialyzability of the
substances observed in the profiles, both in vivo and in vitro.
Protein binding coefficients, mass transfer coefficients, and dialyzer
clearances of the HPLC-solutes were determined in vitro [8].

Ultrafiltrate, obtained from patients dialyzed with permeable
membranes, was used as a blood substitute, in in vitro experiments.
Uremic ultrafiltrate was dialyzed across cuprophane membranes, Gambro
GF80-M and GF180-M, against usual dialysate solution, at 37 degrees
celsius. Transmembrane pressure was kept very close to zero, resulting
in negligible ultrafiltration. Dialyzer clearances ($K_b$), and mass
transfer coefficients ($h_oA$) were calculated. Blood substitute inlet
and outlet, and dialysate outlet concentrations were determined by
HPLC, thus yielding data on 20 solutes characteristic for the uremic
state.
Protein binding levels of the solutes in the HPLC-profiles were determined by ultrafiltration and trichloroacetic acid precipitation.
Dialyzer clearance values for the solutes were recalculated for any
protein binding using the following equation [2]:

$$K = Q_B \frac{1 - \exp\left[\dfrac{h_oA}{Q_B \cdot \mu(1+\rho)} \left(1-\mu(1+\rho)\dfrac{Q_B}{Q_D}\right)\right]}{\mu(1+\rho)\dfrac{Q_B}{Q_D} - \exp\left[\dfrac{h_oA}{Q_B \cdot \mu(1+\rho)}\left(1-\mu(1+\rho)\dfrac{Q_B}{Q_D}\right)\right]}$$

In Fig. 3 dialyzer clearance values as a function of blood flow
are given. Fig. 3a shows $K_b$ values of urea, free indican, and the
recalculated values for indican when 89% protein binding was taken into
account. The latter results in a very flat curve and the solute thus
shows an apparent molecular mass of over 1500 daltons (the in vitro
vitamin $B_{12}$-clearance as determined by the manufacturer of the used
membrane, being 34 ml/min). In Fig. 3b the influence of membrane
surface area on the clearance of urea and the unknown solute UKF3, with
zero protein binding, is shown. Doubling of membrane surface area

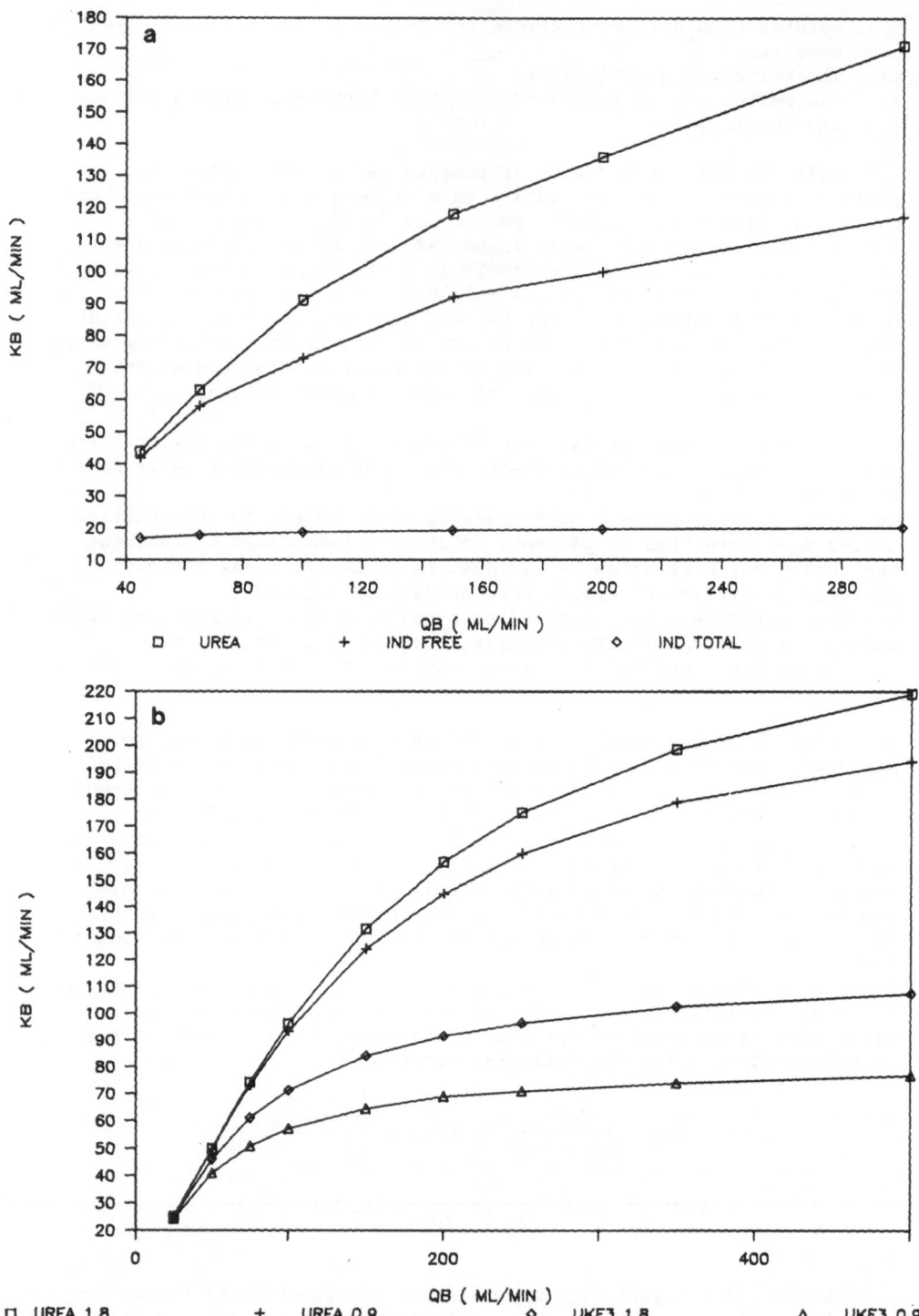

Fig. 3.   a. Clearance values of urea and indican, and indican taking
into account 89% protein binding (lowest curve), as a function
of $Q_b$ in vitro. b. The same for urea and unknown fluorescent
substance UKF3 on membranes of 1.8 $m^2$ (upper curves) and
0.9 $m^2$ (lower curves). UKF3 has no significant protein
binding.

results in 10-15% increase of clearance for urea and 40-50% for UKF3. Thus, indican, UKF3 and a number of other solutes analyzed by HPLC behave in the dialyzer as if they had molecular masses of 500-1500. For some of them this behavior may be attributed to protein binding, while others have zero protein binding.

Summarizing, it may be concluded that besides readily removable solutes as urea, uric acid and creatinine, a number of poorly removable solutes in uremic sera exist, of which the removal will benefit from high surface area dialysis. Moreover, these solutes can be analyzed and quantitated by high performance liquid chromatography simultaneously. Among these solutes are hippuric acid, indican, indole acetic acid, and a number of fluorescent substances. Some of them are unknown and await isolation by preparative HPLC and spectrometric identification.

Apart from publications by Farrell[9,10] relatively little attention has been paid to the question of protein binding in relation to dialyzability during the middle molecule debate, although it was described by Babb and Scribner in 1971[2]. In order to tackle the problem of poor removability due to protein binding Dorson et al [11] described the use of so-called "solutizers" to influence the level of protein binding. In what they call speculative detoxification systems, a solutizer such as ethylalcohol may be added at the arterial side of the dialyzer and removed at the venous side.
It has been reported that heparin releases the enzyme lipoprotein lipase from tissue, which promotes the appearance of free fatty acids in the patients blood, which could act as competitors to protein binding of other solutes.
We observed higher levels of the free fractions of tryptophan and indican postdialysis, compared to predialysis sera[7], in almost all patients studied. This means that heparin already, in a way, may act as a solutizer. This also may relativize a little the very flat curve of clearance versus blood flow (Fig. 3a) for indican and other solutes, as the extent of protein binding decreases during dialysis. The reverse is true for hippuric acid, as protein binding was found to be increasing during dialysis (Table 2), in accordance with data reported by Farrell[10].

Table 2. CHANGE OF PROTEIN BINDING LEVEL DURING DIALYSIS

| Solute | % bound predialysis mean (st.dev.) | % bound postdialysis mean (st.dev.) | p* |
| --- | --- | --- | --- |
| Indican | 89 (5) n=14 | 83 (7) n=14 | 0.002 |
| Hippuric acid | 38 (8) n=14 | 49 (8) n=14 | 0.001 |

*Wilcoxon's test.

In the HPLC analysis of sera of a large number of dialyzed patients, we observed a heterogeneity of the patients' population with respect to the relative concentration of urea, creatinine and a number of poorly removable substances such as hippuric acid, indican, and a number of unidentified fluorescent solutes. Maybe it will prove worthwhile to consider the use of a marker solute, representative for the poorly removable substances, in addition to the well documented kt over V for urea in the National Cooperative Dialysis Study[12]. In

conclusion a statement of Babb et al[13] in 1975 is recalled that may still be topical:
"The succes of a dialysis index depends on the selection of both the proper marker solute and the minimum weekly amount of this solute to be removed either by dialysis alone or by combined contributions of dialysis and residual renal function."

REFERENCES

1. J. Bergstrom, P. Furst, Uraemic toxins in: "Replacement of Renal Function by Dialysis", W. Drukker, F.M. Parsons and J.F. Maher, eds., M. Nijhoff Publ.: 354 (1983).

2. A.L. Babb, R.P. Popovich, T.G. Christopher and B.H. Scribner. The genesis of the square meter-hour hypothesis. Trans. Am. Soc. Artif. Int. Organs, Vol. XVII:81 (1971).

3. A. Schoots, F. Mikkers, C. Cramers, R. De Smet, and S. Ringoir. Uremic toxins and the elusive middle molecules. Nephron 38:1 (1984).

4. P. Furst, J. Bergstrom, A. Gordon, E. Johnsson and L. Zimmerman. Separation of peptides of 'middle' molecular weight from biological fluids of patients with uremia. Kidney Int., 7:272 (1975).

5. J.L. Funck-Brentano, N.K. Man, A. Sausse, J. Zingraff, J. Boudet, A. Becker, G.F. Cueille. Characterization of a 1100-1300 MW uremic neurotoxin. Trans. Am. Soc. Artif. Int. Organs, 22:163 (1976).

6. A.C. Schoots, F.E.P. Mikkers, H.A. Claessens, R. De Smet, N. van Landschoot, and S.M.G. Ringoir. Characterization of uremic 'middle molecular' fractions by gas chromatography, mass spectrometry, isotachophoresis, and liquid chromatography. Clin. Chem., 28(1):45 (1982).

7. A.C. Schoots, H.R. Homan, M.M. Gladdines, C.A. Cramers, R. De Smet and S.M. Ringoir. Screening of UV-absorbing solutes in uremic serum by reversed HPLC. Change of blood levels in different therapies. Clin. Chimica Acta, 146:37 (1985).

8. A. Schoots, R. Vanholder, P. Pieters, M. Piron, R. de Smet, C. Cramers and S. Ringoir. In vitro diffusive mass transfer of UV-absorbing and fluorescent uremic solutes on cuprophane membranes. Submitted for publication (1986).

9. P.C. Farrell, N.L. Gribb, D.L. Fry, R.P. Popovich, J.W. Broviac and A.L. Babb. A comparison of in vitro and in vivo solute-protein binding interactions in normal and uremic subjects. Trans. Am. Soc. Artif. Int. Organs, Vol. XVIII:268 (1972).

10. P.C. Farrell, F.A. Gotch, J.H. Peters, B.J. Berridge and M. Lam. Binding of hippurate in normal plasma and in uremic plasma pre- and postdialysis. Nephron 20:40 (1978).

11. W.J. Dorson, V.B. Pizziconi, C.N. Sizto, R.P. Radnoti, C.J. Zarembinski, and L.M. Aniuk. Solutizers as an adjunct to artificial organ treatment for bound chemicals. Trans. Am. Soc. Artif. Int. Organs, Vol. XXVI:116 (1980).

12. F.A. Gotch and J.A. Sargent. A mechanistic analysis of the National Cooperative Dialysis Study (NCDS). Kidney Int., 28:526 (1985).

13. A.L. Babb, M.J. Strand, D.A. Uvelli, J. Milutinovic and B.H. Scribner. Quantitative description of dialysis treatment: A dialysis index. Kidney Int., 5:23 (1975).

# MIDDLE MOLECULES AND THE 7 C FACTOR

Jonas Bergström, Hans Jörnvall,
and Lena Zimmerman

Karolinska Institute
Department of Renal Medicine
Huddinge University Hospital
S-141 86 Huddinge  Sweden

## Background

Soon after Babb and Scribner in 1972[1] presented The Middle Molecule Hypothesis several investigators performed clinical studies in dialysis patients, designed to increase or decrease the presumed level of middle molecules (MM) in the body fluids relating these changes to the symptomatology of the patients or to the in vitro toxicity of plasma and dialysate (see references in 1). However, most of these studies were inconclusive and no direct determination of MM were performed.

## Earlier results

For the assay of MM in biological fluids, our group developed analytical methods using soft gel chromatography. A MM fraction isolated by gel permeation could be separated into 7-9 fractions by ion exchange chromatography. These fractions were quantified by integration of the peaks on the chromatogram, recorded under standardized conditions[2]. Based on tests with standars of known molecular weight we assumed at that time that the MM fraction which we isolated only contained substances with a molecular weight higher than 1000 daltons, which later on proved to be wrong. Our results showed that MM indeed accumulate in the plasma of severely uremic patients[3]. Furthermore we found that high MM peaks in plasma of individual patients were associated with signs of uremic "sickness" but that there was no strict correlation between symptomatology and accumulation of certain peaks[4,5]. These findings raised the question whether accumulation of MM might be markers of toxicity and "sickness" rather than causally related to toxicity. We studied a small population of patients treated with continuous ambulatory peritoneal dialysis (CAPD) under condition of metabolic balance when they did not take any interfering drugs (see below) and found that there was a good correlation between the appearance rate of three MM fractions estimated from the sum of the elimination in urine and peritoneal dialysates (fraction 7 a, b and c) and the protein intake and urea appearance rate, respectively, indicating that some of the middle molecule fractions are generated by catabolism of proteins[6].

We found that the plasma concentration of fraction 7 c, which was high in some uremic patients, also showed unpredictable variations, which could not be explained by variations in removal rate by dialysis and renal elimination. Considerable time and effort was spent on isolating the main fraction of peak 7 c and mainly through the effort of Lena Zimmerman we were able to show by using HPLC, amino acid analysis, and gas chromatography - mass spectrometry that the 7 c component was glucuronyl-o-hydroxyhippurate (GOHH) a double conjugate of ortho-hydroxy-benzoic (salicyluric) acid with glucuronic acid and glycine having a molecular weight of 371 daltons[7]. Realizing that the compound had a much lower molecular weight than we had earlier thought we had to consider that drugs or conjugates of drugs might interfere in the middle molecule determination. In a retrospective survey of our patient population, we found no apparent influence on the chromatographic pattern due to disturbances from furosemide, allopurinol, hydralazine, methyldopa and beta blockers, i.e. the drugs most commonly prescribed to our uremic patients during the period when our sampling was made[8]. On the other hand, the structure of GOHH suggested that it might be derived from salicylic acid, considering that two of the main known metabolites of salicylic acids are salicyluric acid and salicylic phenolic glucuronide. We therefore performed studies in which we gave acetyl salicylic acid to normal volunteers and uremic patients and could demonstrate that salicylic acid ingestion increased the height of peak 7 c in plasma and urine considerably. We concluded that part of the variation in 7 c concentration in plasma of uremic patients might be due to this drug artifact in spite of our policy being to discourage chronic uremic patients on dialysis from using salisylates as pain relievers due to the risk of bleeding[8].

## New techniques - HPLC

In recent years we have abandoned the soft gel technique and now use high performance liquid chromatography (HPLC) applying the principles of gel permeation, ion exchange and reverse phase HPLC.

In a recent methodological study we explored the possiblity of using gel permeation HPLC and ion exchange HPLC for separating MM and especially to further elucidate the role of GOHH as an exogenous and endogenous metabolite in normal and uremic subjects[9]. We could demonstrate the superiority of HPLC over soft gel for analysis of MM with much better resolution, ion exchange HPLC yielding 20-40 separate peaks instead of 7-9 peaks with soft gel. The reproducability is good. As in other HPLC applications, the difference in peak heights for the same sample injected several times is less than 10%, and the separation time 1 h instead of 4 h with soft gel (Fig. 1).

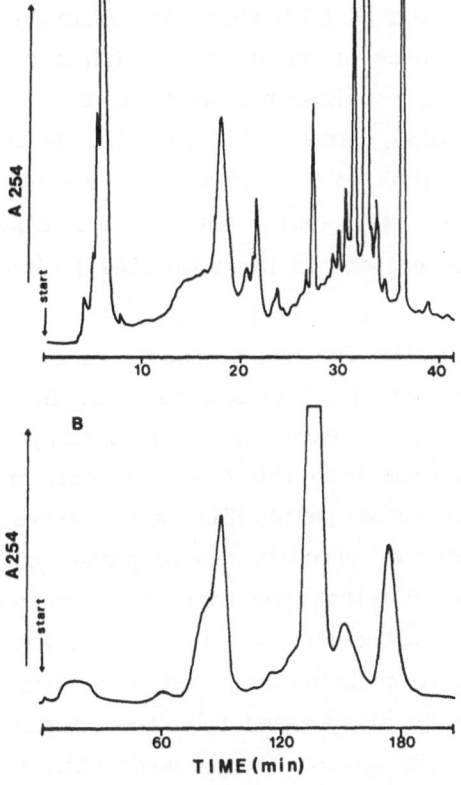

Fig. 1. Ion exchange chromatography of the middle molecule fraction from urine of a uremic patient using: A) HPLC on a TSK 545 DEAE column; B) soft gel chromatography on a Sephadex DEAE A-25 column.

Using gel permeation HPLC we could confirm that ingestion of salicylate for two days yielded a chromatographic peak corresponding to GOHH but in chromatograms of normal urine also a big peak of ortho-hydroxyhippurate (OHH). We could confirm that we were measuring the beta-glucuronide of OHH by treating the urine after salicylate ingestion with beta-glucuronidase after which GOHH disappeared and OHH appeared on the chromatogram. The proportion of OHH to GOHH was much higher (84-88%) in urine from normal subjects than in uremic ultrafiltrate and urine (9-37%). GOHH was searched for in the urine of normal subjects which had not ingested salicylate. A small chromatographic peak was obtained at the same retention volume as GOHH and analyzed for amino acids after hydrolysis. Traces of glycine were detected in all samples but only in samples from two subjects was glycine a major component. The results suggest that small amounts of GOHH may be produced in healthy individuals[9].

So far our investigations have shown that salicylic acid ingestion results in the formation of GOHH which can accumulate in uremic plasma and simulate the presence of an endogenous MM, a possibility not taken into consideration before. However, we believe that salicylate is not the only source of GOHH, since small amounts may be present in normal subjects and the 7 c peak is present to some extent in the majority of the uremic patients. It should, however, be emphasized that other compounds may also be present in the so-called 7 c peak.

We are now searching for new compounds, and especially endogenous glucuronides, in the other MM fractions as well. The so-called 7 a and 7 b fractions isolated by ion exchange chromatography on soft gel represent more acidic compounds than the 7 c compounds. These fractions have been subjected to reverse phase HPLC which reveals that they also contain a large number of chromatographic peaks representing different compounds. We could show that the material corresponding to peak 7 a isolated from uremic ultrafiltrate, i.e. the fraction containing the most acidic compounds, contains at least five different glucuronides. The corresponding peaks disappeared from the original chromatogram and reappeared as new peaks in the chromatogram obtained after hydrolysis with beta-glucuronidase, indicating that all these peaks represent glucuronides formed from endogenous or exogenous sources (Fig. 2). The further characterisation of these compounds as well as other compounds isolated by HPLC may hopefully give us further insight into the molecular basis of uremia.

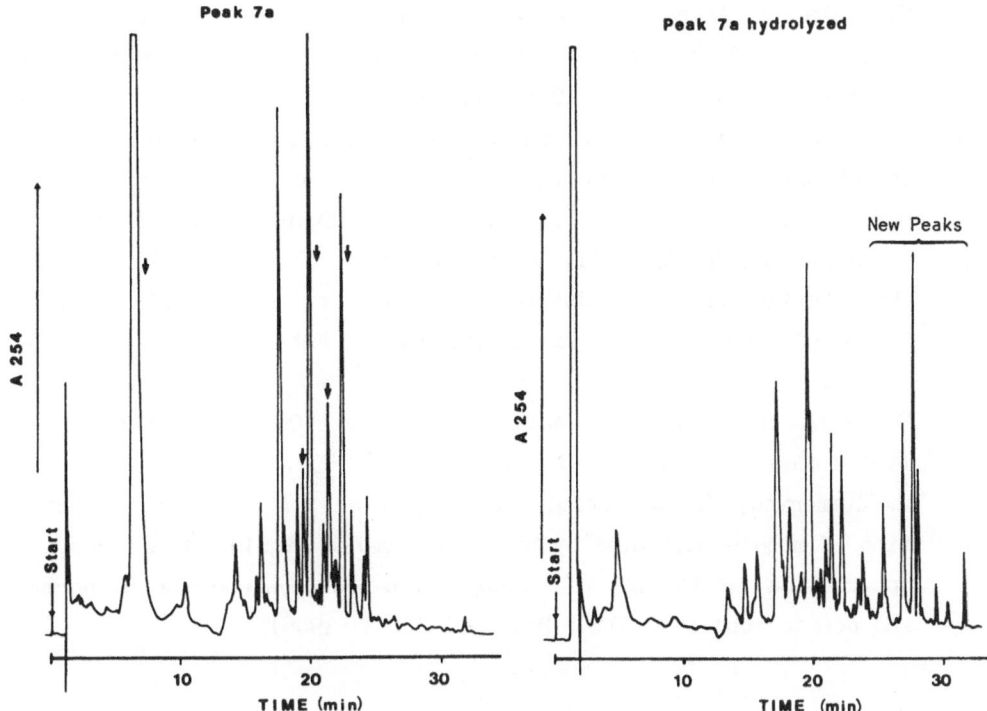

Fig. 2. Reverse phase HPLC on a NOVA-PAC$^{TM}$ C$_{18}$ column of peak 7a isolated from uremic ultrafiltrate before and after hydrolysis with beta-glucuronidase.

We are aware that research on uremic toxins is very complicated and may yield quite unexpected results. Never did we expect that our biochemical approach to uremia would result in the detection and isolation of a new metabolite of salicylic acid which accounts for about 20% of the salicylate metabolites eliminated in the urine.

## References

1. A. L. Babb, P. C. Farrell, D. A. Uvelli, and B.H. Scribner, Hemodialyzer evaluation by examination of solute molecular spectra, Trans. Am. Soc. Artif. Intern. Organs 18:98 (1972).

2. P. Fürst, L. Zimmerman, and J. Bergström, Determination of endogenous middle molecuels in normal and uremic body fluids. Clin. Nephrol. 5:178 (1976).

3. H. Asaba, J. Bergström, P. Fürst, R. Oulés, and L. Zimmerman, Accumulation and excretion of middle molecules. Eur. Dial. Transpl. Assoc. 13:481 (1976).

4. H. Asaba, A. Alvestrand, J. Bergström, P. Fürst, and V. Yahiel, Uremic middle molecules in non-dialyzed azotemic patients: relation to symptoms and clianical biochemistries. Clin. Nephrol. 17:90 (1982).

5. H. Asaba, A. Alvestrand, P. Fürst, and J. Bergström, Clinical implications of uremic middle molecules in regular hemodialysis patients. Clin. Nephrol. 19:179 (1983).

6. J. Bergström, H. Asaba, P. Fürst, and B. Lindholm, Middle molecules in chronic uremic patients treated with continuous ambulatory peritoneal dialysis. Perit. Dial. Bull. 3 (Suppl 1):S7 (1983).

7. L. Zimmerman, H. Jörnvall, J. Bergström, P. Fürst, and J. Sjövall, Characterization of a double conjugate in uremic body fluids. Glucuronidated o-hydroxybenzoylglycine. Febs Letters 129:237 (1981).

8. H. Asaba, L. Zimmerman, and J. Bergström, On drug artifacts in middle molecule analysis. Nephron 39:73 (1985).

9. L. Zimmerman, J. Bergström, and H. Jörnvall, A method for separation of middle molecules by high performance liquid chromatography: application in studies of glucuronyl-o-hydroxyhippurate in normal and uremic subjects. Clin. Nephrol. 25:94 (1986).

# MIDDLE MOLECULES AS A MARKER OF UREMIC TOXINS

Zhao-Guang Wu, Lu-Tan Liao, Zhu-Hui Cai
Zhao-Nian Lu, You-De Cai, Pei-Fang Sheng

Zhong Shan Hospital, Shanghai Med. Univ.
Shanghai, China

In the past decade, toxic uremic solutes in the molecular range of 350-5000 daltons, so called middle molecules (MM), have received much attention, particularly with regard to determination of their chemical structure and biologic toxicity (1-5). Even though substantial proof is still incomplete, uremic toxins with MM have been increasingly identified. In recent years, several new MM compounds have been fractioned and isolated in increased amounts in uremic sera and have been found to be responsible for the clinical manifestations (6-8). The present study is to ascertain the value of quantitative analysis of MM in the evaluation of the efficacy of continuous ambulatory peritoneal dialysis (CAPD), hemodialysis (HD), sequential ultrafiltration and diffusion dialysis (SUD), hemofiltration (HF), hemoperfusion (HP) and reused hollow fiber dialysers.

## MATERIALS AND METHODS

In the past 7 years, MM from plasma or serum, ultrafiltrate, and peritoneal outflow dialysate had been studied in 97 uremic patients by gel chromatography and a group of 32 healthy subjects served as controls.

All blood samples were pretreated with a 40 x 1cm Sephadex G-50 (superfine) column or by a CXA 50 ultrafilter with a cutoff at 50,000 daltons to remove proteins that might interfere with fractionation. Fractionation of MM was performed on a 100 x 1cm Sephadex G 15 column, elution was maintained at 10, 15 or 20 ml/hr. Ultraviolet (U.V.) absorbance (A) was measured at 206 and 225 nm using a Chinese-made 751 G spectrophotometer.

To simplify determinations for routine clinical use. Specific elution fractions such as peak 2 (35-39 ml) and peak 3 (40-45 ml), corresponding to the elution volumes for the MM vitamin B12 (MW 1,355) and Oxytocin (MW 1,007) were collected separately. Absorbances for peak 2 and peak 3 were expressed as mean values.

Fractions in the MM range (peak 2 and 3), as well as macromolecules (peak 1, 1,500-50,000 daltons), were lyophilized separately. Further subfractionation was

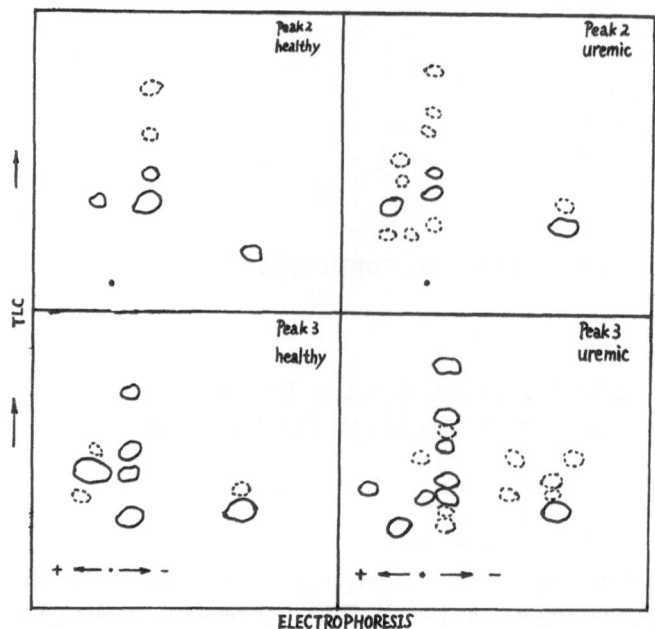

Fig. 1 Thin layer fingerprint chromatograms. Original point. First
dimension: by electrophoresis; Second dimension: by TLC;
Color developed with ninhydrin.

performed by a thin layer fingerprint technique. Fractions obtained from the chromatography column were applied to cellulose-precoated thin layer plastic sheets
(20 x 20 cm, 0.1 cm). The first dimension was by electrophoresis with pyridine:acetic
acid:acetone: water (50:100:375:1875, by volume) buffer, pH 4.4, at 400 volts for 1 hour
The second dimension was by ascending liquid chromatography in a pyridine : n-
butanol:ethanoic acid:water (40:60:12:80 by volume) system, for about 10 hrs at room
temperature, followed by development with 0.5% ninhydrine in ethanol or 0.04%
fluorescamine in acetone (Fig. 1).

RESULTS

Fractionation by both 40x1 cm Sephadex G 50 (superfine) and 100 x1cm Sephadex
G 15 columns was carefully surveyed, using reference substances. Linear regression
between log molecular weight of reference substances and elution volume was satisfactory within a specific molecular weight range. In the Sephadex G 50 column,
the first fraction, eluted at 8-20 ml, contained macromolecules. Later on fractions
at 20-38 ml, a variety of middle as well as small molecules appeared. This portion
was to be further fractionated. Molecular fractionation of different biologic fluids
was carried out in different 100 x 1 cm Sephadex columns, such as G 50 (fine), G 25
(find) and G15. Uremic samples could be separated well into only about 10 peaks,
based upon differences in molecular size, by a Sephadex G 15 column. Macromolecules were eluted in the 27-32 ml fraction (peak 1). Peak with significantly higher
UV absorbance in uremic than in healthy sera appeared in the 35-to-45ml fractions
(peak2 and 3), which corresponded to the elution volumes for the reference substances vitamin B12 and oxytocin.

Reproducibility of gel chromatography was carefully evaluated. A concentrated ultrafiltrate was diluted proportionally with NH4HCO3 buffer and repeatedly eluted in the same Sephadex G 15 column. Mean absorbances for peak 2 and peak 3 were measured. A linear relationship was found for each peak, although their absorbances were different. The mean absorbance of measurements at the same concentration and volume was $0.507 \pm 0.027$, with a coefficient of variation of 5.3%.

The results indicate that the mean normal value of MM in the 32 healthy subjects was $0.123 \pm 0.039$ for peak 2 and $0.144 \pm 0.048$ for peak 3. The mean value of serum MM in the 18 uremic patients before CAPD was $0.337 \pm 0.118$ for peak 2 and $0.484 \pm 0.171$ for peak 3. Mean UV absorbances for peak 2 and peak 3 were consistently higher in uremic patients before treatment than in normal controls ($p < 0.01$). Linear regression equations of UV absorbances for peak 2 and peak 3 were $\hat{Y} = 0.0204 + 1.003\,X$ in healthy subjects and $\hat{Y} = 0.1464 + 1.019\,X$ in uremic patients, with correlation coefficients of 0.812 and 0.713 respectively ($p < 0.01$) (Fig 2).

MM were determined in the plasma of 18 patients before and 20 patients after 1-3 months of treatment with CAPD and their levels were found to be significantly lower compared with initial values. Preliminary results on MM clearance also suggested that CAPD was more effective in the removal of MM than intermittent peritoneal dialysis.

Plasma and ultrafiltrate MM in 23 samples from 12 patients on isolated ultrafiltration and sequential dialysis also displayed consistently higher mean UV absorbances for peak 2 and peak 3 compared with healthy subjects ($p < 0.01$). There was a remarkable increase in the removal of MM from plasma to ultrafiltrate, so the ratio of ultrafiltrate to plasma showed a higher value both for peak 2 ($61.8 \pm 21.7\%$) and peak 3 ($89.5 \pm 27.4\%$).

Serum MM levels were evaluated on 58 occasions in 14 patients before and after hemofiltration (HF). The sieving coefficient was $0.852 \pm 0.314$ for peak 2 and $0.839 \pm 0.306$ for peak 3. MM clearance measured from peak 2 and peak 3 was respectively 42.3 ml/min and 44. 2 ml/min. In contrast, the clearance of urea nitrogen and creatinine was respectively 45.9 ml/min and 40 ml/min. This confirms that HF is much more effective in clearing MM than small molecules.

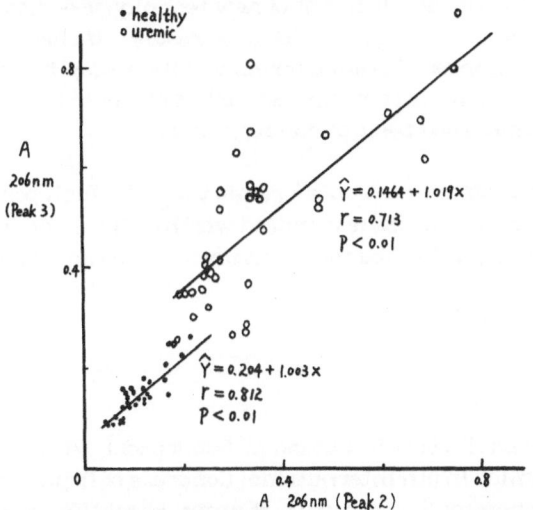

Fig. 2  Correlation and regression for absorbance of peak 2 against that of peak 3.

Erythrocyte membrane sodium and potassium ATPase activity was determined in 37 nondialysis uremic patients, 14 patients treated with HF and 42 normal subjects with a modified Sigström and Whaum method. Results showed that the mean levels of erythrocyte membrane $Na^+ - K^+ - ATPase$ activity were $799\pm62$ u in normal subjects and $335\pm43$ u in nondialysis uremic patients. The mean levels rose almost to normal values in 13 out of 14 patients following HF. The lower the serum MM, the higher was the erythrocyte membrane $Na^+ - K^+ - ATPase$ activity.

Similar changes occurred in MM UV absorbance for peak 2 and peak 3 in 5 patients treated with gelatin-encapsulated activated charcoal hemoperfusion.

Finally, the capacity for clearing MM was one of the parameters used in evaluating the efficacy of a used dialyzer. MM clearance measured from peak 2 and peak 3 on 28 and 38 occasions respectively showed a significant decrease in clearance when reuse exceeded 20 times.

## DISCUSSION

The results of the present study showed that all the uremic patients undergoing CAPD or HF who had markedly improved clinical features of uremia displayed a lower serum MM and relatively higher levels of BUN and creatinine. These facts confirmed our earlier observation that there was a firm correlation between uremic manifestations and MM in uremia (9). Intractable hypertension and large amounts of pericardial effusion in uremic patients could be controlled after a few weeks of treatment with HF or CAPD in our series. It was observed in our HF patients that the red cell membrane $Na^+ - K^+ - ATPase$ activity in 37 uremic patients was very much lower before HF and rose almost to normal values after HF, the lower the serum MM, the higher was the red cell membrane $Na^+ - K^+ - ATPase$ activity, resulting in stabilization of blood pressure. These findings strongly suggest that serial determination of erythrocyte membrane $Na^+ - K^+ - ATPase$ activity, which is suppressed by some uremic toxins with MM (10), may be valuable in elucidating the mechanism of uremic hypertension and metabolic abnormalities in uremic patients.

It is our observation that MM do exist in uremic sera. The MM theory has had considerable impact on the development of new techniques of blood purification in the treatment of chronic renal failure with good results. We feel with certainty, MM could be considered a marker of uremic toxins and its in-depth study would help to shed light on the pathogenesis of uremia as well as evaluate the therapeutic efficacy of various dialyzers and blood purification techniques.

Further work on subfractionation of uremic toxins by high performace liquid chromatography to clarify the issue is indeed worthwhile. In addition, the larger molecules contained in greater abundance in uremic sera should also be given more attention.

## REFERENCES

1. Bergström J., Asaba H., Fürst P., Gordon A., Quadracci L., Zimmerman L., : Middle molecules in uremia. In 6th International Congress of Nephrology, edited by Giovannetti S., Bonomini V., D'Amico G., SKarger, Basel 1976, p. 600.

2. Fürst P., Zimmerman L., Bergström J. : Determination of endogenous middle molecules in normal and uremic body fluids. Clin. Nephrol. 5,178. 1976.

3. Man NK., Uremic neurotoxin in the middle molecular weight range. Artif. Organs 4 : 116 1980.

4. Ringoir S.., Van Landschoot N., de Smet R.; Inhibition of phagocytosis by a middle molecular fraction from ultrafiltrate Clin. Nephrol. 13 : 109, 1980.

5. Dzurik R., Hupkova V., Valovicova E. : Metabolic actions of middle molecules. Artif. Organs 4(Suppl) : 59 1980.

6 Asaba H., Alvestrand A., Bergström J., Fürst P., Yahiel V., Uremic middle molecules in non-dialyzed azotemic patients : relation to symptoms and clinical biochemistries. Clin. Nephrol. 17 : 90 1982.

7. Gallice P., Fournier N., Crevat A., Briot M., Frayssinet R., Murisasco A. : Separation of one uremic middle molecules fraction by high performance liquid chromatography. Kidney Int. 23 : 764 1983.

8. Zimmerman L., Bergström J., Jornvall H. A method for separation of middle molecules by high performance liquid chromatography : application in studies of glucuronyl-o-hydroxyhippurate in normal and uremic subjects. Clin. Nephrol. 25 : 94, 1986

9. Wu ZG, Cai : ZH, Lu ZN, Liao LT, Zhu SQ : Determination of middle molecules in uremic patients. Chinesse Med. J. (Enghlish ed.) 98 : 115 1985.

10. de Wardener HE, Macgregor GA : The relation of a circulating sodium transport inhibitor (the natriuretic hormome?) to hypertension Medicine 62 : 310 1983.

# THE ABIKO FACTOR

Takashi Abiko

Kidney Research Laboratory
Kojinkai
Higashishichiban-cho 84
Sendai 980, Japan

Research on uremic toxins has now been going on for more than 150 years, but despite these efforts the true nature of uremic toxicity has not been fully elucidated yet.

Many substances are known to accumulate in the body fluids of patients with uremic symptoms.

However, none of the peptides have not been identified chemically, and their biological roles are quite unknown.

As you know in cases of uremic patients, variuos kinds of known proteins and peptides are accumulated in their body fluid. Consequently, there is a possibility of the appearance of unknown peptide fragments resulting from unknown enzymatic hydrolysis of proteins and peptides in the course of the metabolism, since the details of the metabolism of these proteins and peptides have not been established yet. Some of these unknown small peptides seem to have toxic activities to patients.

To confirm this hypothesis, we tried to isolate unknown peptides from filtrates and dialysated obtained from uremic patients.

We had continued to study about uremic toxins from 1978 to 1981. During the course of this study, five different kinds of peptides were isolated from filtrates and dialysates of uremic patients and their primary structures were elucidated by us.

(1) H-His-Gly-Lys-OH [1]

1. Patient selection

A uremic patient with following measurement values was selected for study: B P 165/115, BUN 80 mg/dl, creatinine 10.1 mg/dl and a 24 h creatinine of 5.6 ml/min with an urinary output of 600 ml.

2. Separation procedure

```
                    Filtrate (ECUM, 12 1 )
                      │ evapoarated in vacuo
      1.    Amicon Centriflo membrane DM-5
                      │ cut-off at approximately 5000 dalton
      2.    Sephadex G-25
                      │ column size 92.0 X 2.6 cm
                      │ eluate: 1% AcOH
      3.    Sephadex G-15
                      │ column size: 104.0 X 2.6 cm
                      │ eluate: 1% AcOH
      H-His-Gly-Lys-OH
```

Chart 1.  Purification of H-His-Gly-Lys-OH from ECUM filtrate

3. Sequence analysis
     For the sequence analysis of this peptide, manual Edman
degradation procedure was employed.
4. Synthesis of H-His-Gly-Lys-OH
     H-His-Gly-Lys-OH was synthesized by a solution method as
an authentic specimen for identification of the isolated peptide.
5. Physical constants of H-His-Gly-Lys-OH
     $[\alpha]_D^{20}$ + 6.1° (c= 0.5, $H_2O$), amino acid ratios in the AP-M
digest: His 0.89, Gly 0.92, Lys 0.93 (average recovery 78%).
$Rf^1$ 0.01, $Rf^2$ 0.14, single ninhydrin- and Pauly-positive spot.
This peptide was chromatographed on a filter paper, Toyo Roshi
N. 51, at room temperature. $Rf^1$ value refer to the Partridge
system and $Rf^2$ value refer to BuOH-pyridine-AcOH-$H_2O$ (30:20:6:24).
This peptide exhibited a sinlge spot on paper electrophoresis:
Toyo Roshi No. 51 (2 X40 cm), acetate buffer at pH 2.8, mobility
2.1 cm from the origin toward tha anode, after running at potential
gradiant of 60 V / 90 min.

(2) H-His-Pro-Ala-Glu-Asn-Gly-Lys-OH (corresponding to positions
13-19 of $\beta_2$-microglobulin)[2]

1. Patient selection
     A uremic patient with severe neuropathy characterized by
motor paralysis, areflexia, muscular atrophy was selected for
study.
2. Separation procedure
```
                        Filtrate (ECUM, 13 1)
                          │ evaporated in vacuo
          1. Amicon Centriflo membrane DM-5
                          │ cut-off at approximately
                          │ 5000 dalton
          2. Amicon Centriflo membrane UM-05
                          │ cut-off at approximately
                          │ 500 dalton
          3. Sephadex G-25
                          │ column size: 2.6 X 94.0 cm
                          │ eluate: 1% AcOH
          4. Sephadex G-15
                          │ column size: 2.6 X 96.0 cm
                          │ eluate: 1% AcOH
          5. CM-Sephadex C-25
                          │ column size: 2.6 X 96.0 cm
                          │ eluate: $NH_4OAc$ buffer (pH 6.50)
          H-His-Pro-Ala-Glu-Asn-Gly-Lys-OH
```

     Chart 2.  Purification of H-His-Pro-Ala-Glu-Asn-Gly-
               Lys-OH from filtrate

## 3. Sequence analysis

For the sequence analysis of the peptide, manual Edman degradation procedure was employed.

## 4. Synthesis of H-His-Pro-Ala-Glu-Asn-Gly-Lys-OH

The heptapeptide was synthesized by a solution method as authentic specimen for identification of the isolated peptide.

## 5. Physical constants of H-His-Pro-Ala-Glu-Asn-Gly-Lys-OH

$[\alpha]_D^{25}$ - 40.0° (c= 0.8, $H_2O$), $Rf^1$ 0.23, $Rf^2$ 0.33, single ninhydrin- and Pauly-positive spot. Amino acid ratios in the acid hydrolysate: His 0.89, Pro 0.91, Ala 0.91, Glu 0.86, Asp 0.99, Gly 1.01, Lys 1.02 (average recovery 77%). This peptide exhibited a single spot on a paper eletrophoresis: Toyo Roshi No. 51 (2 X 40 cm), acetate buffer at pH 2.8. Mobility 4.6 cm from the origin toward the anode, after running at potential gradiant of 60 V / cm for 120 min.

## (3) Syntheses and immunological activity of the heptapeptide and its analogs

### 1. Purpose of this study

It is well known that cellula immunity is suppressed in patients with chronic renal failure, although there is no certain information about inhibitory effect of uremic toxins on cell-mediated immunity. After primary structure of a peptide isolated from filtrate of a uremic patient was determined to be the heptapeptide, His-Pro-Ala-Glu-Asn-Gly-Lys-OH,[2] we tried to investigate immunological effects of this peptide and its two analogs, [Val3]- and [Gly2]-analog to normal T-cells.

### 2. Syntheses of [Val3]- and [Gly2]-analog

These two analogs were synthesized by a solution method.

### 3. E-rosette formation inhibition test

We investigated the immunological property of these three peptides against E-rosette formation between normal T-cells and sheep erythrocytes. Results of E-rosette formation inhibiting activities of these peptides are shown in Table I.

Table I. Inhibition Activity of E-Rosette Formation by the Heptapeptide and its Analogs

| Dose= mg/ml | H-Gly-Gly-His-OH[a] | His-Pro-Ala-Glu-Asn-Gly-Lys (%) | [Val3]-analog (%) | [Gly2]-analog (%) |
|---|---|---|---|---|
| 0 | 70.0 | 70.0 | 70.0 | 70.0 |
| 1 | 67.0 | 69.0 | 68.0 | 69.0 |
| 3 | 68.0 | 27.1 | 50.7 | 42.0 |
| 5 | 70.0 | 18.8 | 40.0 | 30.0 |

a) Control.

## 4. Results

The heptapeptide showed inhibitory activity on E-rosette formation. [Val³]-analog showed diminished immunological acitivity compared to native heptapeptide. This result indicates that the presence of an amino acid having a bulkier side chain, and which does not take α-helix structure like Val at position 3 induces lower potency. [Gly²]-analog also showed diminshed immunological activity comapred to native heptapeptide. This result also indicates that the difference on the structural feature between Gly and Pro in peptide induces lower potency.

### (4) H-Asp-Leu-Trp-Gln-Lys-OH (corresponding to positions 123-127 of β chain of fibrinogen)[4]

### 1. Patient selection

A uremic patient with following measurement values was selected for study: BUN 98 mg / dl, creatinine 11 mg / dl and a 24 h creatinine clearance of 5.6 ml / min with urenary output of 600 ml.

### 2. Separation procedure

```
                        Filtrate (ECUM, 12 1)
                          ↓   evaporated in vacuo
                1.      Amicon Centriflo membrane DM-5
                          ↓   cut-off at approximately
                          ↓   5000 dalton
                2.      Amicon Centriflo membrane UM-05
                          ↓   cut-off at approximately
                          ↓   500 dalton
                3.      Sephadex G-25
                          ↓   column size; 92.0 X 2.6 cm
                          ↓   eluate; 1% AcOH
                4.      Sephadex G-15
                          ↓   column size; 98.0 X 2.6 cm
                          ↓   eluate; 1% AcOH
                5.      Sephadex G-10
                          ↓   column size; 88.0 X 2.6 cm
                          ↓   eluate; 1% AcOH
                Trp-containing peptide
```

Chart 3. Purification of Trp-containing
pentapeptide from filtrate

### 3. Sequence analysis

For the sequence analysis of the peptide, manual Edman degradation procedure was employed.

### 4. Synthesis of H-Asp-Leu-Trp-Gln-Lys-OH

H-Asp-Leu-Trp-Gln-Lys-OH was synthesized by a solution method as an authentic specimen for identification of the isolated peptide.

### 5. Physical constants of H-Asp-Leu-Trp-Gln-Lys-OH

mp 132-142°C, $[\alpha]_D^{23}$ - 45.1° (c= 1.0, $H_2O$), $Rf^1$ 0.10, $Rf^2$ 0.12, single ninhydrin- and Ehrlich-positive spot. Amino acid ratios in the acid hydrolysate by 4 N methanesulfonic acid containing 0.02% 3-(2-aminoethyl) indole: Leu 0.98, Trp 0.89, Asp 1.02, Glu 0.96, Lys 0.88 (average recovery 80%). Amino acid ratios in the AP-M digest: Leu 1.00, Trp 0.89, Gln 0.87, Asp 0.93, Lys 0.95 (average recovery 79%). This peptide exhibited a single spot on a paper electrophoresis: Toyo Roshi No. 51 (2 X 40 cm), acetate buffer at pH 2.8. Mobility 1.7 cm from the origin toward the anode, after running at a potential gradient of 60 V / cm for 90 min, ninhydrin- and Ehrlich-positive spot.

6. E-rosette formation inhibition test
    This peptide exhibited the inhibition activity by amount
more than 1.0 mg / ml.


Table II. Inhibition activity of E-rosette formation by the
          pentapeptide

| Dose= mg/ml | H-Gly-Gly-His-OH (%) | H-Asp-Leu-Trp-Gln-Lys-OH (%) |
|---|---|---|
| 0 | 77 | 77 |
| 0.1 | 76 | 75 |
| 0.5 | 74 | 75 |
| 1.0 | 76 | 67 |
| 1.5 | 75 | 56 |
| 2.0 | 75 | 49 |

a) Control.


(5) H-Glu-Asp-Gly-OH[5]

1. Patient selection
    Dialysate was obtained from a patient with clinical
neuropathy during the first 2 h of HD trreatment.
2. Separation procedure

    Uremic dialysate (38 1)
1. Amicon Centriflo membrane DM-5
        cut-off approximately
        5000 daltons
2. Sephadex G-15
        column size: 2.6 X 96.0 cm
        elution medium: 1% AcOH
3. Sephadex G-10
        column size: 2.6 X 97.0 cm
        elution medium: 1% AcOH
4. DEAE-Sephadex A-25
        column size: 2.0 X 48.0 cm
        buffer: ammonium acetate
        buffer pH 6.50
5. H-Glu-Asp-Gly-OH

Chart 4. Purification of H-Glu-Asp-
         Gly-OH from Uremic Dialysate

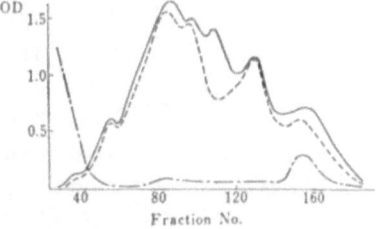

Fig. 1.  Elution Pattern on Sephadex G-15
(230 nm)

———: neuropathic subject.
..........: uremic subject (non-neuropathic).
—·—·—: normal subject.

    As shown in Fig. 1, clear differences were detected between
neurotoxic dialysate and nonneurotoxic dialysate. The fractions
corresponding to tube Nos. 106 through 116 were collected.
3. Sequence analysis
    For sequence analysis of the peptide, manual Edman degradation
procedure was employed.
4. Synthesis of H-Glu-Asp-Gly-OH
    The tripeptide was synthesized by a solution method as an
authentic specimen to confirm the identification of isolated peptide.
5. Physical constants of H-Glu-Asp-Gly-OH
    mp 127-131°C, $[\alpha]_D^{26}$ + 9.6° (c= 1.0 $H_2O$), $Rf^1$ 0.06, $Rf^2$ 0.12,
single ninhydrin-positive spot. Amino acid ratios in the acid
hydrolysate: Gly 1.00, Glu 0.91, Asp 0.94 (average recovery 82%).
Amino acid ratios in the AP-M digest: Gly 1.00, Glu 0.89, Asp
0.95 (average recovery 83%).

6. Inhibitory activity of the tripeptide on LDH
   This peptide inhibited LDH activity as shown in Table III.

Table III. Inhibitory Effect of H-Glu-Asp-Gly-OH on LDH Activity

| Dose= mM | H-Gly-Gly-His-OH[a) (%) | H-Glu-Asp-Gly-OH (%) |
|---|---|---|
| 0.03 | 0 | 5 |
| 0.17 | 0 | 21 |
| 0.33 | 0 | 53 |
| 1.00 | 0 | 100 |

a) Control.

(6) H-Thr-Phe-Gly-Gln-Gly-Thr-Lys-OH[6)

1. Patient selection
   Filtrate by ECUM was obtained from a uremic patient with cell-mediated immunodeficiency.
2. Separation procedure

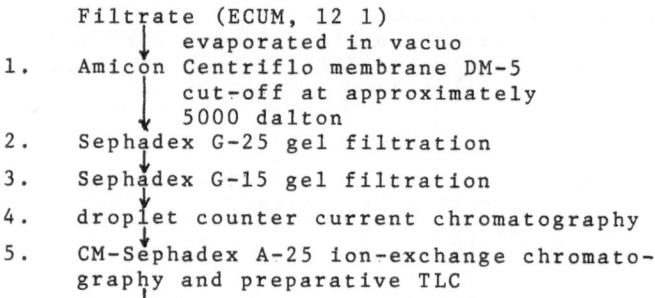

Filtrate (ECUM, 12 1)
   ↓ evaporated in vacuo
1.  Amicon Centriflo membrane DM-5
   ↓ cut-off at approximately
   5000 dalton
2.  Sephadex G-25 gel filtration
3.  Sephadex G-15 gel filtration
4.  droplet counter current chromatography
5.  CM-Sephadex A-25 ion-exchange chromatography and preparative TLC
H-Thr-Phe-Gly-Gln-Gly-Thr-Lys-OH

Chart 5. Purification of the heptapeptide
Thr-Phe-Gly-Gln-Gly-Thr-Lys from hemodialysate

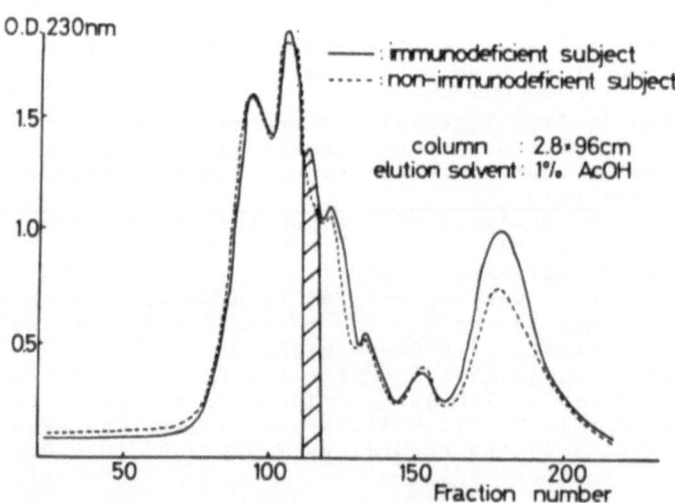

Fig. 2. Elution Pattern on Sephadex G-25

The fractions corresponding to tubes 112 through 115 were present in the filtrate from an immunodeficient patient but were not detectable in nonimmunodeficient filtrate.

3. Sequence analysis

The N-terminal analysis: The N-terminal amino acid of the peptide was determined by the 1-fluoro-2,4-dinitrofluorobenzene method of Sanger as modified by Fraenkel-Conrat et al.

The C-terminal amino acid analysis. The C-terminal amino acid of the peptide was analyzed using carboxypeptidase A according to the procedure of Fraenkel-Contrat et al.

The sequence analysis of the peptide was performed according to the subtractive Edman degradation method.

4. Synthesis of H-Thr-Phe-Gly-Gln-Gly-Thr-Lys-OH

The heptapeptide was synthesized by a solution method as an authentic specimen to confirm the identification of the isolated peptide.

5. Physical constants of H-Thr-Phe-Gly-Gln-Gly-Thr-Lys-OH

mp 127-137°C, $[\alpha]_D^{26}$ - 17.5° (c= 0.3, $H_2O$), $Rf^1$ 0.03, $Rf^2$ 0.20, single ninhydrin- and chlorine-tolidine-positive spot. Amino acid ratios in the acid hydrolysate: Lys 0.83, Glu 0.84, Thr 1.79, Phe 1.00, Gly 2.13 (recovery of Phe 81%). Amino acid ratios in the AP-M digest: Lys 0.84, Gln+Thr 2.69 (Calcd. Gln+Thr as Thr), Phe 1.00, Gly 2.09 (recovery of Phe 80%) (Gln emerged at the same postion as Thr and was calculated as Thr).

6. Inhibitory activity of the heptapeptide on PHA-induced lymphocyte transformation

This peptide inhibited lymphocyte transformation by PHA in a concentration of 100 μg / ml.

Table IV. Influence of the Heptapeptide on the Stimulation of Normal Lymphocytes by PHA

| Concentration of the heptapeptide (μg/ml) | [3H]Thymidine incorporation (cpm) | Concentration of glucagon fragment (1-14)[a] (μg/ml) | [3H] Thymidine incorporation (cpm) |
|---|---|---|---|
| 0 | 49635+1748 | 0 | 46784+1541 |
| 10 | 47428+1536 | 10 | 47289+1642 |
| 100 | 43362+1628 | 100 | 45971+1728 |
| 1000 | 32984+1615 | 1000 | 48362+1489 |

a) Control.

(7) Conclusion

Five different kinds of peptides have been isolated from filtrates and dialysates of uremic patients. Amino acid sequences of them were determined.

Among them, it was found that H-His-Pro-Ala-Glu-Asn-Gly-Lys-OH equals to heptapeptide moiety corresponding to position 13 through 19 of $\beta_2$-microglubulin and H-Asp-Leu-Trp-Gln-Lys-OH equals to pentapeptide moiety corresponding to postion 123 through 127 of pentapeptide of β chain of fibrinogen.

These results seem to suggest that there is a posibility of appearance of unkown peptide fragments resulting from unknown enzymatic hydrolysis of accumulated proteins and peptides in uremic patients.

Some of these unknown peptides may have some toxic activities to patients.

## References

1. T. Abiko, M. Kumikawa, M. Ishizaki, H. Takahashi and H. Sekino, Biochem. Biophys. Res. Commun., 83: 357 (1978).
2. T. Abiko, M. Kumikawa, H. Higuchi and H. Sekino, Biochem. Biophys. Res. Commun., 84: 184 (1978).
3. T. Abiko, M. Kumikawa and H. Sekino, Biochem. Biophys. Res. Commun., 86: 945 (1979).
4. T. Abiko, I. Onodera and H. Sekino, Biochem. Biophys. Res. Commun., 89: 813 (1979).
5. T. Abiko, I. Onodera and H. Sekino, Chem. Pharm. Bull., 28: 1629 (1980).
6. T. Abiko, I. Onodera and H. Sekino, J. Appl. Biochem., 3: 562 (1981).

# POLYAMINES AND UREMIA

Robert A. Campbell

Department of Pediatrics
Oregon Health Sciences University
Portland, Oregon 97201 USA

## INTRODUCTION

In 1678, Antony Van Leeuwenhoek described the deposition of colorless crystals, later to be identified as spermine (SPM) phosphate, in seminal fluid. SPM is one of the aliphatic cations called polyamines (PAs) found in all living things. (See Table 1) The diamine putrescine (PTC) proved to be an essential growth factor for mutant Hemophilus parainfluenza. In mammals, increased biosynthesis of PAs, including spermidine (SPD), was detected during both normal and pathological tissue growth. Increased amount of SPM, SPD, and PTC were found in blood and urine under these circumstances. The nature and regulatory functions of PAs and the enzymes controlling their production and metabolic degradation are beginning to be understood.

## POLYAMINE SYNTHESIS

PAs are derived from the amino acids ornithine and methionine. PTC is formed from ornithine by the enzyme ornithine decarboxylase (ODC). ODC is one of two regulating enzymes for the PA pathways. ODC is vitamin $B_6$ dependent. In appropriate target cells, ODC is activated by all known hormones, several mitogens, drugs, certain amino acids, infecting viruses, osmotic changes and cell membrane injury. A few seconds after plasma membrane stimulation, a rise in ODC activity and cellular PA concentrations occur. These changes are necessary for the rapid receptor and voltage regulated activating calcium fluxes.[1] The half-life of ODC is only a few minutes. Inactivation may be carried out by a PA-induced regulating protein called antizyme.[2] However, sustained ODC biosynthetic activity is seen in cells committed to growth and division. (See Figure 1)

SPD and SPM are formed from PTC through enzymatic additions of

Table 1:  The Polyamines

| Compound | M.W. | Formula |
|----------|------|---------|
| Putrescine | 88.0 | $NH_2(CH_2)_4NH_2$ |
| Cadaverine | 102.0 | $NH_2(CH_2)_5NH_2$ |
| Spermidine | 145.2 | $NH_2(CH_2)_3NH(CH_2)_4NH_2$ |
| Spermine | 202.0 | $NH_2(CH_2)_3NH(CH_2)_4NH(CH_2)_3NH_2$ |

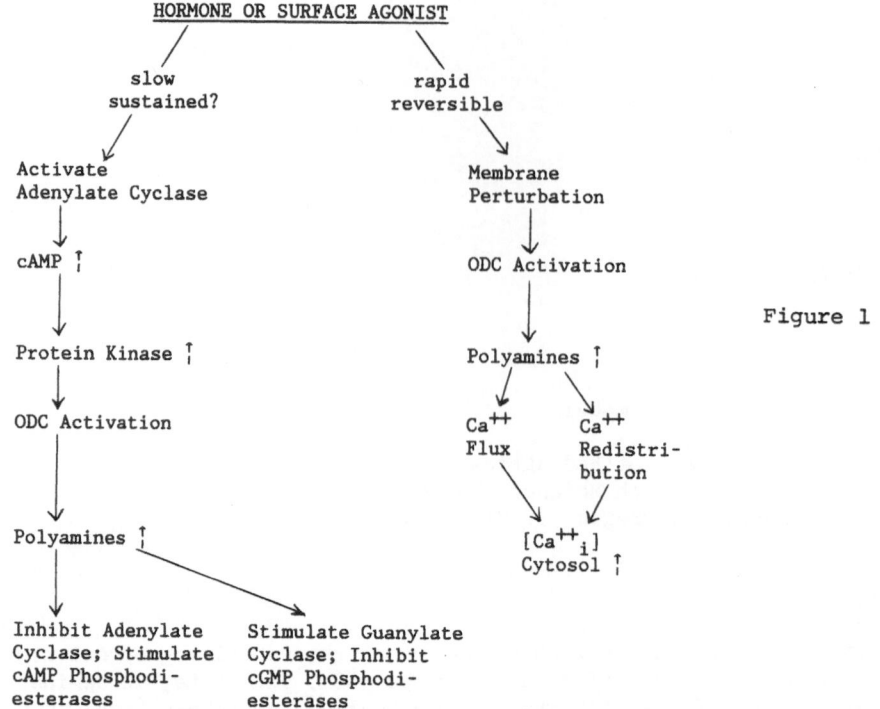

HORMONE OR SURFACE AGONIST

slow
sustained?

rapid
reversible

Activate
Adenylate Cyclase

Membrane
Perturbation

cAMP ↑

ODC Activation

Figure 1

Protein Kinase ↑

Polyamines ↑

ODC Activation

$Ca^{++}$ Flux    $Ca^{++}$ Redistri-bution

Polyamines ↑

$[Ca^{++}_i]$ Cytosol ↑

Inhibit Adenylate Cyclase; Stimulate cAMP Phosphodi-esterases

Stimulate Guanylate Cyclase; Inhibit cGMP Phosphodi-esterases

aminopropyl moieties derived from methionine. PAs participate in stimulating RNA and DNA synthesis and facilitate transcription and translation
by several mechanisms. In vivo and in vitro studies demonstrate that PAs
participate in either the stimulation or inhibition of a variety of
protein kinases. Tight control of intracellular PA concentrations is
maintained for efficient protein synthesis and fidelity of amino acid
sequences.[3] Studies using specific enzyme inhibitors of ODC, such as
alpha-difluoromethylornithine (DFMO), demonstrate the essential nature of
PAs in new product formation, cell differentiation, and proliferation.

POLYAMINE MEASUREMENTS

The choice of measurement of PAs is confounded by the very nature of
PAs, low concentrations, the several PA congeners and body compartmentali-
zation. In addition to free forms, PAs may be ionically complexed or
covalently bound to proteins, peptides and vitamin $B_6$ (Pyridoxal 5'-
phosphate). PAs bind avidly to glass. Up to 50% loss of labeled PAs
stored in glass containers can occur in one hour. Red blood cell (RBC)
hemolysis during sample collection increases plasma and serum values.
Prior to detection, various plasma or serum preparations, including
combinations of acid extraction, hydrolysis and predetection column clean
up, produce quite different results. Because of high PA levels in RBCs,
extracted RBCs have been employed where detection sensitivity may be
limited. However, since PAs accumulate in RBCs in unconjugated form,
dansylation of RBC PAs followed by HPLC and fluorescent detection has
proven to be a very sensitive measure magnifying SPM and SPD changes in
the extracellular space.
Problems may arise during mild PA isolation procedures. PAs resist
dialysis from plasma, serum or tissue extracts. They survive in 10,000 -
30,000 MW fractions during ultrafiltration. They may distribute with
higher MW fractions in resolving and desalting chromatographic procedures.
SPM (202 daltons) contained in tissue extracts or by itself may anomol-
ously chromatograph (Sephadex G-10, G-15, G-25, Bio-Gel P-6) giving a
relative MW of 1400 daltons. Trace amounts may be retarded or lost.[4]

These laboratory observations may be of importance in the in vitro evaluation of biological activity of partially purified samples as they clearly fall in the middle molecule range.

Normal mean serum and plasma levels, depending upon the method used, are in the low nmol/ml range. Packed RBC PA levels may be one to two orders of magnitude higher than those of plasma. Tissue levels vary reflecting the degree of metabolic activity. Pancreatic tissue PA values range from one to two umol/g wet weight whereas skeletal muscle PA values may be one order of magnitude lower. Total daily normal PA excretion in the urine is only about 4-10 mg. Increased urinary PAs occur in a number of conditions including pregnancy, cancer and some genetic disorders of tubular transport. Acetylated PTC and SPD are the predominant urinary forms although a small fraction of unacetylated SPM is present.

In chronic renal failure (CRF) urinary PA excretion is sharply curtailed. Significant elevations of serum PAs begin to be seen when the serum creatinine reaches 6 mg%. In CRF and hemodialysis patients, serum and RBC PAs may be increased two to five times compared to controls. Serum SPD values, in particular, may be increased from four to ten times mean control values. Routine hemodialysis does not reduce serum PA levels to the upper limits of normal. Little reduction occurs in RBCs. A major portion of uremic serum SPD appears to be conjugated, probably by the liver. It is of interest that a basic SPD peptide (Lutz; SPD-peptide) with insulin binding properties has been isolated and characterized in uremic serum. The effects of peritoneal dialysis on serum or tissue PA levels is not known. Finally, in uremic rats SPD concentrations were determined in the following tissues and found to be increased: liver (25%), skeletal muscle (17%) and RBCs (68%).

Based on present knowledge, the liver and kidney share major responsibility for preventing a rapid accumulation of PAs. In addition, total RBC mass has a large PA storage capacity and may serve as a buffer resevoir. In chronic hyperpolyaminemia, oxidative degradation of PAs, based on observed increases of extracellular diamine and polyamine oxidase activities may be substantial. When there are sustained elevations of PAs, as in pregnancy and uremia, plasma diamine and PA oxidase activities gradually increase. In addition, the readily inducible intracellular enzyme SPD/SPM $N^1$-acetyltransferase (SAT) plays a key role in regulating the catabolic limb of intracellular PA oxidation. The acetylated PA, $N^1$-SPD is preferentially metabolized by intracellular PA oxidase. Thus, an enzymatic mechanism is available to reduce concentrations of accumulated SPD. While most tissues contain PA acetylating and oxidizing enzymes, either compromised liver or renal function is associated with significant hyperpolyaminemia.


EXPERIMENTAL EFFECTS OF PA EXCESS

The effects of induced PA excess are qualitatively quite similar to those observed in clinical and experimental uremia. The most prominent involvement is the nervous system. SPM decreased in vitro brain oxygen consumption. In humans intramuscular injection of SPM was followed by persistent vomiting. In acute animal studies, depending on the PA, dose, and route of administration used, adipsia, ataxia, seizures, hind leg paralysis, hypothermia, coma, and death were observed. Parenteral PAs induced release of histamine and catecholamines. Low doses of SPM have a diuretic effect on the kidney. Higher doses produce lesions typical of non-specific acute renal failure. Toxic effects of SPM and SPD on heart muscle include depression of function and myocytolysis. In the lung, PAs rapidly induce pulmonary edema. With respect to anemia, SPM in culture media at concentrations comparable to those seen in uremic serum inhibited development of erythroid colony forming units.[5] The inhibitory effects of either uremic serum or SPM were abolished with the addition of specific anti-polyamine antisera.  In vivo SPM induced anemia has been accomplished using implanted micro-infusion pumps.  PAs can also reversibly inhibit lymphocyte and neutrophil functions (see Immunity).  Most animal studies evaluating PA toxicity have involved acute bolus PA injections.  On this basis, SPM has been shown to be about twenty times as nephrotoxic as SPD, the diamines being relatively non-toxic.  Prolonged administration of low dose SPM also appears to have some proliferative effects (see Mesangial Expansion) without ephrotoxicity.  With respect to potentially adverse effects of PA excess at the biochemical level, PAs are capable of inhibiting $Na^+$-$K^+$ ATPase activity in vitro in several tissues as well as blocking phospholipase $A_2$ and C enzyme activities with membranous substrates.  PAs in concentrations found intracellularly partially inhibit respiration in isolated mitochondria.


CLINICAL PA EXCESS

Several abnormalities are shared by hyperpolyaminemic patients having either uremia or cancer.  Anorexia, weight loss and tissue wasting are common features.  Glucose intolerance and insulin resistance have been demonstrated. The anemias are normocytic normochromic with shortened RBC survival time.  Bone marrow profiles are similar and burr cell formation may be seen in peripheral blood.  Decreased RBC $NA^+$-$K^+$-ATPase activity in the presence of normal ATP levels has been described for both cancer and uremia patients.  Serum inhibitors of mitogenically induced lymphocyte transformation are present. Peripheral neuropathy, disorders of mentation and elevated brain aluminum levels have all been reported. Elevation of aluminum has also been reported in advanced hepatic disease, a hyperpolyaminemic state.  Blood brain barrier (BBBr) abnormalities may be present in these disorders.  In experimental injury to brain microvasculature, inhibition of the PA biosynthetic pathway with DFMO prevents an increase in BBBr permeability.  Thus, the possibility must be entertained that chronic extracellular PA elevations may play a role in BBBr changes. Pyridoxal 5'-phosphate insufficiency is a frequently shared abnormality in all three disorders.  Terminally ill patients have extremely low brain activity of glutamic acid decarboxylase, a $B_6$ dependent enzyme.  Despite clinical similarities, renal function is, by and large, normal in most cancer and liver disease patients.

Other observations may be helpful in suggesting a role for PAs in uremic dysmetabolism. Burn patients may have serum PA levels 4 to 6 times normal. Burn anemia is charaterized by burr cell formation, depressed heme synthesis and shortened RBC survival time. Like uremic RBCs, RBCs from burn patients transfused into normal individuals have a normal survival time. Burn sera inhibits lymphocyte transformation. Burn patients have been shown, as have advanced cancer and uremic patients, to have increased RBC sodium concentrations and depressed $Na^+$-$K^+$ ATPase activity.

Additional clues as to possible mechanisms for uremic glucose intolerance and insulin resistance may also come from non-renal sources. Late pregnancy, a time of rapid increase in blood SPM and SPD levels, is associated with increasing glucose intolerance and insulin resistance. Patients with cirrhosis have similar metabolic lesions.  The administration of growth hormone to animals is a potent stimulus of ODC activity and increased PA production.  Elevated plasma PA levels and increased urinary excretion follows growth hormone administration in humans, reflecting active tissue biosynthesis.  Patients with acromegaly have glucose intolerance, insulin resistance and elevated urinary polyamines.  Prolactin has been recognized as a diabetogenic hormone.  Patients with microprolactinomas are glucose intolerant and insulin resistant.  In animals prolactin stimulates ODC activity in the cells of many tissues.  Both insulin dependent and adult onset diabetes patients have marked increase in total urinary PA excretion. Obesity due to excessive food intake is associated with glucose intolerance and insulin resistance but only so long as excessive food intake is maintained. Increases in PA biosynthetic enzyme activity in many tissues follows the feeding of amino acid mixtures or protein meals to starved experimental animals. Biochemical studies indicating PAs may exert a regulatory role on the glycolytic pathway exist, but are few in number. The effects of chronic PA excess on either glycolysis or gluconeogenesis have not been established. It should be pointed out, however, that sera from patients with cirrhosis, cancer, adult onset diabetes mellitus and uremia all inhibit in vitro phosphofructokinase enzyme activity. While it is unlikely the shared abnormalities of such dissimilar disorders are fortuitous, systematic investigations of the effects of PA excess at all biological levels are needed.

## IMMUNITY

Normal lymphocytes or those from uremic patients have, in the presence of uremic serum, decreased responses to mitogenic stimulation. This is reversed when the cells are washed, resuspended in normal serum and re-stimulated. Other substances reported to be elevated in uremic plasma which can inhibit lymphocyte transformation include methylguanidine, thymosine and cAMP. Levels required are, however, pharmacological. Vitamin $B_6$ insufficiency associated with uremia may contribute to blunted lymphocyte responses. SPD and SPM also inhibit normal lymphocyte responses to mitogenic stimulation if they are enzymatically oxidized to reactive aldehydes. Fetal calf serum, which contains diamine oxidase activity, is necessary to induce the in vitro inhibition of mitogenic stimulation of normal lymphocytes by PAs.[6] Uremic plasma contains both elevated levels of PAs and increased diamine oxidase activity. Likewise during pregnancy, PA levels and PA oxidase enzyme activities increase over time. Lymphocyte transformation can be inhibited by pregnancy serum and the addition of PAs to maternal serum increases its inhibitory potency. Other hyperpolyaminemic states such as hepatic insufficiency, burns and cancer show similar plasma inhibitory activity. With the exception of patients with hepatorenal syndrome, renal function is essentially normal in these patient groups.

Uremic and burn sera contain substances which alter neutrophil behavior. Migration into skin windows is decreased. Responses to chemotaxis in the presence of uremic serum are depressed. However, washed neutrophils from uremic patients show normal chemotactic responses in normal serum. In the presence of either bovine serum (containing diamine oxidase activity) or partially purified PA oxidase activity, SPM and SPD inhibited human neutrophil locomotion. In the presence of human serum albumin, PAs did not.[7] The reversibility of inhibition with washing was similar to that of lymphocytes.

## DOWN REGULATION

Inhibition of cellular responses to elevated circulating hormones is characteristic of uremia. Several PA-related mechanisms promote this unresponsiveness. (See Figure 2) In uremia PAs and their congeners are present in excess in both the extra- and intra-cellular spaces and decreased PA biosynthetic activity is taking place. Antizyme, the endogenous protein inhibitor of ODC activity, could be present due to the sustained cellular PA accumulation of chronic uremia. An uncharacterized cytosolic ODC inhibitor has been demonstrated in the kidney tissue of uremic rats. A second possible inhibitory mechanism involves the documented decrease in cAMP messenger system responses following intracellular PA accumulation in cells.[8] Similar messenger system changes have been reported in uremia. A third mechanism, whereby cellular PA excess might interfere with cellular responses to hormones, involves early PA cell activation. The relatively non-specific membrane receptor activation which stimulates an acute rise in ODC activity and a concurrent rapid but brief rise in PA levels permits transmembrane $Ca^{++}$ fluxes necessary for mediating $Ca^{++}$ dependent cell responses.[1] Excess intracellular PAs might permit an abnormal accumulation and redistribution of intracellular calcium. Excess tissue $Ca^{++}$ is commonly seen in uremia. A fourth possibility rests on the observation that PAs in excess can induce intracellular myeloid body formation in

CHRONIC POLYAMINE EXCESS AND
TARGET CELL HYPORESPONSIVENESS

Possible Adverse Effects:

1) Altered cAMP messenger system activity.
2) Inhibition of receptor recycling by plasma membrane sequestration.
3) Disturbed PA regulation of extracellular $Ca^{++}$ entry and intracellular $Ca^{++}$ redistribution.
4) Altered PA stimulus and inhibition of specific protein kinases.
5) Altered PA concentration-dependent efficiency of protein synthesis with possible misreading.
6) Excess extracellular PA aldehyde generation depressing cell responses to activating stimuli.

heterophagic vacuoles and lysosomes, sequestering quantities of plasma membrane and associated receptors which presumably would otherwise undergo recycling.[9] Finally, it has been shown in vitro that either PA excess or deficiency decreases efficiency of protein synthesis and increases misreading.[2]

MESANGIAL EXPANSION

Renal hypertrophy of remnant tissues and hyperfiltration are seen in CRF. Mesangial expansion leading to ischemia results in obliterative glomerulosclerosis. An undefined humoral factor(s) promotes intrarenal vascular changes and stimulates mesangial cells (MC) to proliferate and to lay down matrix.

Serum obtained from hemodialysis patients stimulates in vitro proliferation of arterial smooth muscle cells (ASMC).[10] When passed through a silica gel column to remove free PAs, proliferative responses are diminished. SPM addition to culture media restored the responses in a dose dependent manner. MCs and ASMCs share common embryonic origin, similar contractile structures, and respond similarly to pharmacological agents such as angiotensin II and vasopressin. MCs might be responding to the PA elevation which occurs in hemodialysis. In renal reduction models, high dietary protein has been shown to accelerate loss of glomeruli. Dietary protein or amino acid mixtures stimulate PA biosynthetic enzyme activities in various body tissues in normal rats.

In a preliminary study of PA effects on glomerular morphology, BalbC mice were treated with low dose i.p. SPM (15 mg/kg/day) for 21 days. No cytotoxic effects were noted in glomeruli or tubules at termination. Changes were noted, however, by morphometric analysis in both the outer cortical (OC) and juxtamedullary (JXT) glomeruli. Comparison of treated and normal JXT glomeruli revealed the glomerular area increased 12%, the mesangial area 69% and the capillary area 28%. The mesangial/glomerular area ratio increased 47%. Glomerular nuclear counts/unit area of glomeruli were unchanged.  SPM produced more pronounced hypertrophy and hyperplasia of JXT than OC glomeruli and the increase in mesangium was disproportionately large.  The greater capillary area was consistent with the hyperfiltration phase of renal insufficiency.  Further studies, extending duration of treatment and employing various combinations of PAs are necessary to reaffirm the correctness of these orginal observations and to determine if PAs can produce the obliterative lesions of end stage renal disease.  It is clear they cause renal hypertrophy.[11]

SUMMARY

PAs are in' racellular regulators of growth and anabolic processes. Toxic properties of PAs are conferred according to the increasing number of cationic charges.  In uremia, PA accumulation occurs both in and outside the cell.  Decreased PA synthesis and blunted PA pathway responses suggest PAs may participate in cellular down regulation.  This would explain the lack of tissue responses to elevated plasma hormones, a feature of uremia.  Such homeostatic control could prevent life-threatening PA toxicity due to an imbalance between production, degradation and excretion.  The concurrent rise in plasma PA oxidative activity, while adversely influencing behavior of lymphocytes and neutrophils reduces the likelihood of direct PA toxicity.  In addition, preliminary observation of increased glomerular size and capillary area of intact kidneys in PA treated mice suggest there could be a PA dependent adaptive enhancement of renal excretion in circumstances of decreased renal mass due to disease, thus, compensatory renal hypertrophy.  It is important to further inspect the non-uremic chronic hyperpolyaminemias as they occur in man.  The continued search for parallel adverse systemic and local effects will be useful in strengthening the base for understanding homeostatic consequences of PA accumulation unobserved by the complex mixture of uremic metabolites.

In conclusion, we have examined some known regulatory and toxic properties of PAs and related these to features of the uremic syndrome. Available information on the various ramifications of PA dysmetabolism in uremia, as with other suspect toxins, is in part circumscribed and indirect.  Caution is to be exercised in evaluating the new and provisional PA related cause and effect relationships suggested here.  Nevertheless, focusing attention on these relationships should stimulate fresh and productive inquiry into the etiopathogenesis of CRF, uremia and dialysis complications.  Insights in to mechanisms of renal growth and compensatory hypertrophy are valuable secondary products of our investigation.

Thanks are in order to J. B. Russi, M. C. Campbell, J. Buskirk, and S. E. Davis for help in manuscript preparation.

Recognition of the fundamental support of the Pediatric Renal Metabolic Laboratory program for Polyamine Research by the Chiles Foundation is made with genuine personal satisfaction.

REFERENCES

1.  Koenig, H., Goldstone, A. D., Iqbal, Z., Fan, C.-C., Lu, C. Y., Trout, J.J., Rapid polyamine synthesis plays a key role in the mediation of receptor- and voltage-regulated calcium fluxes. In Recent Progress in Polyamine Research, L. Selmeci, et al, Eds, Akademiai Kiado, Budapest, 1985, 191-201.

2.  Canellakis, E. S., Heller, J. S., Kyriakidis, D. A., The interation of ornithine decarboxylase with its antizyme. In *Advances in Polyamine Research, Vol. 3*, CM Caldarera et al Eds, Raven Press, New York, 1981, 1-13.

3.  Abraham, A. K., Olsnes, S., Pihl, A., Fidelity of protein synthesis *in vitro* is increased in the presence of spermidine. *FEBS Lett*, 101:93-96, 1979.

4.  Patt, L. M., Barrantes, D. M., Gleisner, J. M., Houck, J. C., Abnormal behavior of polyamines on gel filtration: a cautionary note. *Cell Biology Int Reports*, 5(8):798-803, 1981.

5.  Radtke, H. W., Rege, A. B., LaMarche, M. B., Bartos, D., Bartos, F., Campbell, R. A., Identification of spermine as an inhibitor of erythropoiesis in patients with chronic renal failure. *J Clin Invest*, 67:1623-1629, 1981.

6.  Byrd, W. J., Jacobs, D. M., Amoss, M. S., Influence of synthetic-polyamines on *in vitro* responses of immunocompetent cells. In *Advances in Polyamine Research*, Vol 2, R.A. Campbell, et al, Eds, Raven Press, New York, 1978, 71-83.

7.  Ferrante, A., Inhibition of human neutrophil locomotion by the polyamine oxidase-polyamine system. *Immunology* 54:785-790, 1985.

8.  Caldarera, C. M., Tantini, B., Marmiroli, S., Pignatti, C., Clô, C., Polyamine control of cellular cycle nucleotide response. In *Recent Progress in Polyamine Research*, L. Selmeci, et al, Eds, Akademiai Kiado, Budapest, 1985, 109-118.

9.  Campbell, R. A., LaBerge, T., Campbell-Boswell, M., Brooks, R. E., Talwalkar, Y. B., Myeloid body formation under conditions of polyamine stress. In *Advances in Polyamine Research*, Vol 4, U. Bachrach, et al, Eds, Raven Press, New York, 1983, 107-125.

10. Bagdade, J., Campbell, R., Grettie, D., Bartos, D., Bartos, F., Effects of polyamines on human arterial smooth muscle cells in tissue culture. In *Advances in Polyamine Research*, Vol 2, RA Campbell, et al, Eds, Raven Press, New York, 1978, 345-349.

11. Campbell, R. A. Unpublished observations, 1986.

ROLES OF HIPPURATE AND INDOXYL SULFATE IN THE

IMPAIRED LIGAND BINDING BY AZOTEMIC PLASMA

Paul F. Gulyassy, Elizabeth Jarrard and Linda Stanfel

University of California
UCD Professional Building
4301 X Street
Sacramento, California  95817

## INTRODUCTION

Despite extensive efforts by many investigators the chemical basis of most uremic disorders is not known.  We have been studying a circumscribed disorder, impaired plasma protein binding of acidic drugs and endogenous metabolites, hoping to discover investigative approaches which might be applicable to study of more complex uremic disorders.  Our initial interest in the problem arose from our finding of markedly reduced levels of plasma tryptophan in patients with renal failure, before and after long-term dialysis treatment.  This change appeared to be due to impaired binding of tryptophan by azotemic plasma (1).  The initial hypothesis proposed by others that binding is impaired because plasma albumin is per se abnormal has little support.  Substantial evidence indicates that retained uremic solutes, which bind to plasma proteins, account for this defect (2).  Two aromatic anions, hippurate and indoxyl sulfate, accumulate to substantial levels in uremia and are known to be moderate to strong displacers of several albumin-bound model probes in vitro (3-5).  In addition, Lichten-walner et al have implied that o-hydroxyhippurate may be an important binding inhibitor in azotemic plasma (6).  To evaluate the relative importance of these three aromatic anions  as binding inhibitors, we have developed highly sensitive and specific HPLC methods to measure their concentrations in plasma of patients with a wide range of renal failure.  We have also determined the binding of $^{14}$C-salicylate by normal plasma, azotemic plasma and normal plasma "spiked" with a wide range of hippurate and indoxyl sulfate and correlated the concentrations of these ligands with salicylate binding.  Salicylate was used as a probe because it binds to both of the major binding sites on albumin for aromatic anions.  Finally, we have determined the effect of wide variations in concentrations of pH, calcium and chloride on binding.

## MATERIALS AND METHODS

After obtaining informed consent, heparinized blood was drawn from normal subjects, undialyzed patients with a wide range of renal function and patients receiving regular dialysis treatment.  To evaluate the effects of calcium and chloride on binding we made separate pools of plasma samples from normal subjects and patients with severe hypochloremia or hypocalcemia.

Normal plasma pools were diluted to the same albumin concentration as the azotemic plasma pools with a buffer solution containing 25 mmol $NaHCO_3$ and 110 mmol NaCl per liter. Concentrated stock solutions of $CaCl_2$ or NaCl were added to achieve the wide ranges of these ions seen in clinical circumstances. To test for pH effects (at constant sodium concentration) a range of pH from 6.0 to 8.5 was achieved by addition of concentrated solutions of NaOH, HCl or NaCl.

Hippurate, indoxyl sulfate, p-hydroxyphenylacetate and o-hydroxy-hippurate concentrations were measured as previously described (5,7,8). Concentrated stock solutions of hippurate, indoxyl sulfate or both were added to normal plasma samples to achieve a range of concentrations as found in azotemic plasma. Binding of 6.3 uM $^{14}C$-salicylate was determined by centrifugal ultrafiltration as described (3). Albumin, urea nitrogen, creatinine, chloride, salicylate and calcium were measured by standard methods.

RESULTS

The ranges of BUN and creatinine in the plasma of undialyzed patients with chronic renal failure were 40 to 305 and 3.3 to 42 mg/dl. Over the ranges of 74 to 112 meq/L of chloride, 5.8 to 15.1 mg/dl of calcium and pH 6.5 to 8.5, deviations from normal of these ions in the direction seen in renal failure produced changes in salicylate binding in both normal and azotemic plasma which were either very slight or opposite in direction (increased) to that seen in renal failure. Binding of salicylate was reduced in all the azotemic plasma samples (n = 28), being 61.1 to 97.4% of the binding by the control normal plasma of equal albumin concentration. There was a weak, negative correlation in the azotemic plasma samples between salicylate bound (per cent of control, y) and BUN (mg/dl, x), which was of borderline significance: $y = 92.9 - .076x$, $r = -0.44$, $0.1 > p > 0.05$ but a significant correlation with creatinine: $y = 94.9 - 0.70 x$, $r = -0.69$, $p < 0.01$.

Hippurate and indoxyl sulfate were barely detectible in normal plasma (<10 uM). Increase in hippurate concentration of normal plasma by 25 to 1000 umoles/L progressively but only slightly reduced binding to 99.4% of control at 100 uM and 95.2% at 1000 uM. Addition of 50 to 300 uM indoxyl sulfate also progressively but weakly reduced binding to 99.4% of control at 50 uM and to 96.8% at 300 uM.

The concentrations of both hippurate and indoxyl sulfate were elevated in both undialyzed and regularly dialyzed patient groups, ranging from 18 to 883 uM (mean of 279) and 45 to 202 uM (mean of 137) respectively. There was a weak, non-significant negative correlation between salicylate binding and level of hippurate ($r = -0.34$, $p > 0.1$) but a significant correlation with plasma indoxyl sulfate ($r = -0.73$, $p < 0.001$).

The concentration of o-hydroxyhippurate in twelve of the azotemic samples was <2 uM, the lower level of reliable detection. In nine samples with measurable amounts, there was a very wide range from 8 to 467 uM. Among the latter there was a positive correlation between the concentrations of o-hydroxyhippurate and salicylate.

Finally, we tested the effect of hippurate and indoxyl sulfate combined on salicylate binding. Nine pairs of normal and azotemic plasma samples were studied, each pair of which was made to contain equal concentrations of albumin, hippurate and indoxyl sulfate. The difference from the normal control plasma binding was 17.9 ± 7.6% (S.D.) for the patient samples and 3.2 ± 1.1% for the modified normal plasma (Table 1). Thus these two aromatic anions accounted for only 18% of the binding defect of azotemic plasma.

Table 1. Effects of hippurate and indoxyl sulfate on binding of salicylate by normal and azotemic plasma pairs.

| Pair | Hippurate* (μmol/L) | Indoxyl Sulfate* (μmol/L) | Salicylate Bound (% of control) Normal | Salicylate Bound (% of control) Azotemic+ |
|------|------------|-----------------|--------|----------|
| 1 | 322 | 45 | 97.7 | 87.3 |
| 2 | 118 | 187 | 97.2 | 84.1 |
| 3 | 243 | 114 | 97.1 | 77.4 |
| 4 | 376 | 121 | 96.5 | 86.9 |
| 5 | 107 | 191 | 97.4 | 83.3 |
| 6 | 210 | 196 | 96.3 | 61.1 |
| 7 | 819 | 145 | 94.1 | 77.9 |
| 8 | 301 | 172 | 96.6 | 78.3 |
| 9 | 76 | 158 | 97.9 | 73.1 |

\* Concentrations are in the final mixture of the binding assay - see text.
+ The mean and range for BUN and plasma creatinine of the original azotemic samples (before dilution) were 123 (62-312) and 19.4 (8.2-43) mg/dl.

DISCUSSION

Impairment of salicylate binding correlated with the degree of renal failure over a wide range of plasma creatinine concentration. The abnormalities of plasma pH, calcium and chloride commonly found in uremia had little or no effect on salicylate binding. Although indoxyl sulfate and hippurate concentrations were increased to approach or exceed that of albumin (normally 0.6 mM) these two aromatic anions accounted for only 18% of the binding defect in azotemic plasma. Furthermore, o-hydroxyhippurate is of no significance except perhaps among patients ingesting large amounts of salicylate or aspirin. The present studies indicate that a number of other ligands remain to be identified as the major determinants of this uremic disorder. The recently identified furanoid acids may be among these inhibitors as Collier et al have shown that one of these, 3-carboxy-4-methyl-5-propyl-2-furanpropionic acid, is a very potent inhibitor of binding to albumin (9). An important implication of these studies is that this uremic disorder and perhaps others are due not to a single chemical but to the additive effect of a family of chemicals.

Acknowledgments

This work was supported in part by grant AM-19833 from the National Institutes of Health. Patricia Ramos-Smith provided secretarial assistance.

References

1. A. De Torrente, G. B. Glazer, P. F. Gulyassy, Reduced in vitro binding of tryptophan by plasma in uremia, Kidney Int. 6:222, (1974).
2. P. F. Gulyassy, T. A. Depner, Impaired binding of drugs and endogenous ligands in renal diseases, Am. J. Kidney Dis. II:578, (1983).
3. I. Tavares Almeida, P. F. Gulyassy, T. A. Depner, E. A. Jarrard, Aromatic amino acid metabolites as potential protein binding

inhibitors in human uremic plasma, Biochem. Pharmacol. 34:2431, (1985).

4.  P. F. Gulyassy, A. T. Bottini, L. A. Stanfel, E. A. Jarrard, T. A. Depner, Isolation and chemical identification of inhibitors of plasma ligand binding, Kidney Int. 30:391, (1986).

5.  L. A. Stanfel, P. F. Gulyassy, E. A. Jarrard, Determination of indoxyl sulfate in plasma of patients with renal failure by use of ion-pairing liquid chromatography, Clin. Chem. 32:938, (1986).

6.  D. M. Lichtenwalner, B. Suh, M. R. Lichtenwalner, Isolation and chemical characterization of 2-hydroxybenzoylglycine as a drug binding inhibitor in uremia, J. Clin. Invest. 71:1289, (1983).

7.  P. Igarashi, P. F. Gulyassy, L. A. Stanfel, T. A. Depner, Plasma hippurate in renal failure: HPLC method and clinical application, Submitted for publication, (1986).

8.  P. F. Gulyassy, E. Jarrard, L. A. Stanfel, Contributions of hippurate, indoxyl sulfate and o-hydroxyhippurate to impaired ligand binding by plasma of azotemic humans, Submitted for publication, (1986).

9.  R. Collier, W. E. Lindup, H. M. Liebich, G. Spiteller, Inhibitory effect of the uraemic metabolite 3-carboxy-4-methyl-5-propyl-furanpropionic acid on plasma protein binding, Brit. J. Clin. Pharmacol. 21:610P (1986).

# HIPPURIC ACID AS A MARKER

R. Vanholder, A. Schoots[*], C. Cramers[*], R. De Smet, N. Van Landschoot, V. Wizemann[**], J. Botella[***], and S. Ringoir

Nephrology Department, University Hospital, Ghent, Belgium
[*]Laboratory of Instrumental  Analysis, University of Technology, Eindhoven, The Netherlands
[**]Centre of Internal Medicine, Justus Liebig University, Giessen, FRG
[***]Nephrology Department, Free University, Madrid, Spain

There are some classical markers of uremic solute retention, such as serum urea and creatinine concentration. These parameters are however not always reliable, especially not in dialysis patients, so that it seemed interesting to us to undertake a multifactorial study, in an attempt to define the most suitable marker molecules in uremia.  For this purpose, we used high performance liquid chromatography HPLC as a basic technique.

In fig. 1, the HPLC pattern of uremic serum and of normal non-uremic serum  is illustrated.

As expected, there is more retention in uremia compared to normal. Several peaks have been identified such as creatinine, pseudo-uridine, uric acid, hypoxanthine, indoxylsulphate, tryptophan and hippuric acid. Peaks 4,5 and 6 remain unidentified.  In order to define adequate markers of uremic solute retention, the height of each of these individual peaks was correlated to residual renal function, and also to total HPLC-UV absorbance. Total HPLC-UV absorbance is the cumulated and integrated peak height of all UV-absorbing HPLC peaks together.  We evaluated this parameter as such, but also it's relative evolution before versus after dialysis, and each of these values was compared to the corresponding results obtained for the major individual peaks. Preliminary studies had shown that this total UV-absorbance was directly correlated in a highly significant way to renal function and to overall retention of fluorescent uremic solutes.

Two preliminary studies by our group within this context had suggested a role as a marker for hippuric acid. A first study revealed a significantly better elimination of several substances during hemodiafiltration compared to conventional dialysis (1).  One of these substances was hippuric acid. What was even more intreaguing, if we

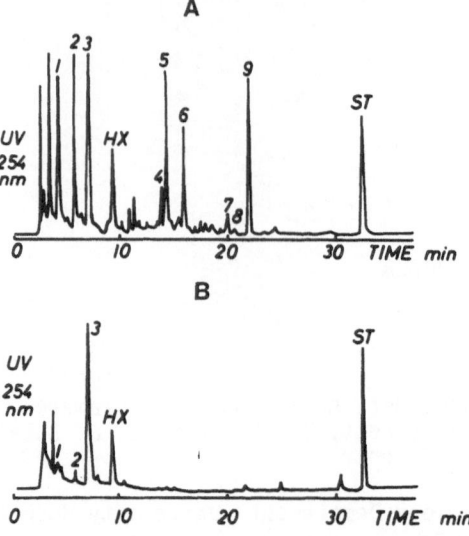

FIG. 1 : HPLC pattern of the ultrafiltrate of uremic serum
(above) and of serum of a healthy person
(below). The following peaks can be recognized :
creatinine (1), pseudo-uridine (2), uric acid
(3), hypoxanthine (Hx), indoxylsulphate (7),
tryptophan (8) and hippuric acid (9). Peaks 4, 5
and 6 remain unidentified. ST is an internal
naphtalene sulfonic standard, allowing the appreciation of the relative height of the peaks.

correlated the pre-dialysis height of the most important observed peaks
to the overall UV absorbance in HPLC, we obtained the most significant
correlation for hippuric acid, together with pseudo-uridine.

A second study was started on a prospective basis in a group of 20
randomly chosen patients where dialysis was shortened from 4 to 2.5
hours (2). One group of patients supported the procedure very well,
whereas the other group showed all signs of clinical intolerance. Pre-
dialysis peak heights of several substances were significantly lower in
the group with good tolerance : creatinine (1.24 vs 1.73), pseudo-
uridine (0.92 vs 1.17), the unidentified peaks 4 and 6 (0.52 vs 1.55 and
0.75 vs 1.49) and hippuric acid (1.37 vs 3.59).

Together with this, it appeared that the group with good tolerance
also had a much better residual renal function, with a creatinine
clearance of 2.6, compared to 0.1 ml/min in the intolerant group. This
suggested a relationship between the pre-dialysis concentration of the
abovementioned substances on one hand, one of them being hippuric acid,
and residual renal function on the other.

These data incited us to investigate more extensively the value of
hippuric acid, a substance with a molecular weight of 179, as a marker
of total uremic solute retention and in the context of uremic toxicity.
This paper is a review of the observations made, and we will try to
answer the following questions :

1) is hippuric acid a useful marker of intra-dialytic solute elimination ?
2) is hippuric acid a useful marker of residual renal function ?
3) what are the sources of hippuric acid ?
4) is hippuric acid toxic ?
5) finally we will make a comparison between the value of HPLC and a colorimetric method in the determination of hippuric acid concentrations.


HIPPURIC ACID AS A MARKER OF INTRADIALYTIC SOLUTE ELIMINATION

In a first protocol, HPLC data were collected from our own centre in Ghent for hemodialysis and hemodiafiltration, from the dialysis centre in Giessen, procured by Dr. Wizemann, for high efficiency hemo-diafiltration and from Prof. Botella's unit in Madrid for conventional dialysis and the so called two chamber technique, which is a combination of hemofiltration and conventional dialysis with a double filter. A total number of 90 patients was studied.  The following parameters were analysed :

1. the pre-dialysis height of different individual  UV- absorbing peaks.

2. the dialysis ratio, i.e. the relation  of pre-dialysis to post-dialysis peak height (pre/post).

3. the extraction, i.e. the relation of pre minus post over pre-dialysis [(pre-post)/pre]

For each of these parameters, the retention and elimination of different individual solutes on HPLC was correlated to the cumulated overall retention and elimination of all U.V.-absorbing compounds together,  starting from the idea that the latter parameter gives a direct estimation of the overall uremic solute retention on one hand and the overall blood purifying capacity of a given dialysis technique on the other.

Table IA shows the degree of correlation (R) of pre-dialysis peak height of different substances towards cumulated overall UV-absorbance. There are enormous differences in the degree of correlation.  Highly significant results were obtained for the unidentified peak 6, pseudo-uridine, uric acid, and hippuric acid; the highest significance was reached for hippurate, with a correlation index of 0.67. The same correlation studies were also repeated for the dialysis ratio (Table IB).

Here statistical significance was reached for all substances under study, but again, the highest significance was observed for hip-puric acid, with a value of 0.75.

These correlation studies were finally repeated for the  post-versus pre-dialysis extraction of solutes, and here again the extraction of individual compounds was compared to the overall extraction of all UV-absorbing compounds together. For hippuric acid, peak 6 and uric acid the correlations are better here than for pre-dialysis peak height and the dialysis ratio.  Here again the best significance is obtained for hippuric acid.

Figure 2 shows the correlation - regression curve that was specifi-cally obtained for hippuric acid.

TABLE I : CORRELATION WITH TOTAL U.V. ABSORBANCE (254 nm)

| | A : PEAK HEIGHT | | B : DIALYSIS RATIO | | C : EXTRACTION | |
|---|---|---|---|---|---|---|
| | R | p | R | p | R | p |
| Hippuric acid | 0.67 | $<0.001$ | 0.75 | $<0.001$ | 0.83 | $<0.001$ |
| Uric acid | 0.53 | $<0.001$ | 0.51 | $<0.001$ | 0.65 | $<0.001$ |
| Pseudo uridine | 0.53 | $<0.001$ | 0.47 | $<0.001$ | 0.54 | $<0.001$ |
| Unidentified peak 6 | 0.46 | $<0.001$ | 0.74 | $<0.001$ | 0.76 | $<0.001$ |
| Unidentified peak 5 | 0.24 | $<0.05$ | 0.62 | $<0.001$ | 0.48 | $<0.001$ |
| Creatinine | 0.15 | N.S. | 0.66 | $<0.001$ | 0.62 | $<0.001$ |
| Indoxylsulfate | 0.10 | N.S. | 0.35 | $<0.001$ | 0.33 | $<0.01$ |
| Unidentified peak 4 | 0.01 | N.S. | 0.62 | $<0.001$ | 0.19 | N.S. |

All calculations were performed on 90 pairs of samples.
R = correlation index; N.S. : non-significant

The correlation is even better than for the two first parameters under study, with a correlation index of 0.83, compared to 0.67 and 0.75 for peak height and dialysis ratio respectively.

Thus, for each of the three formulas that have been studied, the best correlation with total UV absorbance was found for hippuric acid. These data point to hippuric acid as an interesting marker of solute retention and elimination in dialysed patients, whatever the treatment technique.

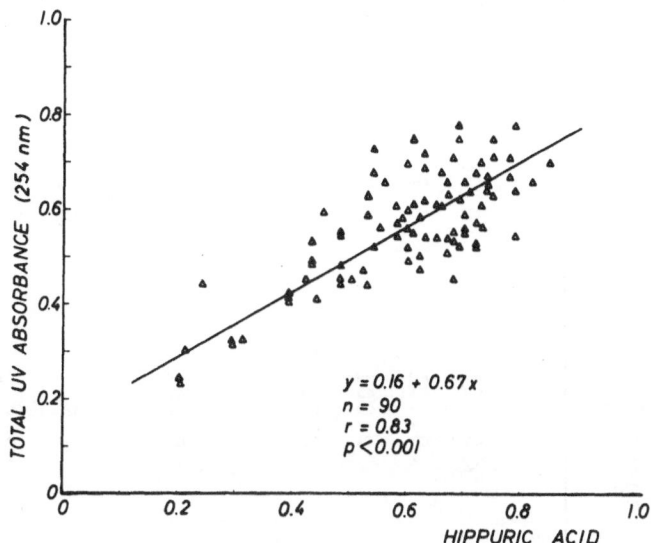

FIG. 2  : Correlation-regression curve between extraction
of total  UV-absorbance and extraction of hippuric acid. The correlation was highly
significant.

## HIPPURIC ACID AS A MARKER OF RESIDUAL RENAL FUNCTION

In another protocol, we evaluated the relationship between peak height and creatinine clearance. Two groups of patients were examined : a group of 32  non-dialysed patients with a creatinine clearance above 5 ml/min and a group of 43 dialysed patients, with a clearance below 5 ml/min. Up to now 4 different parameters have been studied : urea, creatinine, the unknown peak 6 and hippuric acid.  For all these parameters, there was in non dialyzed patients a significant linear correlation with residual renal function (Table IIA).  This correlation became even more significant, when a semi-logarythmic correlation was calculated. This is not unexpected, in view of the hyperbolic interrela-tion of the different points. A similar calculation was performed for dialyzed patients (Table IIB).  When taking into account linear regression, significance is now only reached for hippuric acid and peak 6.  Urea and creatinine are no more related to the residual function. Similar results were obtained from the semi-logarythmic approach, with again an improvement in significance.  The best results were obtained for hippuric acid.

These data indicate that, whereas urea and creatinine are useful indicators of residual renal function before the start of dialysis, this is not the case anymore in dialyzed patients. Here, hippuric acid is the

TABLE II : UREMIC SOLUTE CONCENTRATION VERSUS RESIDUAL RENAL FUNCTION

|  | Linear regression | | semi-log regression | |
|---|---|---|---|---|
|  | R | p | R | p |
| **A. Non-Dialyzed patients (n=32)** | | | | |
| Hippuric acid | 0.74 | < 0.001 | 0.85 | < 0.001 |
| Unidentified peak 6 | 0.50 | < 0.001 | 0.71 | < 0.001 |
| Creatinine | 0.76 | < 0.001 | 0.89 | < 0.001 |
| Urea | 0.76 | < 0.001 | 0.87 | < 0.001 |
| **B. Dialyzed patients (n=45)** | | | | |
| Hippuric acid | 0.57 | < 0.001 | 0.71 | < 0.001 |
| Unidentified peak 6 | 0.40 | < 0.01 | 0.58 | < 0.001 |
| Creatinine | 0.29 | N.S. | 0.15 | N.S. |
| Urea | 0.09 | N.S. | 0.01 | N.S. |

N.S. : non-significant

only well identified compound that has been studied up to now, showing a good correlation with residual function.

In summary, on these data we may assume a strong correlation between hippuric acid concentration on one hand and residual renal function, before and after the start of dialysis, solute retention and solute elimination by hemodialysis on the other.

ORIGINS OF HIPPURIC ACID

The sources of hippuric acid, accumulating in uremic patients, are illustrated in fig. 3.

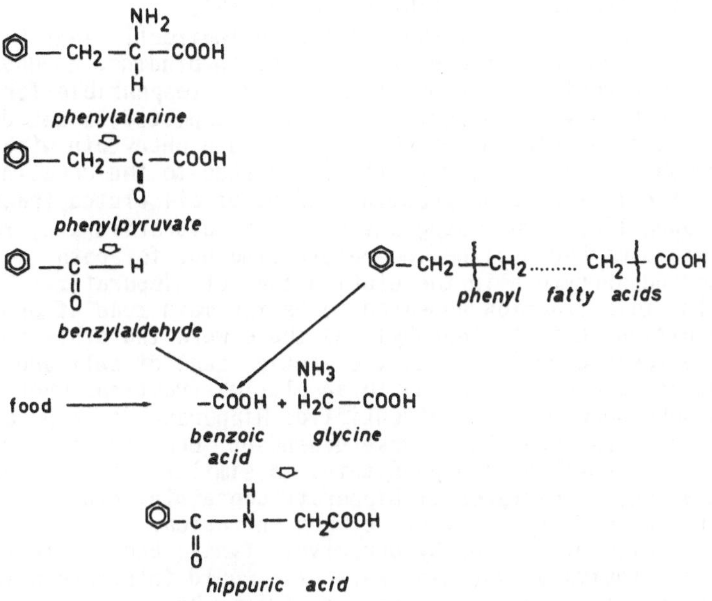

FIG. 3 : Metabolic sources of hippuric acid production.

Hippuric acid is essentially metabolized in the liver from benzoic acid (3), and is then excreted by the kidneys in the urine. The sources of benzoic acid are multiple, the most important source being food intake. In the food, quinic acid is the main precursor. This substance is most frequently found in fruits, to the largest extent in prunes and cranberries (4). There are however other sources, such as tea. Sodium benzoate is also used as a preservative in many processed foods and beverages, jams, jellies, fruit salads, syrups, conserves and deep frozen preparations. Other related food sources are the break-down of phenyl containing fatty acids with an odd number of carbon-atoms, and phenylalanine. Apart from that, the direct production of benzoic acid by the intestinal flora is another source (5). Environmental contact with or ingestion of xylenes and/or toluenes also provokes the production of hippuric acid, and in industrial medicine urinary hippurate is even used as a marker of exaggerated contact with this environmental polluant. Finally, it should be stressed that some brands of heparin contain benzyl-alcohol, and that this may be an important source of hippurate in dialysed patients (6). Thus the sources of hippurate are multiple, and it should be admitted that the presence of such divergent origins as food, beverages, environmental polluants, possible endogenous sources, intestinal flora and/or heparin may bias the presently reported correlation between concentration and renal function. Further studies are needed to evaluate whether alterations in diet or other factors may influence hippurate concentration to such an extent that the relation with renal function becomes disturbed.

## EVENTUAL TOXICITY OF HIPPURIC ACID

It was already shown in 1975 by Boumendil-Podevin of Richet's group in the Journal of Clinical Investigation that hippuric acid interfered with the para-aminohippurate and urate transport at the cortical tubular level (7). These data suggested that hippuric acid might inhibtransport of a variety of organic substances. It was also demonstrated by Porter et al. that hippuric acid, caused net fluid secretion in proximal straight tubules isolated from rabbit kidneys (8).
Data reported by Depner et al. (9), suggested indirectly that hippuric acid might also cause a decrease of drug protein binding. We undertook recently a more direct evaluation of the factors responsible for drug protein binding of several drugs. As might be expected, a net decline was found in both the binding of theophylline and phenytoin with renal failure, and this decline was directly correlated to the creatininemia. Analysis of the influence on protein binding of all eluted fractions of high performance liquid chromatograms of uremic ultrafiltrate, revealed that there was a marked decline in protein binding, for both theophylline and phenytoin in the elution zone of hippurate.
For phenytoin, this fraction appeared to be the main zone of protein binding inhibition, but for theophylline there were two other inhibitory zones, one corresponding mainly to the elution zone of salt and other electrolytes and one corresponding to an elution fraction, where the responsible substance remains unidentified. Hippurate, as a pure substance, caused as such also a dose dependent protein binding inhibition for both theophylline and phenytoin. A similar dose response curve was also obtained, if mixtures of hippurate containing elution fractions were evaluated at different degrees of concentration.
Preliminary results of studies by ourselves, Tanaka and Dzurik (personal communications) indicate that hippurate also would interfere with glucose tolerance and platelet cyclo-oxygenase activity. Recently we also found an inhibition by hippuric acid of the oxygen burst during the phagocytosis of labelled glucose by leucocytes. Thus, it can at least be stated that hippuric acid interferes with at least some biological functions, often in a dose dependent manner, and so it may be considered as a toxin. Further studies are needed on the potential toxicity of this substance.

## HPLC VERSUS COLORIMETRY

A last question that arises, is whether there are simpler and quicker methods available than the presently used HPLC-method for the estimation of hippuric acid concentrations; the latter method indeed turns out to be technically complicated and very labor-intensive. For that reason, we recently tried to correlate hippuric acid concentrations, measured by the HPLC method, to the values obtained by a colorimetric method, first described by Ohmori and coworkers (10).
There was a highly significant correlation between the two methods of concentration estimation, both when taking into account total and free hippuric acid concentration (n = 21, r = 99, p 0.001 for both correlation-regression analyses).
In other words, if hippuric acid concentrations are to be measured as a

marker for overall solute retention and elimination, data can as easily be obtained with the simple and quick colorimetric Ohmori method, as with the HPLC-method.


CONCLUSION

Hippuric acid appears to be related significantly to residual renal function, and to solute retention and elimination. For its determination, there is a colorimetric method available that is as reliable as the more complicated HPLC technique.

REFERENCES

1. A. Schoots, H. Homan, R. De Smet, C. Cramers, R. Vanholder and S. Ringoir : Evaluation of in vivo dialysis efficiency in hemo-dialysis and hemodiafiltration by reversed-phase liquid chromatography. In : First international symposium on single-needle dialysis. Ed. S. Ringoir, R. Vanholder, P. Ivanovich, ISAO Press, Cleveland (1984) p. 151.

2. A. Schoots, R. Vanholder, M. Gladdines, R. De Smet, C. Cramers and S. Ringoir : Hippuric acid and an unidentified compound as possible indicators of residual renal function in dialysed patients. In : Immune and Metabolic Aspects of Therapeutic Blood Purification Systems. Ed. : L.C. Smeby, S. Jorstad, T.E. Wideroe, Karger, Basel (1985) p. 240.

3. M.D. Armstrong, F.C. Chao, V.J. Parker and P. E. Wall : Endogenous formation of hippuric acid. Proc. Soc. Exper. Biol. & Med. 90 : 675 (1955).

4. P.T. Bodel, R. Cotran and E.H. Kass : Cranberry juice and the anti-bacterial action of hippuric acid. J. Lab. Clin. Med., 54 : 881 (1959).

5. E. Bourke, G. Frindt, H. Preuss, E. Rose, M. Weksler and G.E. Schreiner : Studies with uraemic serum on the renal transport of hippurates and tetraethylammonium in the rabbit and rat : effects of oral neomycin. Clin. Science, 38 : 41 (1970).

6. P.C. Farrell, F.A. Gotch, J.H. Peters, B.J. Berridge, and M. Lam : Binding of hippurate in normal plasma and in uremic plasma pre- and postdialysis. Nephron, 20 : 40 (1978).

7. E.F. Boumendil-Podevin, R.A. Podevin and G. Richet : Uricosuric agents in uremic sera. Identification of indoxyl sulfate and hippuric acid. J. Clin. Invest., 55 : 1142 (1975).

8. R.D. Porter, W.F. Cathcart-Rake, S.H. Wan, F. C. Whittier and J.J. Grantham : Secretory activity and aryl acid content of serum, urine, and cerebrospinal fluid in normal and uremic man. J. Lab. Clin. Med., 85 : 723 (1975).

9. T.A. Depner : Suppression of tubular anion transport by an inhibi-tor of serum protein binding in uremia. Kidney Int., 20 : 511 (1981).

10.S. Ohmori, M. Ikeda, S. Kira and M. Ogata : Colorimetric deter-mination of hippuric acid in urine and liver homogenate. Anal. Chem., 49 : 1494 (1977).

# CONCLUSION

## REFERENCES

# SUPPRESSIVE EFFECT OF QUINOLINIC ACID AND HIPPURIC ACID ON BONE MARROW

# ERYTHROID GROWTH AND LYMPHOCYTE BLAST FORMATION IN UREMIA

Yoichiro Kawashima[a], Tsutomu Sanaka[a], Nobuhiro Sugino[a],
Masatomo Takahashi[b] and Hideaki Mizoguchi[b]

[a]Department of Medicine, Kidney Center, [b]Department of
Medicine I, Tokyo Women's Medical College, Tokyo, Japan

## INTRODUCTION

The empirical treatment of chronic renal failure with dialysis therapy
has achieved considerable success. But, there are some problems remaining
unsolved in the management of dialysis therapy. Anemia is one of these
problems.

Numerous studies indicate that inadequate erythropoietin production
and shortened red cell survival contribute to the anemia associated with
CRF. In addition, inhibitors of erythropoiesis in CRF have been thought
to be another important cause of anemia.

It is well known that the bindings of drugs and other substances to
serum albumin are impaired and the major cause of that is believed to be
an accumulation of inhibitory ligands. We call these substances Protein
Binding Inhibitor (abbreviated PB-Ix). Some of these organic acids may be
related to anemia, and at the same time, dysfunction of the immune system.

In this study we studied the role of protein binding inhibitors,
Quinolinic Acid (QA) and Hippuric Acid (HA), whether they may have the
inhibitory action of erythroid colony and lymphocyte blast formation.

## METHOD

Erythroid colony assay: A modification of the methylcellulose technique of Iscove was used for CFU-E. The culture medium, BSA and FCS was
made semi solid with 2.67% methylcellulose. Mallow cells ($3 \times 10^6$/ml) were
cultured in the presence of human urinary erythropoietin (20 U/ml) in plastic tissue culture dishes. Cultures were incubated at 37°C in a highly
humid and 5%$CO_2$95% air condition for 2 days. Aggregates of 7 or more cells
were scored as CFU-E derived colonies. In this study, solutions of QA or
HA were added instead of Iscove medium. Final concentration of QA and HA
were 10, 1, 0.1 mg/dl and 100, 10, 2 mg/dl respectively. To determine the
suppressive mechanism, Kynurenic acid (KA) was added to the culture system,
at a concentration of 6, 1.2, 0.24 mM/L KA. The same study with a calcium
free medium was done.

Lymphocyte blast formation: Lymphocytes were isolated from the same
normal volunteer as previously described. They were incubated in the presence of Phytohemagglutinin and QA or HA for two days. Final concentrations of QA and HA were the same as that in CFU-E.

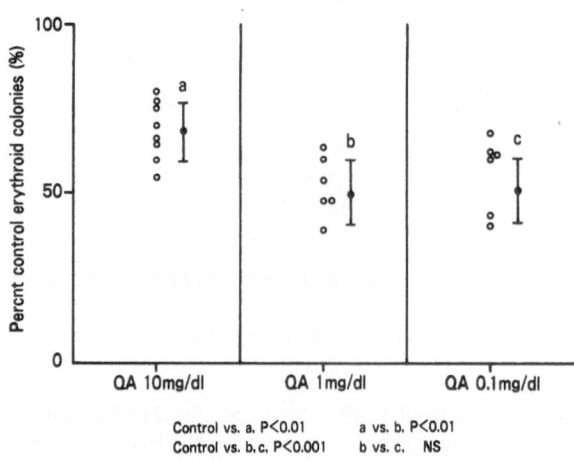

Fig. 1   The effect of Quinolinic Acid (QA)
on erythroid colony formation


Measuring of PB-Ix activity of QA and HA:  Binding of $^{14}$C-phenytoin by
purified bovine serum albumin was measured by centrifugal ultrafiltration
method.   Albumin concentration was 1 g/dl and the result was expressed by
% increase of unbound $^{14}$C-phenytoin at concentration of 10, 1, 0.1 mg/dl QA,
100, 10, 2 mg/dl HA.


RESULT

Erythroid colony assay:  QA suppressed the erythroid colony formation
dose dependentry.  The plasma concentration of QA in uremic patients is
about 0.2 mg/dl at most.  QA had the suppressive effect at the concentration of uremic plasma level. (Fig. 1)  On the other hand, HA had no suppressive effect at the concentration of 100 mg/dl, that is about 10 times of
plasma concentration of CRF patients.  But 10 and 2 mg/dl, HA had slightly
suppressive effect statistically.
KA did not have no improving effect on the suppressed erythroid colony
formation by QA.
With calcium free medium, colony formation was about the same as that
of control.  It is concluded that calcium is not essential for the erythroid
colony formation.
Lymphocyte blast formation:  QA had a strongly suppressive effect at
all concentrations. (Fig. 2)  On the other hand, HA had a suppressive effect
only at 100 mg/dl, and no effect at 10 mg/dl, which is the average plasma
concentration in uremic patients.
PB-Ix activity:  QA and HA inhibited the binding of DPH to BSA.  QA
was about twice as potent as HA. (Table 1)


DISCUSSION

Inhibitors of erythropoiesis in chronic renal failure have been suggested to be important in the pathogenesis of anemia.  In our previous
study, the changes of hematocrit and crude PB-IX activity before and after
CAPD treatment in the same patients indicated that anemia improved parallel
to the decrease of PB-Ix activity.


70

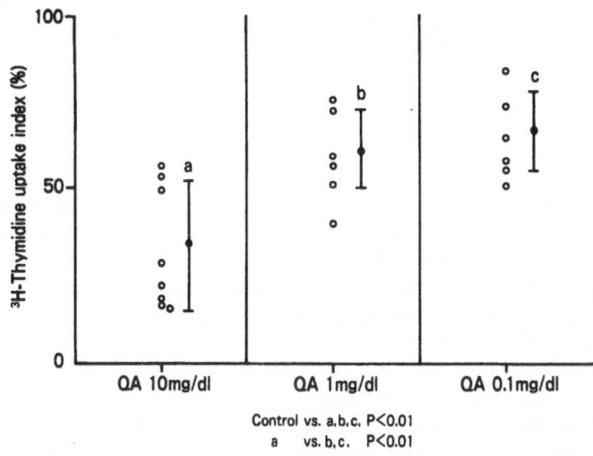

Fig. 2   The effect of Quinolinic Acid (QA)
on lymphocyte blast formation

QA is an organic acid which is thought accumulate in uremic plasma. Koide et al. identified some substances, like QA and Aconitic acid in peak 2a by HPLC and they also reported some of these substances have glycolisis inhibitory action and δ-amino laevulinic acid dehydrase inhibitory action. QA has been suggested to have a role in human neurodegenerative diseases based upon its neuro toxic and convulsant properties.  In this study we showed the suppressive effect of QA and HA on erythroid colony and lymphocyte blast formation.

As for lymphocyte reaction we have already reported that crude PB-Ix binds the PHA receptor onto the lymphocyte surface competitively may block the signal to synthesize DNA.

The mechanism of suppression on erythroid cell remains unclear in this our study.  Erythropoietin is a glycoprotein, M.W. approximately 39,000, which acts on the surface receptor of erythroid progenitor cell to induce hemoglobin synthesis.  Erythropoietin is essential to the erythroid proliferation system and some substances like cAMP and some amino acid may work cooperatively to accelerate the action of erythropoietin.  QA is a tryptophan metabolite and may interact with erythropoietin.  We added Kynurenic Acid which reportedly acts as an antagonist of QA in the nerve system.  Buy KA had no improving effect on the suppressed erythroid colony formation by QA.  On the other hand, judging from the structure formula, QA might have the chelate action of calcium.  So the study with calcium free medium was also done.  Since the colony formation was about the same as that of the control, we could not clarify the suppressive mechanism.

Table 1   Inhibitory effect of uremic metabolites on
the binding of DPH to bovine serum albumin

| Inhibitor | % Increase in unbound $^{14}$C-DPH at inhibitor concentration (mg/dl) | | | | | |
|---|---|---|---|---|---|---|
| | 100 | 50 | 10 | 2 | 1 | 0.1 |
| Quinolinic Acid | 63 | 41 | 31 | — | 19 | 16 |
| Hippuric Acid | 38 | — | 19 | 11 | — | — |

But a possible interaction with amino acid are thought to be as one of the causes. From this point of view, we must do more experiments using other antagonists of QA.

REFERENCES

1. S. F. Wallner and R. M. Vautrin: Evidence that inhibition of erythropoiesis is important in the anemia of chronic renal failure. J Lab Clin Med 97:170 (1981)
2. H. W. Radtke, A. B. Rege, et al.: Identification of Spermine as an Inhibitor of erythropoiesis in Patients with chronic renal failure. J Clin Invest 67:1623 (1981)
3. N. N. Iscove, F. Sieber and K. H. Winterhalter: Erythroid colony formation in cultures of mouse and human bone marrow; Analysis of the requirement for erythropoietin by gel filtration and affinity chromatography on agarose-concanavalin A. J Cell Physiol 83:309 (1974)
4. K. Koide, et al.: Study of Middle molecules. Jap J Nephrol 26:537 (1984)
5. T. Sanaka, Y. Kawashima, et al.: Suppressive effect of uremic protein binding inhibitor on cultivated lymphocyte. Jap J Nephrol 25:1063 (1983)
6. A. H. Gonong and C. W. Cotman: Kynurenic acid and Quinolinic acid act at N Methyl-D-Aspartate receptors in the rat hippocampus. J Pharm Exper 236:293 (1985)

# PURINE METABOLITES IN URAEMIA

H.A. Simmonds, J.S. Cameron, G.S. Morris,
L.D. Fairbanks, and P.M. Davies

Purine Research Laboratory and Renal Unit
United Medical Schools of Guy's and St. Thomas'
London, U.K.

Many investigators have described the large number of unknown solutes of low molecular weight, within the purine range, which accumulate in the plasma during progressive reduction in renal function (1). Their identification is important both in terms of understanding the toxicity which occurs in uraemia and in devising new therapeutic approaches, particularly for dialysis. The question is whether some of these could be purines and if so what would the significance of this be? Elevated plasma cAMP levels which correlated with plasma creatinine concentrations were the first purines to be implicated in renal failure (1). Since all cells require a balanced supply of purines for growth and survival it is clear that there may be other abnormalities.

## Importance of purines in cellular metabolism

Purines are effectively anchored inside the cell as the nucleotide by attachment to ribose-phosphate. The vital role of purine nucleotides in intracellular metabolism as the basis of energy (ATP) and genes (DNA), as well as in membrane signal transduction, translation and protein synthesis (GTP, cAMP/cGMP, RNA) has been well documented (2). Purine metabolism (Figure 1) is concerned with the efficient recycling of nucleosides and bases derived from these nucleotides during the wear and tear of daily life (wound healing, muscle work, erythrocyte senescence, providing essential nourishment for the brain, etc.). The small amount of purine not recycled (2-4mmol/day on a low purine diet) is excreted in the form of uric acid. However, it has now been recognised that purine metabolism has two additional, and hitherto unrecognised, roles. The first concerns the removal of adenine, the end-product of the polyamine pathway (3), the second is removal of adenosine, the end-product of S-methylation reactions, which is essential for the normal functioning of this pathway (Figure 1).

The latter route is involved in creatinine production and is estimated to be a major metabolic source of adenosine and the only source of homocysteine (4). These pathways have assumed considerable significance recently, because attention has now been focused on the extracellular functions of purine nucleosides and bases. A number of physiological roles have been proposed for adenosine in the regulation of cardiac, neurological, muscle, respiratory and renal function (5). The source of this adenosine is not established with certainty, but the S-methylation pathway appears a likely candidate (4). The bases hypoxanthine and

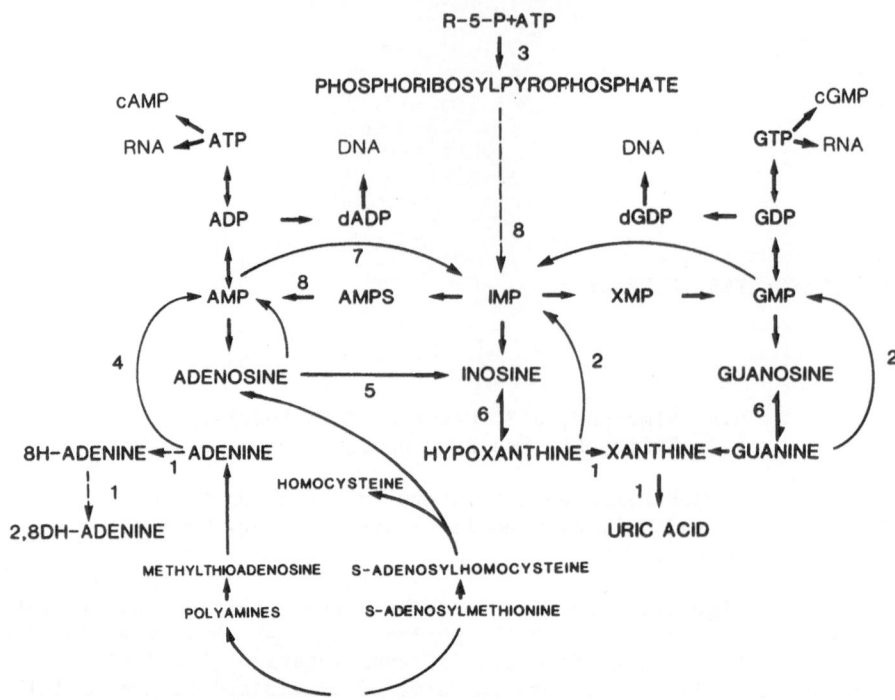

Figure 1. Pathways of purine metabolism showing the routes of de novo
synthesis from phosphoribosylpyrophosphate (PP-ribose-P), compared with
salvage from the base – or nucleoside in the case of adenosine. The figure
also indicates the importance of these pathways for the removal of
adenosine and adenine, the end-products of the S-methylation and polyamine
pathways, respectively. Dotted lines indicate the alternative route of
metabolism when the salvage enzyme is absent or defective (2H-adenine =
2-hydroxyadenine, 2,8DH-adenine = 2,8-dihydroxyadenine).
Numbers 1 to 8 indicate the inherited enzyme defects now known (1=Xanthine
oxidase, 2=hypoxanthine-guanine phosphoribosyltransferase, 3=PP-ribose-P
synthetase, 4=adenine phosphoribosyltransferase, 5=adenosine deaminase,
6=purine nucleoside phosphorylase, 7=myoadenylate deaminase,
8=adenylosuccinate lyase).

xanthine have also been implicated as modulators of neurotransmission
through acting as natural ligands for the benzodiazepine receptors in
brain (6). Positive correlations have been demonstrated between CSF
xanthine concentrations and poor appetite plus weight loss in patients
with depressive illnesses.

## Importance of the kidney in controlling blood levels of purine metabolites

How then might any of the above processes be affected in the uraemic
state? The first possibility relates to potential inhibitory effects of
compounds accumulating in uraemia on the enzymes involved in this normal
recycling of cellular waste. The second is the fact that the kidney plays
a vital role in controlling the levels of purine metabolites in the body.

Table 1  Plasma concentrations, clearances and urine solubilities of purine bases and their analogues in healthy subjects.

| Parameter | Xanthine | Hypoxanthine | 2,8-DHA | Uric acid[**] | Adenine | Oxipurinol |
|---|---|---|---|---|---|---|
| Plasma level (μmol/l) | <0.1 | <5.0 | – | m(0.15–0.30) f(0.13–0.26) c(0.10–0.23) | 0.07 | <30.0 |
| Clearance (%GFR) | 130 | 110 | 394 | m  8.0 f  14.0 c  20.0 | 160 | 30 |
| Solubility* (mmol/l) pH 5.0 pH 8.0 | 0.5 0.86 | 10.3 11.0 | 0.016 0.030 | 0.9 11.9 | 6.0 ? | 1.3 4.6 |

– not a normal metabolite.  * Values for human urine, except for adenine where values for water only were available.  ** mmol/l.

Plasma levels are mean values for normal subjects (m=male,f=female,c=child); oxipurinol concentrations are maximal values for normal subjects on 300mg allopurinol/day. Clearances are also mean values and, with the exception of uric acid, have been obtained in subjects either given the appropriate purine load (or purine analogue), or in patients with a metabolic defect where high levels of the relevent purine metabolite accumulates (6,12). Results are expressed as a percentage of the GFR.

In normal man most nucleosides and bases are unlikely to accumulate because of further rapid metabolism (Figure 1). An additional protective mechanism, which ensures that plasma and extracellular fluid nucleoside and base levels are generally at the limits of detection, is that (even when accumulating in consequence of an inherited metabolic defect) these metabolites are cleared at, or in excess of the GFR (Table 1). The nucleosides inosine, guanosine, and their deoxyanalogues also undergo net secretion (unpublished observations). The exceptions are adenosine, which is seemingly actively reabsorbed in the proximal tubule (7), and uric acid which is handled in a complex fashion involving bidirectional transport (8), resulting in net reabsorption and excretion of as little as 10% of the filtered load depending on age and sex (Table 1). Obviously, in uraemia accumulation of other nucleosides and bases could also result, particularly in inherited enzyme defects. Alternatively, accumulating metabolites (purine or non-purine) could simulate such defects by acting as enzyme inhibitors or exert their effects by acting extracellularly as receptor agonists, antagonists or uptake blockers. Whether any of these putative effects could be implicated in the cellular toxicity observed in the uraemic state, requires more detailed investigation.

Purine metabolites accumulating in uraemia

The situation with regard to uric acid has been well documented. It is not currently considered as a uraemic toxin. The relationship betwen uric acid and renal function is shown in Figure 2. As can be seen, the plasma urate level rarely doubles, even in uraemia. (A high urate in excess of 600 μmol/l is indicative of a metabolic defect and should be investigated).

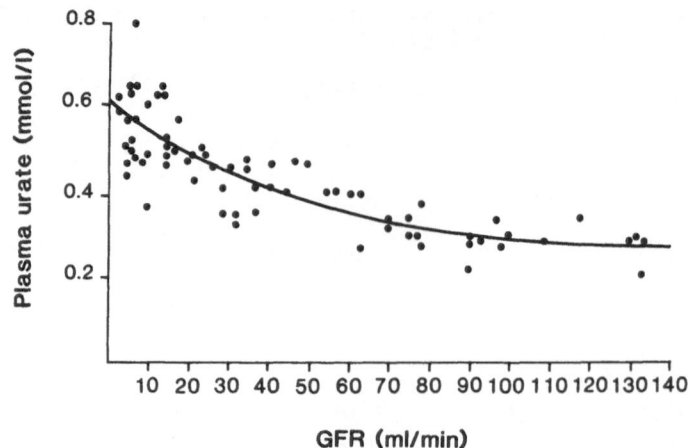

Figure 2. Relationship between GFR and plasma uric acid in non-gouty patients with renal disease showing that plasma urate remains essentially unchanged until the GFR is below 60ml/min and even when less than 10ml/min plasma urate levels rarely double.

Three factors have been implicated: increased fractional excretion by the remaining nephrons, increased extra-renal excretion and thirdly decreased production (8). Increased fractional excretion - up to 80% of the filtered load in terminal uraemia - is apparently due to a combination of decreased reabsorption and increased secretion. Normally 1/3 of the uric acid lost daily is considered to be excreted via the gut, but in renal failure this may increase to 2/3. However, the latter data are derived from studies in a very few patients and it is conceivable that inhibition of the S-methylation pathway, the major route of creatinine production in the liver (eg by homocysteine which is known to accumulate in renal failure (9), or creatinine itself (10)), could also explain both the purine and creatinine deficit (11) sometimes noted in uraemia.

Another important fact which is not generally appreciated is that purine bases are extremely insoluble (Table 1) and as such can themselves be nephrotoxic, leading to permanent renal damage and terminal uraemia. 2,8-dihydroxyadenine (2,8-DHA) and xanthine are the most nephrotoxic, due to their extremely low solubility, and unlike uric acid this cannot be improved by alkalinisation of the urine (3,12). Patients with uric acid, xanthine or 2,8-DHA crystal-induced nephropathies have frequently presented in coma requiring maintenance dialysis. In one patient with 2,8-DHA nephropathy an adenosine-type compound was found in the urine on several such occasions (13). Whether this compound was itself toxic, or alternatively adenine (shown to be toxic in vitro, but only at millimolar levels (14)) could have accumulated, has not been established because plasma prior to dialysis was not available. Adenine is also an endogenous ligand for the benzodiazepine receptor. Alternatively, adenine accumulation in quantity could lead to product inhibition of the polyamine pathway. The effect of polyamine accumulation in uraemia has been discussed elsewhere at this symposium. Certainly product inhibition is a potentially toxic factor for many methylation reactions, since adenosine as well as homocysteine must be removed for such reactions to proceed (4). Adenosine analogues could obviously have a similar effect.

It is equally noteworthy that grossly elevated plasma xanthine levels have been found in patients requiring dialysis for xanthine nephropathy. Concentrations of around 6–700μmol/l (normal values even in allopurinol treated patients with normal renal function <10μmol/l) were noted in patients developing acute renal failure when given allopurinol concomitantly to reduce the urate load during agressive therapy for different malignancies (12). They have also been noted in a xanthinuric patient in terminal uraemia. Whether or not CSF xanthine levels would be raised correspondingly is not known, but in view of the potential for interaction with benzodiazepine receptors and the effects shown for xanthine on appetite, is a possibility which merits consideration (6). Patients with severe hypoxanthine-guanine phosphoribosyltransferase (HGPRT) deficiency and purine overproduction in consequence of the defect are equally sensitive to allopurinol (12) and can develop xanthine stones or xanthine nephropathy; they are also much less neurologically disturbed at lower allopurinol doses and hence lower xanthine levels (unpublished observations).

## Purine analogues as uraemic toxins

Another uraemic toxin which accumulates secondary to treatment for hyperuricemia and gout is the xanthine analogue oxipurinol, the active metabolite of allopurinol (8). Oxipurinol is handled by the kidney in a manner similar to uric acid and is likewise retained in renal failure, particularly by patients with familial gouty nephropathy with an inherited defect in urate and hence oxipurinol clearance (12). In such patients, even with a relatively mild reduction in renal function, oxipurinol levels 3 times the recommended level of 100μmol/l or less, have been recorded on only 200 mg allopurinol per day. Consequently, the dose of allopurinol must be carefully monitored in renal failure and reduced to as little as 100 mg on alternate days in uraemic patients on dialysis, to avoid the risks of severe bone marrow depression, or dermatitis, frequently reported (8,12). Even greater care should be taken in patients given thiazide diuretics (or purine analogues) concomitantly, where oxipurinol retention and consequently the above side effects, will be greatly potentiated.

## Altered erythrocyte purine nucleotide concentrations as potential toxins in the uraemic state

The above findings may well be important in another context. It is significant that in all the inherited purine disorders so far recognised (Figure 1) which involve cellular toxicity, accumulation of abnormal nucleotides, or altered levels of normal nucleotides, have been identified in the patient's erythrocytes (15). This has been related in turn to the accumulation of abnormal purine metabolites in body fluids (16). Up to fourty UV absorbing components have been found in uraemic plasma, compared with less than 10 for normal plasma (17). One of these is 5-amino-4-imidazole carboxamide (1), which has been shown to accumulate as the triphosphate in the Lesch-Nyhan erythrocyte. It could be anticipated therefore that the red cell might have a similar capacity to form abnormal nucleotides from any unusual purine, or related compound, accumulating in uraemia. Erythrocyte nucleotides have indeed been studied in uraemia and high levels of ATP and GTP have been reported by a number of groups (18-20). Increased diphosphate as well as IMP, IDP and ITP levels have also been noted and there has been some controversy as to the explanation for these raised nucleotide levels (19).

Some have attributed this to an increased young red cell population, which have much higher ATP and GTP levels. Raised APRT activity, also characteristic of the young red cell, has been demonstrated in uraemia in

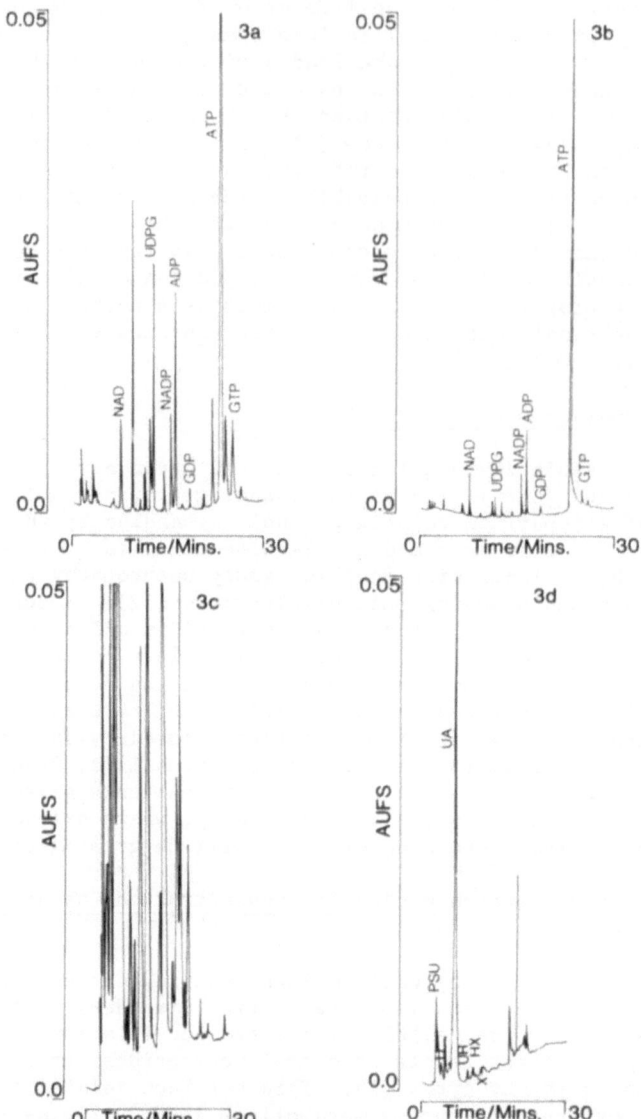

Figure 3. HPLC traces recorded at 254nm, 0.05 absorbance units full-scale (AUFS) and a flow rate of 1ml/min, following injection of 5ul of erythrocyte extract from a uraemic patient (3a), compared with a healthy control (3b) using an anion exchange system (15). Note the large number of additional peaks in 3a compared with 3b where there are only 7 main peaks, namely ATP, ADP, GTP, GDP, UDPG, NADP and NAD. 3c and 3d show plasma levels of UV absorbing components from the same subjects (50ul injections) recorded using a reversed phase HPLC system (16) confirming the large number of additional peaks in uraemic plasma (17) compared with the control, where the principal components are pseudouridine (Psu), uric acid (UA), hypoxanthine (Hx) and xanthine (X).

support of this argument (18). Alternatively, the hyperphosphataemia of uraemia has been invoked to explain these raised nucleotide levels (20). However, a recent study showed first and most importantly that although ATP and GTP were reduced only transiently by dialysis, both showed a rapid reduction to normal post transplantation. Moreover, both rose rapidly in association with episodes of acute rejection (19). Since the return to normal occurred at a time of increasing reticulocyte numbers, and in some patients the high levels were noted in the presence of normal plasma phosphate, there is obviously no consistent relationship between red cell ATP, red cell age or plasma phosphate levels in uraemia. This was supported by studies using 51 Cr which confirmed that the fall in ATP and GTP could not be accounted for by a new population of metabolically normal cells.

Our own studies (unpublished observations) have shown the presence of other as yet unidentified nucleotides in the uraemic erythrocyte (Figure 3a), in addition to the raised ATP and GTP levels found by others. Different high pressure liquid chromatographic (HPLC) systems were however necessary to demonstrate this. These findings may be equally important in establishing which metabolites are responsible for the uraemic state. Specifically, whether any correlation exists between the chromatographic characteristics of these nucleotides and any unidentified UV-absorbing components accumulating in uraemic plasma (Figure 3b). High levels of 4-amino-5-imidazole carboxamide have been found in uraemic plasma (1). This could lead to intracellular accumulation of the corresponding nucleotide triphosphate, as well as inhibition of de novo purine synthesis through an effect on the bifunctional adenylosuccinase enzyme, followed by a reduction in the ATP pool (21). Since the concentration of many putative uraemic toxins (e.g.methylguanidines, polyamines) is higher in the erythrocyte than in the plasma (1), resolution of the number of non-phosphorylated UV absorbing components, which elute together in the front during HPLC of erythrocyte extracts, may be equally useful.

In summary, further investigation of the UV absorbing components in the uraemic erythrocyte, as well as plasma, is warranted. Such studies may well assist in determining, for instance, whether abnormal pyrimidine metabolism is directly associated with membrane instability leading to the accelerated haemolysis of chronic renal failure (20). Alternatively, whether this is an effect of altered ATP levels per se, since ATP is known to be important for retaining the discoid shape and hence the flexibility of the erythrocyte which enables it to squeeze through tiny vessels. It also protects the phospholipid membrane from attack by extracellular phospholipases and hence accelerated lysis (22). Identification of these unusual nucleotides should also enable clarification of the origin and significance of at least some of the many components which have been demonstrated to accumulate in uraemic plasma (17). Such knowledge must assist our understanding of their potential to act as uraemic toxins, either directly, or via their phosphorylated derivatives. More importantly they should establish whether any of them are indeed purine metabolites. Almost certainly many metabolic pathways, and not one single pathway, is going to be involved in the toxicity associated with the uraemic state.

REFERENCES

1. Bergstrom J, Furst P. Uraemic Toxins. Chapter 19. In: Drukker W, Parsons FM, Maher JF, eds. Replacement of renal function by dialysis. Boston:Martinus Nijhoff, 1983;2nd Edition:354-390.
2. Henderson JF, Paterson ARP. Nucleotide Metabolism: an introduction. NY: Academic Press 1973.

3. Simmonds HA, Van Acker KJ. Adenine phosphoribosyltransferase deficiency: 2,8-dihydroxyadenine lithiasis. Chapter 52. In: Stanbury JB, Wyngaarden JB, Fredricksen DA, Goldstein JL, Brown MS, eds. The metabolic basis of inherited disease. NY: McGraw-Hill, 1983;5th edition:1144-1156.

4. Hershfield MS. S-adenosylhomocysteine hydrolase as a target in genetic and drug-induced deficiency of adenosine deaminase. In: Berne RM, Rall TW, Rubio R, eds. Regulatory function of adenosine. The Hague: Martinus Nijhoff, 1983: 171-177.

5. Various authors. Ibid.

6. Niklasson F. Experimental and clinical studies on human purine metabolism. Abstracts of Uppsala Dissertations from the Faculty of Medicine. Uppsala: Fyrus Tryk, 1983.

7. Thompson CI, Sparks HV, Spielman WS. Renal handling and production of plasma and urinary adenosine. Am J physiol 1985;248:F545-F551.

8. Cameron JS, Simmonds HA. Uric acid, gout and the kidney. J Clin Pathol 1981;34:1245-1254.

9. Cohen BD. Uraemia: pillar of salt or column of nitrogen? Editorial. Dialysis & Transplantation 1980;9:535-538.

10. Mikami H, Orita Y, Ando A, Fujii M, Kikuchi T, Yoshihara K, Okada A, Abe H. Metabolic pathway of guanidino compounds in chronic renal failure. In:Lowenthal A, Mori A, Marescau B, eds. Urea cycle disorders. NY:Plenum Press 1983:449-458.

11. Jones JD, Burnett PC. Creatinine metabolism in humans with decreased renal function:creatinine deficit. Clin Chem 1974;20:1204-1212.

12. Simmonds HA, Cameron JS, Morris GS, Davies PM. Allopurinol in renal failure and the tumour lysis syndrome. Clin Chim Acta 1986;in press.

13. Greenwood MC, Dillon MJ, Simmonds HA, Barratt TM, Pincott JR, Metreweli C. Renal failure due to 2,8-dihydroxyadenine urolithiasis. Eur J Paediatr 1982;138:346-349.

14. Kishi T, Kittaka E, Hyodo S, Kashiwa H, Karakawa T, Suzawa T, Sakura N, Sakano T, Usui T. Inhibition by adenine of in vitro immunological functions of normal and adenine phosphoribosyltransferase deficient human lymphocytes. Immunopathol. 1985;10:157-162.

15. Simmonds HA, Watson AR, WEbster DR, Sahota A, Perrett D. GTP depletion and other erythrocyte abnormalities in inherited PNP deficiency. Biochem Pharmacol 1982;31:941-946.

16. Morris GS, Simmonds HA. A single system for the evaluation of purine and pyrimidine nucleosides and bases together with their analogues in biological fluids. Adv Exp Med Biol 1986;195A:593-601.

17. Schoots AdC, Homan HR, Gladdines MM, Cramers CAMG, de Smet R, Ringoir SMG. Screening of UV-absorbing solutes in uremic serum by reversed phase HPLC - change of blood levels in different therapies. Clin Chim Acta 1985;146:37-51.

18. Becher HJ, Weise HJ, Volkermann U, Schollmeyer P. Enhanced purine nucleotide synthesis in erythrocytes of uraemic patients. Klin Wochenschr 1980;58:1243-1250.

19. Rejman ASM, Mansell MA, Grimes AJ, Joekes AM. Rapid correction of red cell nucleotide abnormalities following successful renal transplantation. Br. J Haematol 1985;61:433-443.

20. Angle CR, Swanson MS, Stohs SJ, Markin RS. Abnormal erythrocyte pyrimidine nucleotides in uraemic subjects. Nephron 1985;39:169-174.

21. Sabina RL, Holmes EW, Becker MA, The enzymatic synthesis of 5-amino-4-imidazolecarboxamide riboside triphosphate (ZTP). Science 1984;223:1193-1195.

22. Gazitt Y, Ohad I, Loyter A. Changes in phospolipid susceptibility toward phosphlipases induced by ATP depletion in avian and amphibian erythrocyte membranes. Biochim Biophys Acta 1975;382:65-72.

# THE ROLE OF LIPIDS IN PROGRESSIVE GLOMERULAR DISEASE

William F. Keane, Bertram L. Kasiske,
and Michael P. O'Donnell

Department of Medicine, Hennepin County Medical Center
University of Minnesota, 701 Park Avenue
Minneapolis, Minnesota 55415

## ABSTRACT

The role of lipids in the pathogenesis of focal glomerulosclerosis (FGS) was evaluated using two chemically different lipid lowering agents, clofibric acid and mevinolin. Pharmacologically, these two agents have different mechanisms of action. Clofibric acid affects both cholesterol and triglyceride metabolism, while mevinolin inhibits 3-hydroxy-3 methyl-glutaryl coenzyme A reductase, the rate limiting enzyme in cellular cholesterol synthesis. In two different models of FGS in which hyperlipidemia occurs, the obese Zucker rat and the 5/6 nephrectomy model, both agents significantly reduced FGS and albuminuria. Since glomerular hemodynamic function is normal in obese Zucker rats, these results suggested that lipids are an independent factor in the pathogenesis of FGS. Moreover, in the 5/6 nephrectomy model, the beneficial effects on glomerular structure of reducing serum lipids occurred despite persistent systemic and glomerular hypertension. Thus, we postulated that a synergistic interaction between lipids and hypertension might exist in the pathogenesis of FGS.

## INTRODUCTION

The notion that lipids participate in kidney injury has intrigued investigators for nearly a century. Whether perturbations in lipid metabolism directly contribute to renal injury or are only a frequent complication of kidney disease is unknown. Virchow commented on the "fatty degeneration of the renal epithelium in Bright's Disease" and stated that the only way "clear notions upon the subject can be obtained consists in reexamining whether the condition of fatty degeneration is a primary or a secondary one, whether it sets in as soon as the disturbance can be perceived, or whether it does not occur until some other perceptible disturbance has gone before" (1).

Histologically, lipid material has been observed in both the glomerulus and tubular cells in various renal diseases. In addition, foam cells, believed to be lipid-laden macrophages, have been identified in kidney vessels and interstitium, as well as in glomeruli of patients with renal disease. The extensive glomerular and tubular lipid deposition in young patients with the nephrotic syndrome led Munk, in 1916, to coin the term "lipoid nephrosis" (2). Subsequently, other investigators noted that a variety of different renal diseases were also characterized by deposits of lipids and/or the presence of foam cells. In 1951, Wilens, intrigued by the association of intercapillary glomerulosclerosis and "glomerular lipidosis" in hypertensive diabetic patients, concluded that "a combination of hyperlipidemia and elevated intraglomerular blood pressure might be responsible for the penetration of lipid containing materials into the intercapillary substance of the tufts" (3). Histological evidence of lipid deposition in diseased glomeruli has been confirmed by biochemical studies of whole glomeruli and preparations of purified glomerular basement membranes (4).

It has been established that diet-induced hypercholesterolemia can lead to glomerular injury and progressive nephron destruction in several species of laboratory animals. Moreover, in immune- and nonimmune-mediated models of glomerular disease, diets high in cholesterol content aggravate glomerular injury (5). The mechanism(s) whereby diet-induced hyperlipidemia exaggerated glomerular injury is unknown. These associations between lipids and glomerular disease led Moorhead and colleagues to speculate that progression of chronic kidney disease in man might be mediated by abnormal lipid metabolism (6). Although several dietary studies have demonstrated a relationship between lipid abnormalities and glomerular injury, there is a paucity of experimental data to support the contention that endogenous lipid abnormalities cause progressive renal disease.

EXPERIMENTAL DESIGN AND RESULTS

Hyperlipidemia and Renal Disease in Obese Zucker Rats: In our laboratory, the role of hyperlipidemia in modulating glomerular injury was suggested by results of studies in the obese Zucker rat (7,8). Obesity

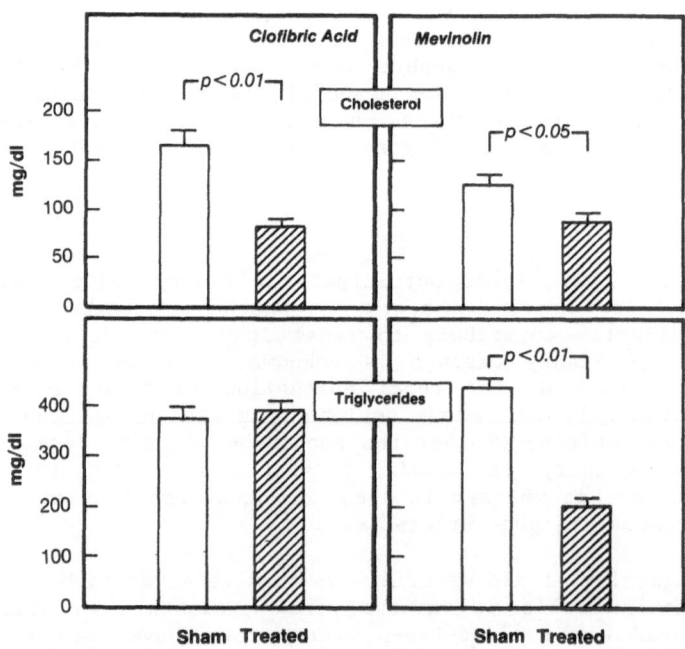

Figure 1. Fasting serum cholesterol and triglyceride levels in male obese Zucker rats treated until approximately six months of age with clofibric acid or mevinolin. Similar differences in lipid levels were noted throughout the study.

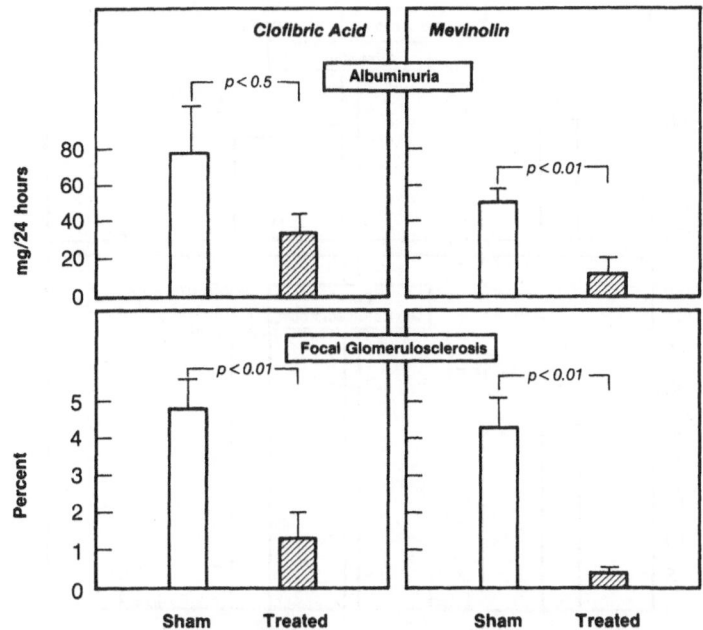

Figure 2. Twenty-four hour urine albumin excretion rates in obese Zucker rats treated until approximately six months of age with clofibric acid or mevinolin. Extent of focal glomerulosclerosis is depicted in clofibric acid or mevinolin treated rats after approximately six months of treatment.

in the Zucker rat is an autosomal recessive trait. Obese Zucker rats exhibit hyperphagia, hyperlipidemia, hyperinsulinemia and peripheral insulin resistance. Thus, these obese rats have metabolic abnormalities that resemble those seen in human obesity and type II diabetes mellitus.

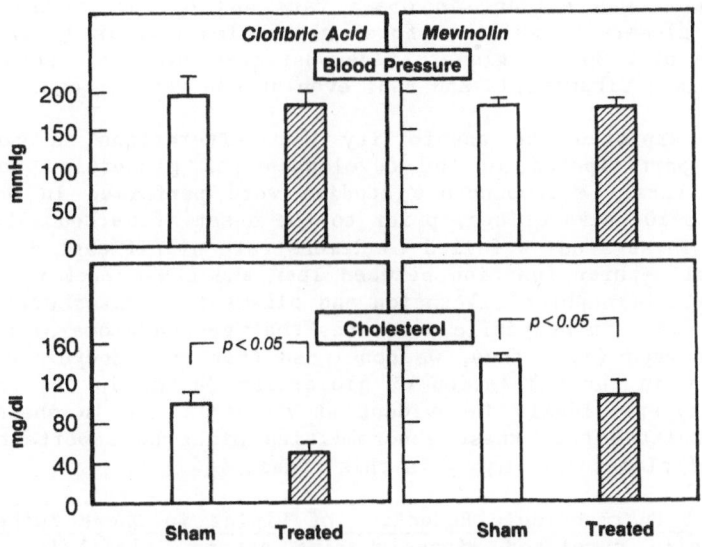

Figure 3. Tail cuff blood pressures after twelve weeks of treatment with clofibric acid or mevinolin in 5/6 nephrectomy rats. Blood cholesterol levels after twelve weeks of therapy are depicted in lower panel.

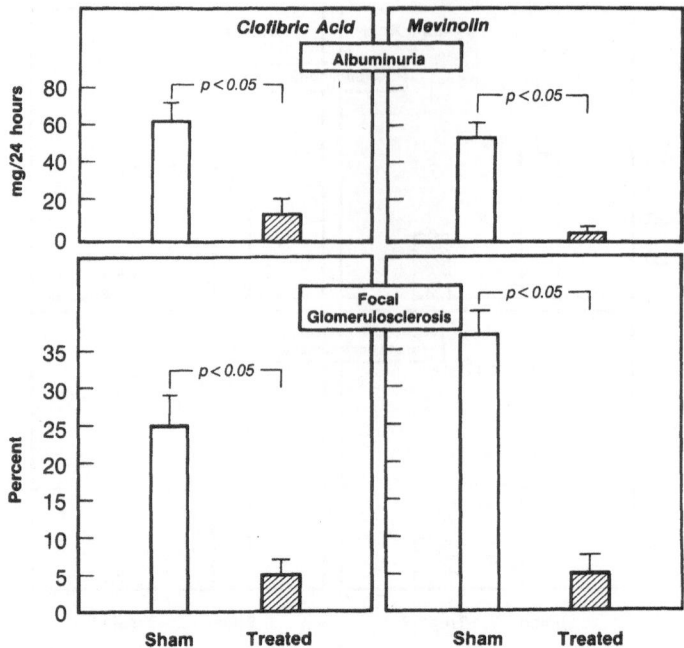

Figure 4.  Twenty-four hour urine albumin excretion rates in rats with 5/6 nephrectomy after twelve weeks of clofibric acid or mevinolin therapy in 5/6 nephrectomy rats is presented in lower panel.

Obese Zucker rats demonstrate elevated cholesterol and triglyceride levels at an early age.  These lipid abnormalities were associated with development of mild albuminuria (~10 mg/24 hours) by twelve weeks of age, and by six months of age albuminuria increased markedly (~100 mg/24 hours) (7).  No significant differences in glomerular architecture between obese and lean littermates were evident at six weeks of age by light microscopy.  However, by twelve weeks of age, expansion of the mesangial matrix was evident in obese rats and by six months of age, nearly 5% of glomeruli exhibited focal glomerulosclerosis (FGS).  By 14 months of age over 30% of glomeruli demonstrated FGS (7).  Lean littermates had minimal albuminuria and FGS, even at one year of age.

We also explored the possibility that alterations in glomerular hemodynamics participated in the development of glomerular injury in obese Zucker rats.  Micropuncture studies were performed in obese and lean rats at 8-10 weeks of age, prior to the onset of marked albuminuria or FGS (8).  These studies failed to demonstrate significant differences in superficial nephron function between lean and obese Zucker rats.  In particular, single nephron filtration and plasma flow rates were similar and no differences in hydraulic pressures that govern glomerular filtration were observed (8).  Thus, we concluded that nonhemodynamic factors were important in the initiation of glomerular damage in this model of FGS.  Since hyperlipidemia is evident at an early age in obese Zucker rats, we postulated that these abnormalities might be important in the development of glomerular injury in this model.

Effect of Pharmacologic Reduction of Lipids in Obese Zucker Rats: Two structurally unrelated lipid-lowering agents, clofibric acid and mevinolin were used to reduce serum lipid levels in obese Zucker rats (9).  Clofibric acid, the pharmacologically active form of clofibrate effects both cholesterol and triglyceride levels.  Mevinolin inhibits

3-hydroxyl-3 methylglutaryl coenzyme A reductase, the rate limiting enzyme of cellular cholesterol synthesis. This results in increased cell surface lipoprotein receptors and reductions in serum cholesterol.

Obese Zucker rats were treated daily with either clofibric acid (200 mg/kg body weight subcutaneously), or vehicle. In another experiment, obese rats were given 4 mg/kg body weight of mevinolin, or vehicle, subcutaneously. Treatment was initiated at six weeks of age and continued until rats were approximately 6 months old. Body weight, food intake, blood pressure, urine albumin excretion and serum cholesterol and triglycerides were regularly monitored. Both clofibric acid and mevinolin significantly lowered cholesterol levels. In addition, mevinolin significantly reduced triglyceride levels (Figure 1). Significant reductions in albuminuria and the frequency of FGS were also observed in obese rats treated with either lipid lowering agent (Figure 2). Neither clofibric acid or mevinolin caused significant changes in body weight, food intake or tail cuff blood pressures. These experimental results suggested that endogenous hyperlipidemia may contribute to progressive glomerular injury independent of glomerular hemodynamic factors.

Effect of Pharmacologic Reduction of Lipids in Rats with Reduced Renal Mass. Surgical reduction of renal mass (5/6 nephrectomy) in normal rats results in anatomical hypertrophy and increased function of the remaining nephrons. This adaptation to nephron loss is associated with the development of marked systemic hypertension and progressive destruction of the residual nephron population. It has been proposed that glomerular hemodynamic adaptations that occur after renal ablation, particularly glomerular hypertension, are critical to the initiation and progression of FGS in this model (10). In this regard, reduction in systemic and glomerular pressures by angiotensin converting enzyme inhibition ameliorated glomerular damage (11). Since hyperlipidemia occurs soon after nephron ablation, we postulated that lipid abnormalities might also contribute to nephron injury. To test this hypothesis, we treated Sprague-Dawley rats with 5/6 nephrectomy with either clofibric acid or mevinolin for twelve weeks (12).

Both clofibric acid and mevinolin significantly reduced serum cholesterol levels in 5/6 nephrectomy rats without affecting food intake or body weight (Figure 3). Rats with 5/6 nephrectomy were hypertensive and neither agent significantly lowered systemic blood pressure (Figure 3). However, both clofibric acid and mevinolin significantly reduced albuminuria and the incidence of FGS (Figure 4).

Micropuncture studies were also performed in 5/6 nephrectomy rats after 3-4 weeks of clofibric acid treatment and prior to development of FGS and marked albuminuria. These studies demonstrated that clofibric acid treatment did not reduce single nephron function or glomerular capillary pressures (13). Thus, these results suggested that hyperlipidemia that accompanies reduction in functioning nephron mass may be an independent risk factor in the development of FGS.

DISCUSSION

Our data support the concept that lipids are an important factor in progressive glomerular injury. The studies in the Zucker rat model of obesity demonstrated that FGS developed in the absence of nephron hyperfiltration or glomerular hypertension. Both clofibric acid and mevinolin significantly reduced serum lipid levels and ameliorated glomerular injury in obese Zucker rats, lending credence to the role of lipid abnormalities in the pathogenesis of FGS. In addition, amelioration of glomerular injury and albuminuria by clofibric acid or mevinolin in 5/6 nephrectomy rats occurred independently of changes in systemic hypertension or hyperfunctioning of remaining nephrons. However, it has been shown that reduction in both systemic and glomerular hypertension is also associated with reduction in FGS in 5/6 nephrectomy rats (11).

Collectively, studies in the 5/6 nephrectomy model suggest that alterations in lipid metabolism and glomerular hypertension could interact synergistically in the pathogenesis of FGS. Indeed, an analagous relationship between these factors has been shown to be important in the development of atherosclerosis. The glomerulus has many structural features which resemble arteries commonly involved in atherosclerosis. Mesangial cells, for example, are structurally similar to arterial smooth muscle cells, important in the pathogenesis of atherosclerosis. Lipid-laden macrophages, or foam cells, are frequently found in both early atherosclerosis lesions, as well as in human and experimental models of FGS. Thus, factors important in the pathogenesis of FGS may be similar to those that influence the development of atherosclerosis. The relative importance of lipids versus glomerular hypertension in the initiation and progression of glomerular disease is presently unknown. In addition, the role of other mechanisms, such as coagulation factors is still incompletely understood. Nonetheless, the studies reviewed herein underscore the complexity of the process that leads to chronic glomerular injury and progressive loss of renal function.

ACKNOWLEDGEMENTS:

Part of this work was supported by grants from the National Institutes of Health, RO1-AM37396 and R23-AM37112 and Hennepin Faculty Associates. Donna Boehmer and Diane Erickson helped prepare the manuscript.

REFERENCES

1.  Virchow R. 1860. A more precise account of fatty metamorphosis. In Cellular Pathology, translated from the second edition by F. Chance. 350.
2.  Munk F. Die Nephrosen. Die Lipoidnephrose. Medizinische Klinik 12:1047, 1916.
3.  Wilens SL, Elster SK, Baker JP. Glomerular lipidosis in intercapillary glomerulosclerosis. Ann Int Med 34:592, 1951.
4.  Misra RP. 1972. The glomerular basement membrane. In Renal Disease. Sir Douglas Black, ed. 3rd edition Blackwell Scientific Publications, Oxford. 187.
5.  Klahr S, Buerkert J, Purkerson ML. Role of dietary factors in the progression of chronic renal disease. Kidney Int 24:579, 1983.
6.  Moorhead JF, El-Nakos M, Chan MK, Varghese Z. Lipid nephrotoxicity in chronic progressive glomerular and tubulo-interstitial disease. Lancet ii:1309, 1982.
7.  Kasiske BL, Cleary MP, O'Donnell MP, Keane WF. Effects of genetic obesity on renal structure and function in the Zucker rat. J Lab Clin Med 106:598, 1985.
8.  O'Donnell MP, Kasiske BL, Cleary MP, Keane WF: Effects of genetic obesity on renal structure and function in the Zucker rat. II. Micropuncture studies. J Lab Clin Med 106:605, 1985.
9.  Kasiske BL, O'Donnell MP, Daniels F, Keane WF: The lipid lowering agent clofibric acid reduces glomerular injury in obese Zucker rats. Clin. Res. 34:600A, 1986.
10. Hostetter TH, Olson JL, Rennke HG, Venkatachalam MA, Brenner BM. Hyperfiltration in remnant nephrons: a potentially adverse response to renal ablation. Am J Physiol 241:F85, 1981.
11. Anderson S, Meyer TW, Rennke HG, Brenner BM. Control of glomerular hypertension limits glomerular injury in rats with reduced renal

mass.   J Clin Invest 76:612, 1985.

12.  Kasiske BL, O'Donnell MP, Daniels F, Keane WF:  The lipid lowering
     agent clofibric acid ameliorates renal injury in the 5/6 nephrectomy
     model of chronic renal failure.  Clin Res 33:488A, 1985.

13.  O'Donnell MP, Kasiske BL, Daniels F, Keane WF.  Clofibric acid does
     not affect glomerular hemodynamics in rats with 5/6 nephrectomy.
     Kidney Int 29:324, 1986.

# ALUMINUM AN UREMIC TOXIN

Patrick C. D'Haese, Frank L. Van de Vyver, Ludwig V. Lamberts, and Marc E. De Broe

University of Antwerp, University Hospital Antwerp, Dept. of Nephrology-Hypertension, Wilrijkstraat 10, B-2520 Edegem (Antwerpen), Belgium

## INTRODUCTION

Aluminum has historically been regarded as non essential, since up to now no physiological function could be ascribed to it. Environmental exposure to aluminum is virtually universal as aluminum constitutes a substantial part of the earth's crust (8%) and is found in food, medicine and cosmetics. Besides this aluminum has a lot of industrial applications. Until recently, aluminum was generally considered to be a relatively non-toxic metal, which is reflected by the scarce and often wrong information presented in the literature dealing with the toxicology of trace elements. However, since it was found that aluminum is the causative factor in some dialysis-related diseases, the issue of the origin and physio-pathology of aluminum accumulation-toxicity in these patients has received the interest it deserves.

## METABOLISM AND PHARMACOKINETICS OF ALUMINUM

Notwithstanding its ubiquity, serum, blood and tissue aluminum levels in healthy individuals are quite low[1]. This is due to the presence of a strong gastrointestinal barrier and by an efficient renal elimination of the metal.

Since direct investigation of intestinal absorption has been impossible due to the lack of suitable radionuclides, the amount of aluminum absorbed by the gastrointestinal tract has been estimated by making the sum of the urinary excretion rate and the probable rate of aluminum accumulation in the tissues[2]. The latter can, according to the body content, be estimates to maximally 2 µg/day[2]. The normal urinary aluminum excretion is found to be 20 to 50 µg/day. According to these data the gastrointestinal uptake fraction for a normal aluminum intake of 20 mg is supposed to be 0.1 to 0.5%[2,3]. Almost all (90%) of this fraction is subsequently excreted by the kidney. Indeed, the kidney plays an essential role in the metabolism of aluminum in man. The majority (80%) of the aluminum in the serum is bound to transferrin which is the most important or even the sole carrier protein of this element in plasma[4,5]. Unbound aluminum is filtered by the glomerulus, reabsorbed and secreted (exocytosis) by the proximal tubular cells. It is obvous that in patinents with severe renal failure aluminum can rapidly increase in serum and tissues and may thus behave as an uremic toxin.

ALUMINUM IN HUMANS

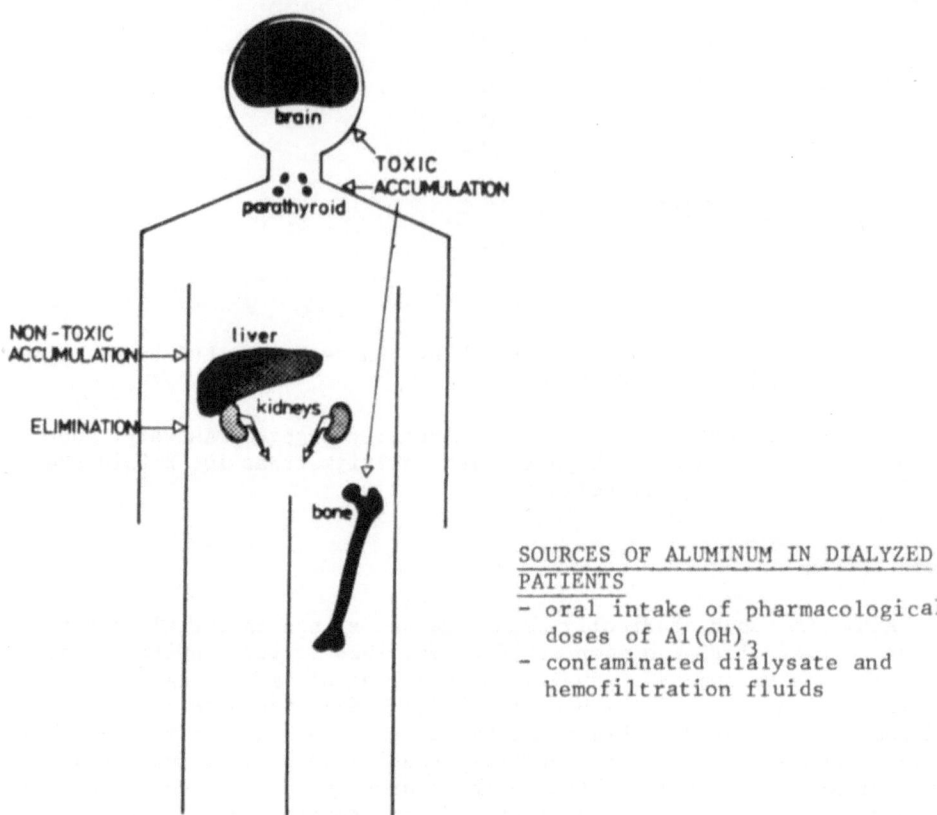

SOURCES OF ALUMINUM IN DIALYZED
PATIENTS
- oral intake of pharmacological
  doses of $Al(OH)_3$
- contaminated dialysate and
  hemofiltration fluids

SOURCES OF ALUMINUM ACCUMULATION

Patients with severe renal failure (creatinine clearance below 10 ml/min) and those already admitted to a chronic dialysis program, ingesting pharmacological doses of $Al(OH)_3$ are at considerable risk for aluminum accumulation-toxicity. Moreover in chronic dialysis the aluminum present in the dialysate and parenteral fluids circumvenes the natural barriers and may present an even greater hazard for aluminum absorption [1], [6], [7].

It has been suggested that individual aluminum absorption rates vary and that some patients may absorb excessive amounts. A number of factors have been suggested to modulate aluminum absorption from the gastrointestinal tract[8]. These are parathyroid hormone (PTH), pH, vitamin D, cumulative amount of aluminum intake, size of the aluminum species and fluoride. Recent data[9] show that children with ESRF are highly susceptible to aluminum accumulation originating from aluminum hydroxide intake.

## TOXICITY OF ALUMINUM

Extensive reviews of aluminum exposure and health issues have concluded that in patients with normal renal function, environmental exposure to aluminum and its compounds presents no risk to human health[10]. In employees of aluminum processing industries plumonary lesions and encephalopathy due to inhalation of aluminum are described and an increased risk for lung cancer was suggested repeatedly[11]. Phosphate depletion due to the therapeutic intake of aluminum hydroxide (e.g. gastrointestinal ulcer) has also been observed. However, the prototype patient for aluminum accumulation-toxicity is the patient with end stage renal failure treated by one or another type of dialysis. In these patients the clinical expresson of an exaggerated aluminum body burden include vitamin D resistant osteomalacia[12], dialysis encephalopathy[13] and to a lesser extent, microcytic anemia[14].

## ALUMINUM LEVELS AND TOXICITY

Compared to a normal population aluminum levels in ESRF patients treated by dialysis are significantly elevated[1]. Moreover the element is distributed heterogeneously in the body and is particularly stored in bone and liver[1,7,15]. It appears however that the relationship between tissue concentration and toxicity is rather complex (Figure 1)[7]. While bone and liver both show elevated aluminum levels, toxicity has only been established in bone. On the contrary, aluminum toxicity has been demonstrated in the brain although brain levels are relatively low (Table 1). This suggests that toxicity of aluminum is directly related to an ultrastructural critical localization.

Table 1.

| ALUMINUM | LEVEL | TOXICITY |
|---|---|---|
| Bone | high | established |
| Liver | high | absent |
| Brain | relatively low | established |
| Parathyroid | relatively low | interference with PTH release (?) |

The subcellular localization of aluminum in the lysosomes and the Küppfer cells of the liver can be considered as an important mechanism for neutralizing the metals toxic effect. Among the different localizations of aluminum in bone, the osteoid/calcified bone boundary (OCBB) seems to be a critical one. Indeed, patients with aluminum induced osteomalacia have the element localized at the OCBB. A significant correlation was observed between the histomorphological measure of stainable aluminum at the OCBB and the bone aluminum concentrations (figure 2). All cases with aluminum levels below 30 μg/g showed none or negligible stainable aluminum (Al < 1%).

It is worth noting that aluminum and iron have often been found at the same ultrastructural sites (e.g. in liver lysosomes, the OCBB, bone marrow cells, and parathyroid chief cells)[7]. Since iron may substantially interfere with the aluminon staining[16], attention must be paid when results of histochemical investigation are evaluated in thes dialyzed patients prone to develop iron overload.

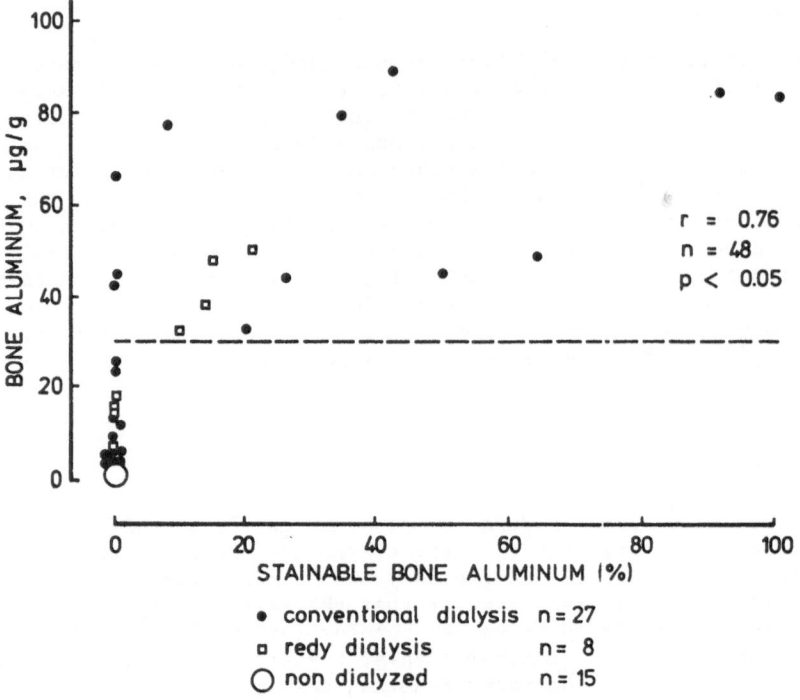

Fig. 2. Relationship between the histochemical measure of aluminum present at the OCBB and the bone aluminum concentrations.

Furthermore the occurrence of aluminum in the secretory granules of the parathyroid cells[17] may inhibit PTH release resulting in the documented relatively low serum PTH levels found in dialysis patients with aluminum induced osteomalacia.

Biochemical and epidemiological data have shown that aluminum is also strongly involved in the pathogenesis of a neurological syndrome called dialysis encephalopathy. The disease is characterized by speech difficulties, seizures and progressive dementia terminating in death when dialysis is continued. Although it appears that in these encephalopatic patients aluminum is particularly stored in the gray brain matter, very little is known concerning its ultrastructural localization in brain tissue. How far aluminum is the causative or contributing factor in the development of Alzheimer's disease, a progressive form of presenile dementia, is not clear.

DIAGNOSIS AND THERAPY OF ALUMINUM OVERLOAD

Since only a small fraction of aluminum is present in the blood one can reasonably state that basal serum aluminum has only a limited value for assessing the total aluminum body burden and will particularly reflect recent exposure (table 2).

Table 2. Value of serum aluminum

- Parameter of acute exposure
- Limited correlation with total body aluminum
- Useful indicator for toxicity when determined regularly
  e.g. every 4 months
- Monitoring as soon as exposure exists

| Serum aluminum level | Implications |
|---|---|
| - Less than 50 µg/L | - Aluminum-related bone disease unlikely but possible |
| - 50 - 100 µg/L | - Aluminum-related bone disease quite possible, especially if serum PTH levels are low or low-normal, hyper-calcemia is present, or the DFO test is positive |
| - Greater than 100 µg/L | - Aluminum-related bone disease very probable, but not invariably present expecially if serum PTH levels are high. |

Until now, bone aluminum and histology of undecalcified bone specimen have proven to be the best parameters for the diagnosis of aluminum accumulation. It has been suggested recently[7,18] that the increment of the serum aluminum level after chelation with desferrioxamine (DFO) is a helpful parameter for assessing the aluminum body burden. However, since with the DFO test false negatives as well as false positives are frequently observed[7,9], results of the DFO test are only indicative for abnormal aluminum body burden and bone biopsy should therefore be performed for definite diagnosis.

Since the DFO-aluminum complex is easily removable by dialysis (molecular weight +/- 650) and since the chelator can be used (maximally 2 g/week) over a longer period "long-term DFO therapy" is possible.

Although the aluminum removed after DFO administration is rather low, the clinical improvement is remarkebly rapid, especially in cases of aluminum induced bone disease. It is clear that in order to allow a correct diagnosis of the DFO test as well as evaluation of therapy standaridized protocols should be followed[15]. Tentative alogrithms for monitoring aluminum levels and evaluation of therapy are shown in Figure 3.

Since successful therapy of aluminum accumulation in CRF patients depends greatly on the ability to remove the DFO-aluminum complex the use of a hemoperfusion column (ALUKART) in combination with a conventional dialyzer is recommended since with the latter device the DFO-aluminum clearance is increased substantially (figure 4).

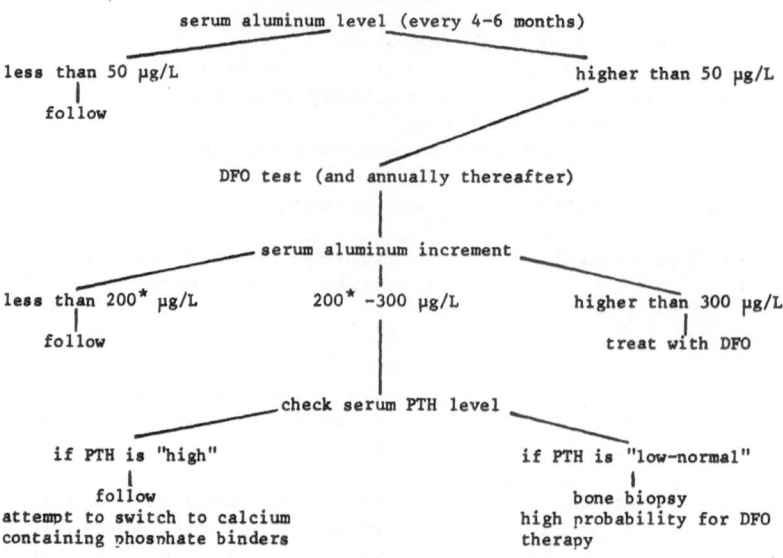

**I. ASYMPTOMATIC PATIENTS**

serum aluminum level (every 4-6 months)

- less than 50 μg/L → follow
- higher than 50 μg/L

DFO test (and annually thereafter)

serum aluminum increment

- less than 200* μg/L → follow
- 200* -300 μg/L
- higher than 300 μg/L → treat with DFO

check serum PTH level

- if PTH is "high" → follow, attempt to switch to calcium containing phosphate binders
- if PTH is "low-normal" → bone biopsy, high probability for DFO therapy

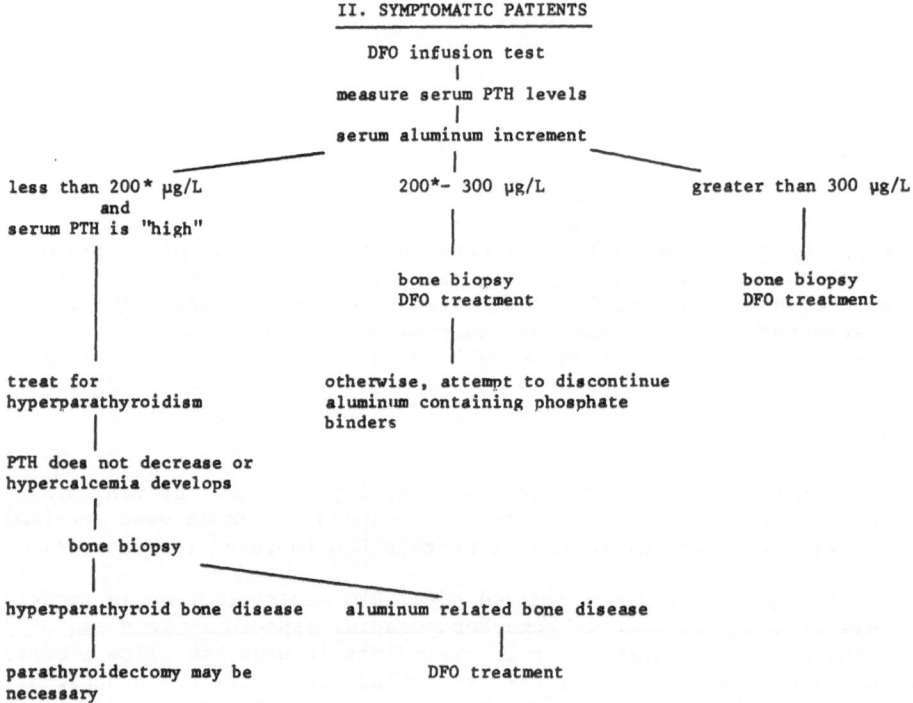

**II. SYMPTOMATIC PATIENTS**

DFO infusion test

measure serum PTH levels

serum aluminum increment

- less than 200* μg/L and serum PTH is "high" → treat for hyperparathyroidism → PTH does not decrease or hypercalcemia develops → bone biopsy
  - hyperparathyroid bone disease → parathyroidectomy may be necessary
  - aluminum related bone disease → DFO treatment
- 200*- 300 μg/L → bone biopsy, DFO treatment → otherwise, attempt to discontinue aluminum containing phosphate binders
- greater than 300 μg/L → bone biopsy, DFO treatment

* We advocate a lower limit of normal for the DFO test of 100 μg/L although 200 μg/L is the usual figure quoted.

Fig. 3.   Shema for aluminum monitoring and therapy.
(Published with the permission of T.S. Ing et al[15]).

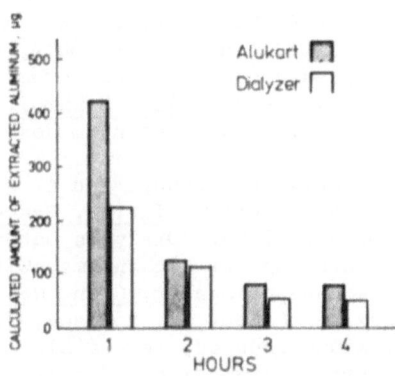

Fig. 4.  Extraction of aluminum by the ALUKART and the dialyzer
         during a dialysis session

CONCLUSION

    Aluminum can be considered as an uremic toxin.  The dialyzed patient
is the predilected target for a toxic accumulation of aluminum.  However,
recent intensive research in that field has allowed the clinician to
monitor carefully and to detect early signs of toxicity.  Furthermore,
secondary prevention has become possible by identification of risk factors
such as : children, diabetic nephropathy and parathyroidectomy.

REFERENCES

1. F.L. Van de Vyver, A.B. Bekaert, P.C. D'Haese, H. Kellinghaus,
     U. Graefe and M.E. De Broe, Serum, blood, bone and liver aluminium
     levels in chronic renal failure, Trace Elem. Med. 3:52 (1986).
2. P.O. Ganrot, Metabolism and possible health effects of aluminum,
     Env. Hlth. Med. 3: 52 (1986).
3. W.D. Kaehny, A.P. Hegg, and A.C. Alfrey, Gastrointestinal absorption
     of aluminum from aluminum-containing antacids, New Engl. J. Med.
     296:1389 (1977).
4. S.W. King, M.R. Wills, and J. Savory, Serum binding of aluminum,
     Res. Comm. Chem. Pathol. & Pharmacol. 26:161 (1979).
5. G.A. Trap, Plasma aluminum is bound to transferrin, Life Sci. 33:311
     (1983).
6. F.L. Van de Vyver, J.P. Van Waeleghem, M.E. De Broe, P.C. D'Haese,
     and A. Heyndrickx, Water treatment and dialysis dementia, Lancet 4:
     1106 (1982).
7. F.L. Van de Vyver, F.J. E Silva, P.C. D'Haese, A.H. Verbueken, and
     M.E. De Broe, Aluminum toxicity in dialysis patients, Contr. Nephrol.
     55:198 (1986).
8. A.C. Alfrey, Gastrointestinal absorption of aluminium, Clin. Nephrol.
     24:84 (1985).

9. A.M. Roodhooft, F.L. Van de Vyver, P.C. D'Haese, K. Van Acker, W.J. Visser, and M.E. De Broe, Aluminum accumulation in children on chronic dialysis : predictive value of serum aluminum levels and desferrioxamine infusion test, Clin. Nephrol. (1986, submitted).

10. H.R. Skalsky, and R.A. Carchman, Aluminium homeostasis in man, J. Am. Coll. of Toxicol. 2:405 (1983).

11. A. Andersen, B.E. Dahlberg, K. Magnus, and A. Wannag, Risk of cancer in the Norwegian aluminum industry, Int. J. Cancer 29:295 (1982).

12. T. Drueke, and G. Cournot-Witmer, Dialysis osteomalacia : clinical aspects and physiopathological mechanisms, Clin. Nephrol. 24:26 (1985).

13. A.C. Alfrey, Dialysis encephalopathy, Clin. Nephrol. 24:15 (1985).

14. M. Touam, F. Martinez, B. Lacour, R. Bourdon, J. Zingraff, S. Di Giulio, and T. Drueke, Aluminum-induced, reversible microcytic anemia in chronic renal failure : clinical and experimental studies, Clin. Nephrol. 19:295 (1983).

15. M.E. De Broe, P.C. D'Haese, and F.L. Van de Vyver, Aluminum toxicity, in: "Handbook of Dialysis," T.S. Ing and T.S. Daugirdas, eds., Little Brown & Co (1987, in press).

16. A.H. Verbueken, F.L. Van de Vyver, W.J. Visser, R.E. Van Grieken, and M.E. De Broe, Laser microprobe mass analysis (LAMMA) to verify the aluminon staining of bone, Stain Technol. 61:287 (1986).

17. A.H. Verbueken, F.L. Van de Vyver, W.J. Visser, F. Roels, R.E. Van Grieken and M.E. De Broe, Use of laser microprobe mass analysis (LAMMA) for localizing multiple elements in soft and hard tissues, Biol. Trace Elem. Res. (1986, in press).

19. A. Fournier, P. Fohrer, P. Leflon, P. Moriniere, M. Tolani, G. Lambrey, R. Demontis, J.L. Sebert, F.L. Van de Vyver, and M.E. De Broe, The desferrioxamine test predicts bone aluminum burden induced by $Al(OH)_3$ in uraemic patients but not mild histological osteomalacia, in: "Proc. EDTA-ERA," A.M. Davison, ed., Pitman Med., London, 21:371 (1984).

# THE INSULIN-RESISTANCE INDUCING FACTOR ASSOCIATED WITH UREMIA

Dean H. Lockwood, Gary R. Hayes, and Michael L. McCaleb
Endocrine-Metabolism Unit, Department of Medicine
University of Rochester School of Medicine and Dentistry
Box 693 Rochester, NY 14642

Insulin resistance is a prominent feature of several clinical conditions including obesity, non-insulin dependent diabetes mellitus (Type II), fasting, glucocorticoid excess, myotonic dystrophy and uremia. Evidence for insulin resistance in chronic renal disease is provided by the demonstration of elevated circulating levels of insulin and a blunted glucose response to an injection of the hormone[1]. More direct evidence of insulin resistance is provided by a reduction in insulin-stimulated glucose uptake by skeletal muscle when measured by the forearm perfusion technique[2] or by the euglycemic insulin clamp[3]. In _vitro_ studies in liver, muscle and fat from uremic rats have shown resistance of at least some parameters of insulin action[4].

Our laboratory has been interested in insulin action at the cellular and subcellular levels in normal and insulin-resistance states including uremia. Much of our understanding of the cellular lesions in insulin resistance has resulted from new knowledge of the mechanisms of insulin action. Insulin initiates its actions by interacting with specific glycoprotein receptors on the surface of the plasma membrane. In ways not completely understood, a signal is generated and transmitted coupling insulin binding to insulin action. A likely participant in coupling is the insulin-receptor kinase which is activated by hormonal binding and stimulates phosphorylation of tyrosine residues. Other candidates for signalling are a group of small molecular weight compounds which presumably are generated in, or released from, the plasma membrane in response to insulin binding, and traverse the cell to influence various well known anabolic reactions. These second messengers probably stimulate protein kinases and phosphatases which, in turn, activate or inhibit certain insulin-regulated enzymes. Examples of such enzymes are pyruvate dehydrogenase, glycogen synthase and fatty acid synthetase. The mechanism by which insulin stimulates glucose transport in muscle and fat has recently been partially clarified. In response to insulin, glucose transporters from intracellular locations migrate to the plasma membrane where they contribute to increased glucose uptake.

To better understand the cellular alteration(s) responsible for insulin resistance of uremia, we have extensively investigated insulin action in fat cells from 80% nephrectomized rats. This animal model is characterized by a 4-5 fold increase in blood levels of urea nitrogen and creatinine and a doubling of the circulating insulin levels. Levels of plasma glucose and the counterregulatory hormones glucagon, corticosterone and growth hormone remain unchanged[5].

Adipocytes from uremic animals have a normal number of insulin receptors and there is no alteration in insulin binding characteristics[5]. The insulin receptor tyrosine kinase activity is also unchanged by uremia. This parameter of insulin action was investigated in vitro utilizing partially purified insulin receptors and a synthetic substrate as well as in situ by studying the effect of insulin on $^{32}$P incorporation into the insulin receptor itself.

In contrast, the glucose transport system is significantly impaired by uremia. Uptake of the non-metabolizable glucose analog 3-0-methyl-glucose or 2-deoxy-D-glucose is markedly reduced in uremic adipocytes in the absence and presence of insulin[5]. Preliminary studies in our laboratory evaluating further the mechanism of reduced glucose uptake indicate that the number of glucose transporters available in the cell are diminished by about 20% in uremia. Additional studies of insulin action in uremia have revealed reduced glucose metabolism and decreased production or activation of the "second messenger" that stimulates pyruvate dehydrogenase[6]. Collectively, these results clearly indicate that the insulin resistance in this animal model system resides beyond the insulin receptor (Fig. 1).

Although tissues from uremic humans have not been extensively investigated, the existing studies support the concept that this resistance is also due to a post-receptor defect(s). For example, Smith and DeFronzo[3] demonstrated reduced insulin action on glucose utilization in peripheral tissues of uremic patients without a concomitant alteration in insulin binding to their circulating monocytes.

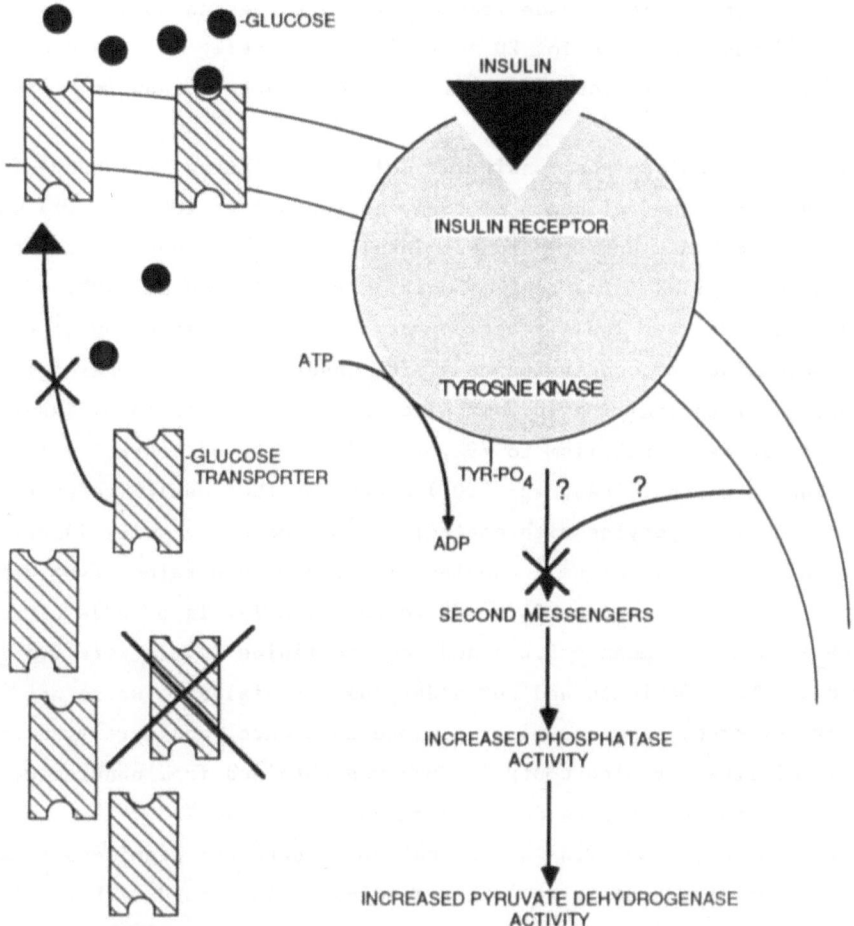

Fig. 1. Schematic representation of impaired insulin action in uremia.
The sites of the observed decreases in "second messenger"
production and glucose transporter number are indicated by X.
The latter may lead to decreased movement of transporters to
the plasma membrane thus decreasing glucose uptake.

We were particularly interested in the insulin resistance of uremia
in view of the evidence that a circulating factor may play a causal role.
In uremic man, Hampers and colleagues reported an improvement in glucose
intolerance in patients undergoing dialysis therapy[7].  Following dialysis,
there is amelioration of insulin resistance as demonstrated by  infusions
of glucose and insulin.  Furthermore, the uptake of glucose in the absence
of insulin is reduced when a variety of tissues from normal animals is
incubated with partially purified serum from uremic patients[8].  Our

observation that adipose tissue from uremic rats regains normal insulin responsivity when cultured for 20 hours in a chemically defined medium[5] further supports a role for circulating factors in this insulin resistant state.

To further explore the possibility of a circulating resistance factor, we have evaluated the influence of dialyzed (cutoff at 1000 mol wt) whole or fractionated human uremic serum on insulin action in nonuremic rat adipocytes maintained for 3 to 5 ½ hours in Parkers' Medium 199[9]. In our standard bioassay, insulin action was assessed by measuring $[^{14}C]$ glucose oxidation and conversion to lipids under conditions where glucose transport was rate limiting (0.2 mM glucose). An inhibition of insulin-stimulated glucose metabolism to >9%, i.e. 2.31 S.D. of control value, was considered a significant (p< 0.01) change of the insulin response. Pretreatment of adipocytes with control serum produced results identical to those of Medium 199 alone. Samples of serum were obtained from uremic patients ranging in age from 21 to 88 years. The levels of urea nitrogen were 109 ± 42 mg/dl (mean ± S.D.) and the creatinine levels were 7.3 ± 3.2 mg/dl. These patients had not undergone any dialysis, where not receiving any drugs known to alter glucose tolerance, and were free of diabetes mellitus. Pooled control serum was obtained from nonuremic, nondiabetic subjects ranging in age from 20 to 70 years[9].

Approximately two-thirds of the patients contained significant amounts of insulin-resistance activity. This finding is in accord with previous in vivo studies demonstrating that not all uremic patients are glucose intolerant. However, most of our patients with more severe uremia (BUN >90 mg/dl and creatinine >4 mg/dl) had the insulin-resistance inducing factor in their serum. The ability of uremic serum to inhibit insulin action was time- and concentration-dependent. If present, the resistance factor manifested itself between 3 to 5½ hours of incubation and did not induce a greater degree of resistance beyond the latter time[10]. Concentration studies revealed a relatively large concentration of serum (40% to 50%) was necessary to demonstrate insulin resistance in our bioassay. In additional investigations, the resistant serum recreated those cellular alterations observed in adipocytes obtained from partially nephrectomized rats. The serum inhibited basal and insulin-stimulated 2-deoxy-D-glucose transport and glucose metabolism without altering either insulin sensitivity or insulin binding[9]. To date, we have not investigated the influence of uremic serum on insulin receptor tyrosine kinase or on the glucose transporter, translocation process.

The insulin-resistance factor present in uremic serum has been partially characterized and purified[10]. As shown in Table I, the insulin-resistance activity is heat stable, but is completely distroyed by digestion of serum with trypsin indicating that the factor has a protein component. This factor is also free of asparagine-linked carbohydrate moities, is stable within the pH range of 5-9, and has an apparent isoelectric point (pI) between 6 and 7. Several lines of evidence indicate that the insulin-resistance activity is not a known counterregulatory hormone or substrate known to antagonize insulin action. The levels of insulin, parathyroid hormone and growth hormone in the sera of uremic patients do not correlate with the degree of insulin-resistance activity. The addition of hormones (insulin, glucagon, parathyroid hormone, growth hormone and cortisol) known to be elevated in uremia do not cause insulin resistance in our bioassay system. Finally, a partially purified fraction of uremic serum that possesses the insulin-resistance activity, does not contain any detectable amounts of the hormones listed above, or the neurotransmitters, dopamine, norepinephrine, epinephrine or free fatty acids which antagonize insulin action[10]. The prolonged time dependency for induction of insulin resistance suggested that protein synthesis is an obligatory step. This was confirmed by the demonstration that the addition of the protein synthesis inhibitor, cycloheximide, to the bioassay system prevented the expression of resistance by uremic serum, but did not influence insulin action in the presence of control serum[10].

TABLE 1

Characteristics fo Insulin Resistance Factor
For Uremic Human Serum

| Treatment | Result |
|---|---|
| A) Heat: 100ºC, 5 min (N=5) | Stable |
| B) Trypsin: 1 mg/ml, 37ºC, 1 hr (3) | Inactivated |
| C) Cortisol, parathyroid hormone, glucagon, free fatty acids, norepinephrine, epinephrine | Not present |
| D) Concanavalin A Column   (6) | Not absorbed |
| E) pH change:   3   (2) | Inactivated |
|     5 to 9   (4) | Stable |
|     12   (2) | Inactivated |
| F) Chromatofocusing   (4) | Apparent pI 6 to 7 |
| G) Cycloheximide (0.1µM) during bioassay   (4) | Resistance not expressed |

The insulin-resistance factor has been purified to $\sim$ 200,000-fold through the sequential employment of heating (100° C, 5 min), gel chromatography (Sephadex G-25) and ion exchange chromotography (DEAE cellulose). Preliminary experiments using HPLC indicate that this partially purified fraction is not near homogeneity.  Because the resistance activity did not elute in the total included volume during gel chromatography, the molecular size of insulin resistance activity was estimated using dialysis tubing with different molecular weight cutoffs.  The resistance activity in whole serum was retained within dialysis tubing of 3,500 mol wt.  However, dialysis under these conditions following heat treatment resulted in a loss of activity suggesting that the inhibitory factor is released from a larger molecular weight substance during this initial step of purification. The insulin-resistance activity was retained when the heat treated supernatant was dialyzed in tubing of 1,000 mol wt cutoff and lost when tubing of 2,000 mol wt cutoff was employed[10].  From these studies, we conclude that the factor has a mol wt characeristic of "middle molecules" which are, of course, unique to renal failure.

To further explore the possibility that the insulin-resistance factor is a uremic toxin, we evaluated the ability of dialyzed and partially purified serum from nonuremic patients to induce insulin resistance (Table II).  Using the bioassay system, there was no evidence of resistance activity in sera from patients with the insulin-resistant states of uncomplicated obesity, type II diabetes mellitus, fasting or aging[10]. In addition, sera from a few subjects with elevated circulating levels of glucocorticoids and one subject with acromegly did not induce resistance. Furthermore, as predicted by in vivo studies, insulin-resistance activity was markedly reduced in serum after patients underwent chronic hemodialysis therapy[10].

In summary, a circulating small molecular weight peptide, unique to uremia, is apparently capable of inducing insulin resistance in target cells by protein synthesis-dependent mechanisms.  This acidic "uremic toxin" appears to be distinct from the basic peptide present in uremic serum that alters basal metabolism of glucose in insulin-responsive and nonresponsive tissues[8].  This factor, described by Dzurik, also behaves differently on gel chromatography.  Our current studies are concerned with extending our knowledge of the mechanism by which this factor induces a post-receptor defect.  Future directions include further attempts at purification of the factor.  To accomplish this, greater amounts of toxin are needed.  Urine from uremic patients and filtrates from hemodialysis will be examined for the presence of insulin-resistance activity.  From

TABLE 2

## Status of Uremic Factor in Other Conditions of Insulin Resistance

| Subjects | Age | IBW | Glucose | Insulin | Insulin-Stimulated Total Glucose Utilization | |
|---|---|---|---|---|---|---|
| | | | | | Dialyzed Serum | Purified Serum |
| | (yr) | (%) | (mg/dl) | ( μU/ml) | $(nmol \cdot 10^5 cells \cdot 90 min)$ | |
| Control (17)[†] Buffer | 26+4[§] | 98+10[§] | 89+14[§] | 15+3[§] | 10.2+ 1.6[Π] 10.5+ 1.6 | 9.1+ 0.2[Π] 9.1+ 0.5 |
| Obese (7) Control | 44+9 | 179+25 | 110+35 | 29+9 | 11.1+ 0.8 11.2+ 0.6 | 11.0+ 1.3 9.8+ 1.1 |
| Type II DM (6) Control | 55+6 | 138+28 | 188+83 | 30+16 | 8.3+ 0.7 8.6+ 0.8 | 10.7+ 1.0 9.8+ 0.9 |
| Fasting(3) Control | 29+1 | 108+7 | 71+8 | 9+2 | 9.9+ 1.7 9.9+ 1.9 | 9.4+ 1.0 9.2+ 1.3 |
| Aged (6) Control | 74+2 | 110+15 | 80+11 | 16+6 | 8.7+ 1.0 9.2+ 1.3 | ND[¶] |

[†] Number of subjects are given in parentheses.   [§] Data in this column are mean + S.D.
[Π] Data in this column are mean + S.E.  All values of insulin-stimulated total glucose utilization are not significantly different from paired control experiments.
[¶] Not determined. Reproduced by permission from reference 10.

a practical standpoint, a bioassay which is more sensitive to the insulin-resistance factor is needed.  Preliminary experiments indicate that dialyzed uremic serum diluted 100-fold inhibits insulin-stimulated lipogenesis in cultured hepatocytes.  Finally, development of an antibody to the insulin-resistant factor would be advantageous for several reasons.  The availability of an antibody could aid in further purification of the factor, help identify tissues of origin, and perhaps be clinically useful through establishment of a radioimmunoassay.

## References

1.  R. A. DeFronzo, R. Andres, P. Edgar, and W. G. Walker, Carbohydrate metabolism in uremia: a review, Medicine 52:469 (1973).
2.  F. B Westervelt, Abnormal carbohydrate metabolism in uremia, Am. J. Clin. Nutr. 21:423 (1968).
3.  R. A. DeFronzo, J. D. Tobin, J. W. Rowe, and R. Andres, Glucose intolerance in uremia, J. Clin. Invest. 62:425 (1978).
4.  J. M. Amatruda, J. N. Livingston, and D. H. Lockwood, Cellular mechanisms in selected states of insulin resistance: human obesity, glucocorticoid excess, and chronic renal failure, Diabetes/Metabolism Reviews 1:293 (1985).
5.  B. L. Maloff, M. L. McCaleb, and D. H. Lockwood, Cellular basis of insulin resistance in chronic uremia, Am. J. Physiol. 245:E178 (1983).
6.  J. F. Caro, F. Flli, M. K. Sinha, and D. Brancaccio, Insulin resistance in uremia: in vitro model to study mechanism(s) in the liver, Diabetes 33(Suppl 1):138A, (1984).
7.  C. L. Hampers, J. S. Soeldner, P. B. Doak, and J. P. Merrill, Effect of chronic renal failure and hemodialysis on carbohydrate metabolism, J. Clin. Invest. 45:1719 (1966).
8.  R. Dzurik, Metabolic alterations caused by uremia, Proc. Eur. Dial. Transplant Assoc. 17:577 (1980).
9.  M. L. McCaleb, R. Mevorach, R. B. Freeman, M. S. Izzo, and D. H. Lockwood, Induction of insulin resistance in normal adipose tissue by uremic human serum, Kidney Int. 25:416 (1984).
10. M. L. McCaleb, M. S. Izzo, and D. H. Lockwood, Characterization and partial purification of a factor from uremic human serum that induces insulin resistance. J. Clin. Invest. 75:391 (1985).

## Acknowledgements

The investigations described were supported by United States Public Health Service grant, AM20129. Dr. McCaleb's current address is: Ayerst Laboratories, Department of Biochemistry, CN 8000, Princeton, NJ 08540. We wish to thank Mrs. Elizabeth Skelton and Shirley Thomas for their skillful secretarial assistance.

# PATHOGENESIS AND CONSEQUENCES OF THE ALTERATION OF GLUCOSE METABOLISM IN RENAL INSUFFICIENCY

Rastislav Dzúrik, Viera Spustová, and Mária Geryková

Medical Bionics Research Institute
Bratislava, Czechoslovakia

## ABNORMALITIES OF CARBOHYDRATE METABOLISM

### Abnormal Glycogen Metabolism

The decreased hyperglycemic response of patients in renal insufficiency (RI) to the applied glucagon or epinephrine and abnormal galactose tolerance test pointed to the abnormal liver glycogen metabolism and its decreased liver concentration (Cohen, 1962; Westervelt and Schreiner, 1962). However, later on the glucagon and epinephrine studies have not been confirmed and normal liver glycogen concentration was found in a group of patients with normal caloric intakte in our laboratory (Dzúrik and Brixová, 1968). The liver glycogen concentration was normal even in patients with abnormal glucose tolerance test. Similar findings were published on muscle glycogen by Bergström (Bergström and Hultman, 1969). It appears now that the glycogen concentration and metabolism depend primarily on nutritional state and not on renal insufficiency. Consequently, adequate caloric intake is the best prevention of this abnormality.

### Impaired Gluconeogensis

Both decreased and increased gluconeogenesis have been described. Gluconeogenesis is stimulated by the accumulated lactate, alanine and probably also serine, which are taken up and used for glucose and glycogen synthesis by the liver. On the other hand gluconeogenesis is inhibited by an inhibitor of the apparent m.w. over 10 000, inhibiting gluconeogenesis and the P-enolpyruvate carboxykinase (Lamberts et al., 1976). It was isolated from the sera of the uremic subjects and from urine of RI and healthy subjects (Dzúrik et al., 1980). The inhibition is probably responsible for the fasting hypoglycemia and increased sensitivity to sulfonylurea antidiabetics in RI patients. However, the fasting hypoglycemia is present in less than 10% of RI patients.

## Abnormal Glucose Utilization

Up to half the patients in RI suffer from fasting hyperglycemia and about two thirds from abnormal glucose tolerance
test. Fasting hyperinsulinemia and increased insulin response
to glucose load suggest the insulin resistance. This has been
definitely proved by the sophisticated glucose and insulin
clamp studies by De Fronzo (De Fronzo et al., 1983). The inhibition of glucose utilization is localized at the postreceptor
level, though there are cell types, such as red blood cells
with the decreased insulin receptor binding.

## INHIBITORS OF GLUCOSE UTILIZATION

If insulin resistance is only a consequence of the post-
receptor defect, what is the responsible inhibitor? One of the
suggested inhibitors could be parathormone (PTH) but also
other hormonal disturbances were described in RI. Moreover, a
great number of various inhibitors could participate.

Rat tissue slices or human blood cells incubated in the
diluted uremic sera utilize less glucose than the matched ones
incubated in sera obtained from healthy volunteers. The inhibition of glucose utilization was found in various tissues or
cells: striated muscle, kidney cortex, brain and liver slices
and human red blood cells and blood platelets. The deproteina-
tion of the sera did not remove the inhibitory activity and by
a tedious 9-step isolation procedure an inhibitor of glucose
utilization (IGU) (Dzúrik et al., 1973) was isolated. IGU
inhibited glucose utilization at the level of P-fructokinase
noncompetitively, which was in a good accordance with the data
on postreceptor defect of glucose utilization in RI. The
complete review of extensive studies with IGU was published
recently (Dzúrik and Spustová, 1984). However, data indicating
the presence of an additional inhibitor of glucose utilization
accumulated and it was obligatory to return to the systematic
studies with the uremic sera.

Deproteinized uremic serum fractioned by gel permeation
chromatography on a column of Sephadex G-15 eluted with 0.03
mol/l ammonium acetate was separated to a series of crude
fractions. The inhibitory activity was eluted in a marked
region Figure 1. where IGU was eluted. However, just simple
change on pH by substituting ammonium acetate with acetic acid
separated the inhibitory activity to two regions the original
one and the second one eluted later on. It was hypothetized
that an additional inhibitor was present in the uremic serum
and it should have been of low molecular weight and strongly
anionic nature. This inhibitor was completely purified by
means of combination of anion-exchange and cation-exchange
chromatography with reverse phase chromatography. Its structure
was defined by means of elementar analysis, mass spectrometry
and NMR spectroscopy and to our great surprise it was found to
be a simple hippurate, i.e. benzoyl glycine.

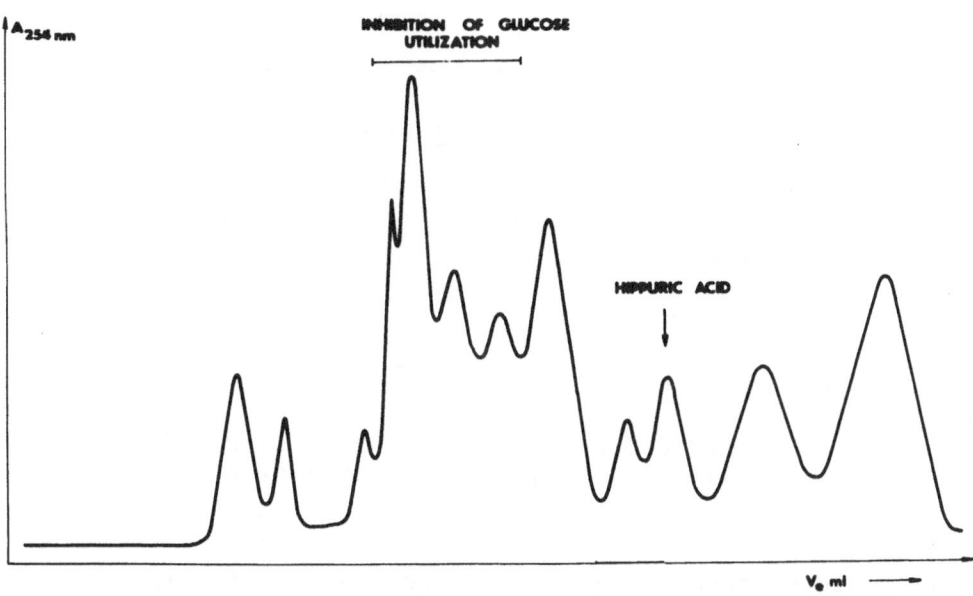

Fig. 1. Gel permeation chromatography of serum deproteinate on a column of Sephadex G-15.

## Hipurate

Hippurate serum concentration in healthy subjects varies around 0.01 mmol/l. In RI its concentration rises markedly, to about 1 mmol/l but values up to 5 mmol/l were found in RI patients after the administration of benzoate without apparent signs of toxicity (Mitch and Brusilow, 1982). The 1 mmol/l hippurate concentration was used for the screening of its effects in glucose utilization.

Synthetic hippurate inhibits glucose utilization in striated muscle and to a lesser degree also in rat kidney cortex slices, but not in other tested tissues, red blood cells or platelets Figure 2. Thus, while similar to IGU in muscle and kidney cortex it differs from IGU by the absence of action in other tissues or blood cells. The predominant action in striated muscle could participate in muscle weakness in a part of RI patients. However, even its renal action could be of physiological importance because of its accumulation in proximal tubular cells and nephron lumen.

When comparing other hippurates with benzoyl glycine it was found that hippurate inhibitory activity was maximal. The smaller inhibition was caused with o-OH hippurate, i.e. salicylurate and no activity with p-aminohippurate or o-OH hippuryl--ß-glucuronide at the same concentrations. However, even the action of salicylurate is of academic interest because of minimal salicyluric synthesis in the organism. The only exception would be the administration of salicylates to the RI patients.

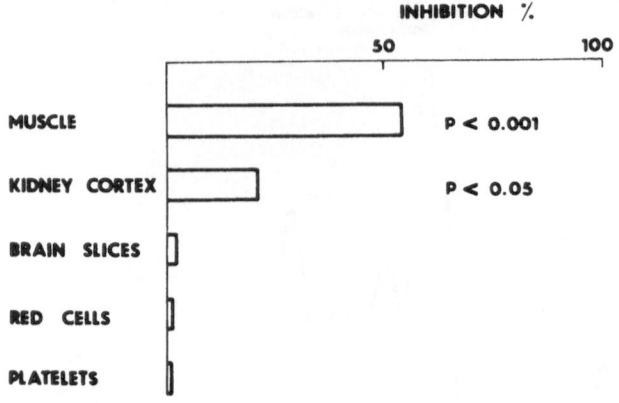

Fig. 2. The effect of hippurate on the glucose
utilization in various tissues or cells.


It could be summarized: 1) The most important defect of
abnormal carbohydrate metabolism in RI appears to be the
inhibited glucose utilization. 2) The inhibition is caused by
an insulin postreceptor defect. 3) Uremic blood contains at
least two inhibitors of glucose utilization. One of them, i.e.
hippurate was already structurally defined. 4) Hippurate
inhibits glucose utilization in muscle and kidney cortex. The
inhibitory activity could participate in muscular weakness of
RI patients.

REFERENCES

Bergström, J., and Hultman, H., 1969, Glycogen contet of
        skeletal muscle in patients with renal failure,
        Acta Med Scand., 186:177.
Cohen, B. D., 1962, Abnormal carbohydrate metabolism in
        renal disease, Ann Int Med., 57:204.
De Fronzo, R. A., Smith, D., and Alvestrand, A., 1983,
        Insulin action in uremia, Kidney Int., 16:S102.
Dzúrik, R., and Brixová, E., 1968, Liver glycogen
        concentration in patients with chronic uremia,
        Experientia, 24:552.
Dzúrik, R., Hupková, V., Černáček, P., Valovičová, E.,
        and Niederland, T. R., 1973, The isolation of an
        inhibitor of glucose utilization from the serum
        of uremic subjects, Clin Chim Acta, 46:77.
Dzúrik, R., Spustová, V., and Černáček, P., 1980,
        Inhibitor of renal gluconeogenesis (IGN): Additio-
        nal physiological modulator?, Int J Biochem, 12:103.
Dzúrik, R., and Spustová, V., 1984, Pathogenesis of the
        metabolic alterations in renal insufficiency,
        Rev Czech Med., 7:207.
Lamberts, B., Brunner, H., Ochs, H. G., Spellerberg, P.,
        and Heintz, R., 1976, Effect of urine metabolites
        from healthy and uremic subjects on gluconeogenesis
        in slices of rat kidney cortex and liver, Clin
        Nephrol, 6:465.

Mitch, W. E., and Brusilow, S., 1982, Benzoate-induced changes in glycine and urea metabolism in patients with chronic renal failure, J Pharmacol Exp Therap., 222:572.

Westervelt, Jr., B. F., and Schreiner, G. E., 1962, The carbohydrate intolerance of uremic patients, Ann Int Med., 57:266.

SEPARATION OF PEPTIDIC INHIBITOR OF

ERYTHROPOIESIS IN UREMIA

Akira Saito, Hiroshi Ogawa, Tai Gi Chung and
Tomomitsu Hotta

The Bio-Dynamics Research Institute
1-3-2  Tamamizu-cho, Mizuho-ku, Nagoya 467 Japan

INTRODUCTION

The cause of anemia in patients with chronic renal failure(CRF) is multifactorial such as lack of erythropoietin[1], hemolysis[2], poor nutritional status and an existence of inhibitor of erythropoiesis[3].

It is well known that anemia of CRF patients improves with CAPD[4]. However, we have reported that the anemia also improves with protein-permeable hemodialysis and hemodiafiltration[5,6,7,8]. The point in common for these techniques is the increased removal of middle molecules which are not easily removed with conventional hemodialysis(HD). Therefore, the presence of inhibitor of erythropoiesis in the middle molecules was suspected and the analytical evaluation was performed.

MATERIALS AND METHODS

Sixty liters of ultrafiltrate collected from 16 HD Patients with severe anemia was used as the sample.  For the filters, KF-101-15C of Kuraray Co., Ltd., and TAF-120S of Terumo Co., Ltd., were used as high-performance filter.

Each 10L of the ultrafiltrate was applied for Amberlite XAD-4 column(2L) and fractions eluted with 10L of 30% 2-propanol was pooled. The sample was lyophilized and dissolved in water.  After that, gel chromatography was performed by Sephadex G-50 column(3.5 X 40cm) and potassium phosphate buffer for the fractionation of molecules from 1,000 to 10,000 daltons.  The sample was further separated by ion-exchange chromatography with DEAE-Sephadex A-25 and each 10cc was collected.  For further analysis, reversed-phase chromatography on Shimazu Simpack C-18 column(0.6 X 15cm) was performed using 0.1% trifluoroacetic acid(TFA) in water as solvent A and o.1% TFA in methanol as solvent B.  Two linear steps were consisted for the gradient elution program.

Inhibitory effect on erythroid colony(CFU-E) formation was examined for each step using mice bone marrow cell culture. The mice bone marrow cell culture was conducted following the method by Maeda et al[9].

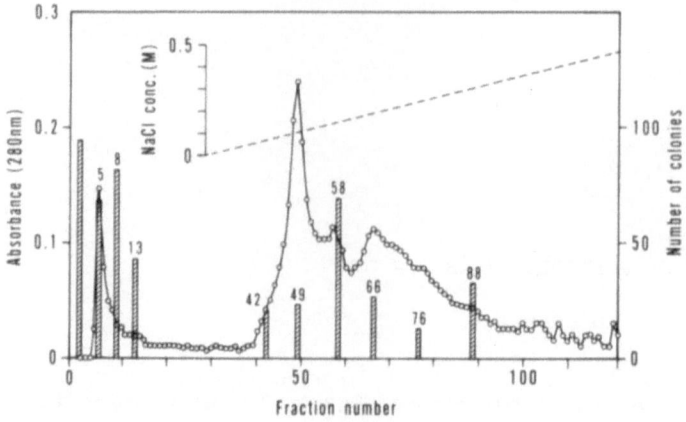

Fig. 1 Inhibitory effects of fractions obtained by DEAE-Sephadex
A-25 chromatography on erythroid colony formation in mice
bone marrow cell culture.


The 8 fractions obtained by chromatography on DEAE-Sephadex A-25
column were lyophilized and dissolved in 1 ml of 0.1 ml/L borate buffer
(PH 8.0).  In the solution, 4 ml of pronase was added.  The digestion
was conducted at 37°C with gentle shaking for 15 h.  The digest was
dialyzed against water with a 1,000 daltons exclusion limit membrane.
Finally, the sample was lyophilized and dissolved in 2 cc of water and
assayed for CFU-E.

RESULTS

30% 2-propanol solution obtained by Amberlite XAD-4 column
chromatography was assayed for CFU-E.  Colony formation of the sample
was 53% compared with control.  The sample was separated with gel
chromatography on Sephadex G-50 and pooled the fractions of 1,000 to
10,000 daltons showing inhibition of CFU-E formation.  After
lyophilization, the sample was separated with DEAE-Sephadex A-25 column
chromatography.  Chromatogram and the number of erythroid colonies of
each fraction is shown in Figure 1, fraction Fr.B, No.40-50, and
fraction Fr.C, No.66-88, were collected.

The results of the study of CFU-E inhibition for Fr.B and Fr.C
before and after proteolytic digestion were as follows:77±10 colonies
before treatment, 204±8 colonies after treatment for Fr.B and 2±0.5
colonies before treatment and 98±10 colonies after treatment for Fr.C.
The inhibitory effect decreased significantly to the control level for
both fractions.

In order to more precisely examine inhibition of CFU-E formation of
Fr.B and Fr.C, their addition into the culture was made at 5 different
concentrations.  Although Fr.C exhibited dose-related inhibition of
CFU-E formation.  Fr.B did not show such inhibition.

For the next step, a relationship between inhibition of CFU-E
formation of Fr.C and erythropoietin was examined. Fr.C and samples of
spermine were respectively added to bone marrow cell cultures containing
6 different concentrations of erythropoietin. Although inhibition of
CFU-E formation of Fr.C decreased with an increase in the erythropoietin
concentration, the number of colonies was always lower than in control.

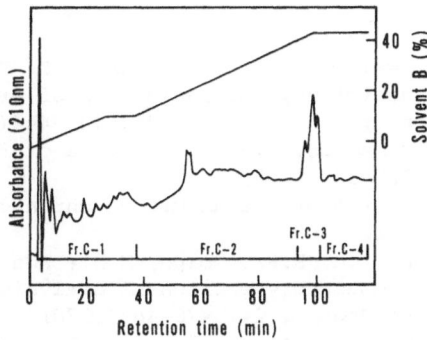

Fig. 2. HPLC pattern of Fr.C. Fr.C was 41% of the colony
formation compared with control.

The action of Fr.C in the erythropoietin dose-response was similar to
the inhibitory effect of spermine.

The chromatogram of Fr.C sample obtained by reversed-phase
chromatography on Simpack C-18 column is shown in Figure 2. The
inhibition was seen in the eluent with methanol concentration of 39-42.5%.
This is presented as Fr.C-3 in Figure 2.


DISCUSSION

The present study was undertaken on account of the following
factors: 1) Inhibition of CFU-E formation in normal human bone marrow
cell culture is seen in fractions of molecules greater than 1,500
daltons and of small molecules[10], 2) this inhibitor(of the molecules
greater than 1,500 daltons) is removed by high-performance filters and
3) the inhibitor is adsorbed on Amberlite XAD-4 resin. The treatment
of a large quantity of the sample with Amberlite XAD-4 is considerably
an effective means in this analysis.

Fractions exhibiting inhibition of CFU-E formation were separated
with chromatography on Sephadex G-50 or DEAE-Sephadex A-25 respectively.
Althouth 2 fractions, Fr.B and Fr.C, obtained by chromatography on
DEAE-Sephadex A-25 showed inhibition, only the latter exhibited a
dose-related inhibition.

In order to clarify the characteristics of the fraction, inhibition
of CFU-E formation before and after proteolytic digestion was examined.
Since the inhibitory effect reduced to the control level after the
digestion, the active site of the factor was composed of peptide.

In addition, change in inhibition of CFU-E formation of the
inhibitory fraction was examined by changing concentration of
erythropoietin in the culture. The inhibitory effect did not disappear
despite an increase in erythropoietin. The fact made us to surmise
that the inhibitory effect of the factor directly acts on erythroid
colony forming cells rather than that it acts as an inhibitor of
erythropoietin. We will continue to purify, identify and evaluate the
inhibitory factor more intensively.

REFERENCES

1. McGonigle, R.J.S., Wallen, J.D., Shadduck, R.K. Fisher, J.K.:
   Erythropoietin deficiency and inhibition of erythropoiesis in
   renal insufficiency.  Kidney Int. 25: 437-444(1984)
2. Desforges, J.F. and Dawson, J.P.: The anemia of renal failure. Arch.
   Int. Med. 101: 326-332(1958)
3. Fischer, J.W.: Mechanism of the anemia of chronic renal failure.
   Nephron. 25: 106-111(1980)
4. Moncrief, J.W., Popovich, R.P., Nolph, K.D.: Additional experience
   with continuous ambulatory peritoneal dialysis(CAPD). Tran. Am.
   Soc. Artif Intern Organs. 24: 476-483(1978)
5. Saito, A., Chung, T.G., Kanazawa, I., Oda, O. and Ohta, K.: Clinical
   evaluation of protein-permeating hemofilter and analysis of middle
   molecules in ultrafiltrate. Jpn. J. Artif Organs. 6: 907-910(1981)
6. Saito, A., Chung, T.G., Kanazawa, I., Oda, O. and Ohta, K.: Middle
   and large molecule removal of protein-permeating hemodiafiltration.
   In: Proc. Int. Sympo. Hemoperf. Artif. Organs. Vol. 4: 42-46(1982)
7. Saito, A., Naito, H. and Hirohata, M.: Dialytic removal of middle
   molecules and low molecular weight proteins. In: Progress in
   Artificial Organs-1983, 412-416(1984)
8. Saito, A., Ogawa, H., Chung, T.G. and Maeda, K.: CAPD and protein-
   permeating hemodialysis: A clinical comparison. In: Frontiers in
   peritoneal dialysis. 307-311(1986)
9. Maeda, H., Hotta, T. and Yamada, H.: Murine erythroid colony-promoting
   factor derived from thymocyte conditioned media. Exp. Hematol.
   11: 835-840(1983)
10. Maeda, H., Hotta, T., Saito, A., Chung, T.G.: Inhibitors of
    erythropoiesis in serum of patients with chronic renal failure.
    Abst. Jpn. Soc. Nephrol. 235(1983) (in Japanese)

# UREMIC TOXINS INHIBIT THE PLATELET MALONYLDIALDEHYDE PRODUCTION RATE

H. Tanaka, S. Itoh, S. Yamagami, T. Kishimoto, M. Maekawa,
and S. Ringoir*

Department of Urology, Osaka City University Medical School
1-5-7 Asahimachi, Abenoku, Osaka 545, Japan
*Renal Division, University Hospital
De Pintelaan 185, B-9000 Ghent, Belgium

## INTRODUCTION

Platelet cyclo-oxygenase activity is known to be low in uremic
patients[1,2], but it is still lower in maintenance hemodialysis patients[3].
Malonyldialdehyde (MDA) is one of the terminal substances from arachidonic
acid, which is converted to prostaglandins by cyclo-oxygenase.  Although
some MDA are produced through other pathways, the platelet MDA production
rate is used to evaluate platelet cyclo-oxygenase activity and is a
guideline in detecting the abnormality of prostaglandin metabolism.  The
aim of this study is to find out the inhibiting factor of this platelet
MDA production rate in hemodialysis patients.  We observed the level of
inhibition of the activity of normal platelets incubated with several
kinds of substances which are suspected to be uremic toxins.

## MATERIALS AND METHODS

Twenty-three healthy men, aged 25 to 36 years, and 10 maintenance
hemodialysis patients (3 males and 7 females), aged 37 to 69 years, were
selected for this study.

Normal platelet rich plasma (PRP) from the blood of a healthy man was
obtained with 1/10 volume of 3.8 % sodium citrate by centrifugation at
200 G for 10 mins (KR-40 and KR-702, Kubota, Tokyo, Japan).  PRP was
divided by volume and centrifuged at 600 G for 20 mins after adding 1/10
volume of 77 mM EDTA.  In this way platelet pellets with the same platelet
count and volume distribution were prepared.  Each platelet pellet was
incubated with uremic plasma or normal plasma at 37°C for 30 mins.  The
uremic plasma used for incubation was that of before and after hemodialysis.  Uremic toxins suspected to be the inhibiting factor were urea
(50, 100, 200, 400 mg/dl as urea nitrogen), creatinine (5, 10, 20,
40 mg/dl), guadininosaccinic acid (GSA; 50, 100, 400, 1600 μg/dl) and
methylguanidine (MG; 25, 50, 100, 400 μg/dl), and they were solved with
10 mM Tris-HCl buffer pH 7.4 containing 140 mM NaCl, 4 mM EDTA and 0.3 mM
glucose and used for incubation in the same way as the uremic plasma.
Lyophilized fractions of the ultrafiltrate from dialysis patients separated by large scale liquid chromatography (Sephadex 15) were also

studied. The ultrafiltrate was obtained from patients on dialysis with the cuprophane membrane and on hemodiafiltration with the PMMA membrane and PAN membrane (Asahi). They were solved with 0.15 M phosphate buffer pH 7.4 and incubated with the normal platelets as well. The platelet MDA production rate of the incubated platelets was measured by the TBA method[4],[5]. The optical density was read to be at 532 nm with a spectro-photometer (Uvidex-60, Jasco Medical Instrument, Inc., Tokyo, Japan and 220A, Hitachi Ltd., Tokyo, Japan).

The level of inhibition of the platelet MDA production rate was expressed as the inhibition index (I.I.; %) and calculated as follows:

for uremic plasma,
I.I. = (MDA production rate of normal platelets incubated with uremic
       plasma) ÷ (MDA production rate of normal platelets incubated with
       normal plasma obtained from a healthy man) X 100

for uremic toxins,
I.I. = (MDA production rate of normal platelets incubated with buffer
       containing uremic toxins) ÷ (MDA production rate of normal
       platelets incubated with buffer) X 100

The results were expressed as mean ± SD, and statistical analysis was done by the Student's t-test.


RESULTS

I.I. for the uremic plasma before and after hemodialysis was 90.3 ± 4.0 % (N=10) and 99.8 ± 4.5 % (N=10), respectively, as shown in Fig. 1, and there was a significant difference between the two (P<0.001). The uremic plasma before hemodialysis significantly inhibited the platelet MDA production rate compared with that after hemodialysis. I.I. for urea,

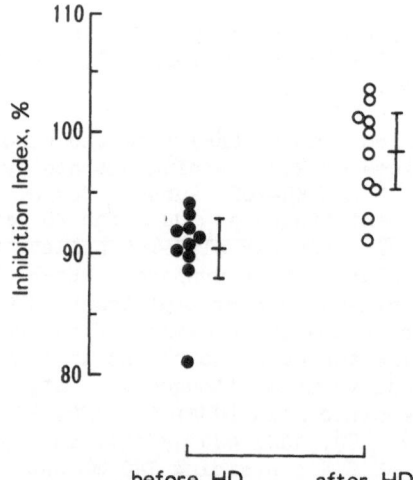

Fig. 1.   Inhibition index for uremic plasma before and after hemodialysis.
          There was a significant difference between the two (P<0.001).

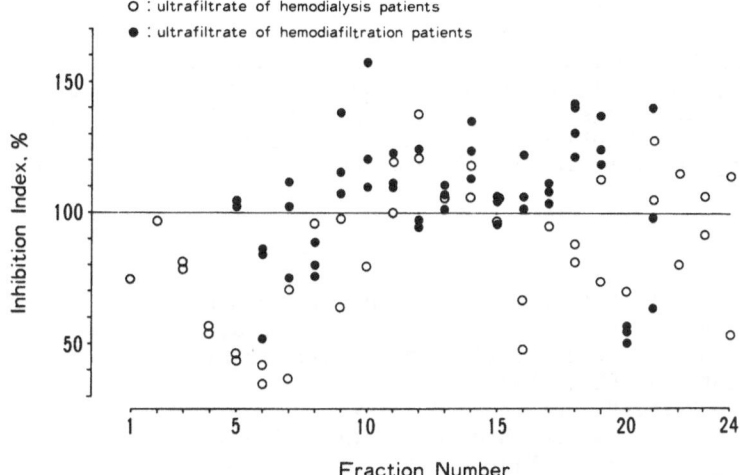

Fig. 2.  Inhibition index for lyophilized fractions of the ultrafiltrate from dialysis patients separated by large scale liquid chromatography.

creatinine, GSA and MG was nearly 100 % in all of the concentrations, and none of them inhibited the platelet MDA production rate.  I.I. for fraction 3 to 7, 16 and 18 of the ultrafiltrate from the patients dialyzed with the cuprophane membrane was low, and that for fraction 6, 8 and 20 of the ultrafiltrate from the patients on hemodiafiltration with PMMA and PAN membranes was also low (Fig. 2).


DISCUSSION

Research on uremic toxins is often done by incubating normal platelets with uremic plasma[6,7].  In this study we measured the MDA production rate of normal platelets incubated with several kinds of substances which are suspected to be uremic toxins and evaluated the level of inhibition. When the platelet pellets were prepared, PRP was divided by volume so that each pellet had the same platelet count and volume distribution.  This is important for minimizing errors, because the platelet has different functions according to its size and age[8,9].

The inhibiting factor of the platelet MDA production rate was found in uremic plasma before hemodialysis, although it disappeared after hemodialysis.  This suggests that the inhibiting factor is a relatively small molecular substance and is dialyzable.  To find out this small molecular substance in uremic plasma, uremic toxins including urea, creatinine, GSA and MG were selected and studied, but they did not inhibit the platelet MDA production rate in any of the concentrations.  However, some of the lyophilized fractions of the ultrafiltrate from dialysis patients separated by large scale liquid chromatography did inhibit the platelet MDA production rate.  Some fractions contained a high concentration of NaCl and had high osmolarity by using this chromatographic technique, which might be taken for uremic toxins.  What's more, the samples were not pure and contained many substances.  More studies have to be made to clarify this point and to identify the inhibiting factor of the platelet MDA production rate which can be called the uremic toxin.

## CONCLUSION

The inhibiting factor of the platelet MDA production rate was present in pre-dialysis uremic plasma of maintenance hemodialysis patients. However, the substance was not urea, creatinine, GSA nor MG, and might be contained in the fractions of the ultrafiltrate from hemodialysis patients separated by large scale liquid chromatography using Sephadex G 15.

## REFERENCES

1. G. Remuzzi, A. Benigni, P. Dodesini, A. Schieppati, M. Livio, G. De Gaetano, J. S. Day, W. L. Smith, E. Pinca, P. Patrignani, and C. Patrono, Reduced platelet thromboxane formation in uremia, J. Clin. Invest. 71:762 (1983).
2. O. Ylikorkala, K. Huttunen, J. Jarvi, and L. Viinikka, Prostacyclin and thromboxane in chronic uremia: effect of hemodialysis, Clin. Nephrol. 18:83 (1981).
3. H. Tanaka, K. Umimoto, N. Izumi, K. Nishimoto, T. Maekawa, T. Kishimoto, and M. Maekawa, Platelet life span in uraemia, Proc. Euro. Dial. Transpl. Assoc. 21:306 (1984).
4. M. Okuma, M. Steiner, and M. Baldini, Studies on lipid peroxides in platelets, J. Lab. Clin. Med. 75:283 (1970).
5. H. Tanaka, K. Umimoto, N. Izumi, K. Nishimoto, H. Yoshihara, Y. Katoh, T. Kishimoto, and M. Maekawa, Platelet dysfunction and its life span in uremia, Trans. Am. Soc. Artif. Intern. Organs 30:207 (1984).
6. G. Remuzzi, D. Marchesi, M. Livio, A. Schieppati, G. Mecca, M. B. Donati, and G. De Gaetano, Prostaglandins, plasma factors, and hemostasis in uremia, in: "Hemostasis, Prostaglandins, and Renal Disease," G. Remuzzi, G. Mecca, G. De Gaetano, eds., Raven Press, New York (1980).
7. M. C. Smith and M. J. Dunn, Impaired platelet thromboxane production in renal failure, Nephron 29:133 (1981).
8. S. Karpatkin, Heterogeneity of human platelets. VI. Correlation of platelet function with platelet volume, Blood 51:307 (1978).
9. M. Kraytman, Platelet size in thrombocytopenias and thrombocytosis of various origin, Blood 41:587 (1973).

# A POSSIBLE REGULATORY SYSTEM OF MICROTUBULE

# FORMATION AMONG UREMIC TOXINS

Diane Braguer*, Philippe Gallice*, Jean-Pierre Monti*,
Claude Durand**, Antoine Murisasco**, and Aimé Crevat*

*Laboratoire de Biophysique, Faculté de Pharmacie
 27 Bd Jean Moulin, 13385 Marseille Cedex 5, France
**Service de Néphrologie, Hôpital Sainte-Marguerite
 Marseille, France

## INTRODUCTION

Tubulin is an intracellular protein, whose polymerization leads to the formation of microtubules (MT). MT are an essential component of axons of nerve cells, and the rate of axon regeneration depends on the synthesis or polymerization of tubulin [1]. Uremic plasma inhibits the growth of nervous fibers in chicken embryo. Moreover, in uremic neuropathy, axonal degeneration is frequently observed [2]. Thus, in uremic patients, MT formation from tubulin polymerization may be inhibited. If this effect also occurs in vivo, uremic toxins (UT) might be involved in the pathogenesis of uremic neuropathy.

In the present paper, we studied in vitro the action of uremic toxins on MT formation. This work has been performed using either pure tubulin (tubulin 6S) or tubulin accompanied by microtubule-associated proteins (MAPs), both in microtubular protein (MTP) form.

## MATERIAL AND METHOD

### Isolation of uremic toxins

Uremic toxins (UT) were obtained using urine from healthy subjects after simple filtration, or ultrafiltrate of plasma of uremic patients with chronic renal failure treated by hemodialysis. UT were isolated by using a semi preparative method previously described [3]. The first step is a gel permeation yielding 8 fractions. Fraction 2 (containing UT) is submitted to an anionic exchange. Six subfractions were obtained. The 2-5 fraction is divided into 10 pure compounds (2-5-1 to 2-5-10) by HPLC.

---

This work was supported by a grant of INSERM n° 851007

## Preparation of tubulin

Microtubular protein (MTP) was obtained from fresh pig brains using Shelanski's method [4]. Tubulin 6S was purified according to Weingarten's method [5].

## Polymerization study

The polymerization of MTP or tubulin was started by a jump of temperature from 4°C to 37°C in the presence of guanosine 5'triphosphate. Then we recorded the increase in turbidity at 340 nm as previously described [6]. The morphology of MTP obtained after 30 minutes was controlled by electron microscopy. The polymerization in the presence of UT was expressed as a percentage of the control value without UT.

## RESULTS

### 2-5 fraction

The 2-5 fraction inhibits MTP and tubulin 6S polymerization (Fig. 1) in a dose dependent manner (Table 1).

### 2-5-3 component

In table 2, we indicate the percentage of inhibition of tubulin 6S polymerization by 2-5-3 compound. This effect is dose dependent and about ten fold higher than that of 2-5 fraction.

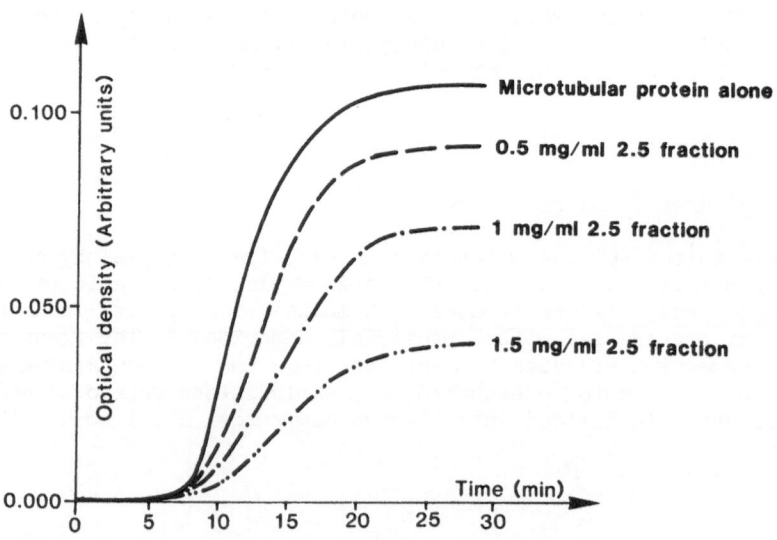

Figure 1. Action of 2-5 fraction on the polymerization of microtubular protein.

Table 1. Inhibition of MTP and tubulin 6S polymerization by the 2-5 fraction.

| 2-5 Fraction | | Inhibition MTP | Inhibition Tubulin 6S |
|---|---|---|---|
| Normal urine | 0.5 mg/ml | 17 % | 26 % |
| Patient Sou. | 1 mg/ml | 30 % | 42 % |
| Patient R. | 1 mg/ml | 10 % | 22 % |
| Patient Sol. | 0.5 mg/ml | 15 % | 20 % |
| Patient Sol. | 1 mg/ml | 30 % | 40 % |

DISCUSSION

Our results show that the 2-5 fraction from uremic patients inhibits microtubule formation from MTP and pure tubulin as previously reported [6]. Whatever its origine, normal urine or uremic plasma, 2-5 fraction has an inhibitory effect, so we can consider the active molecule as an endogenous compound present in normal plasma in low quantity and accumulating in uremic patient.

Among HPLC components, the 2-5-3 compound inhibits in vitro the tubulin polymerization in a dose dependent manner. If this inhibitory effect occurs in vivo as well as in vitro it would lead to an axonal degeneration. We could then consider that the 2-5-3 component as being involved in uremic neuropathy. The structure of 2-5-3 component is not yet determined but preliminary results show that it is quite different from that of myoinositol which is thought to be related to the neural dysfunction in uremic patients [7]. Consequently, even if 2-5-3 compound is correlated with uremic neuropathy, these findings show that several compounds might be implicated in this pathology.

For most patients, a low nervous conduction velocity (NCV) corresponds to a high inhibitory effect of their 2-5 fraction and to a high level of 2-5-3 compound in their 2-5 fraction. But some uremic patients with a high level of 2-5-3 component exhibit a normal NCV and their 2-5 fraction induces an abnormally low inhibition of MTP formation (Table 3).

Table 2. Inhibitory of tubulin 6S polymerization by 2-5-3 compound.

| Dose 2-5-3 (µg/ml) | 11 | 12 | 20 | 22 | 110 | 170 |
|---|---|---|---|---|---|---|
| % Inhibition | 10 | 11 | 21 | 24 | 48 | 62 |

Table 3. Comparison between |2-5-9|/|2-5-3| ratio and inbibition of microtubule formation by 2-5 fraction.

| Patient | NCV | 2-5-3 in 2-5 fraction | Inhibition of MT formation by 2-5 fraction | $\frac{|2-5-9|}{|2-5-3|} \times 100$ |
|---------|-----|------------------------|---------------------------------------------|----------------------------------------|
| S. | 30 | 15 % | 70 % | $\simeq 0$ |
| G. | 50 | 33 % | 30% | 1.4 |
| M. | 52 | 14 % | 12 % | 7.1 |

So we hypothesize that there might be, in the 2-5 fraction of these patients, a compound which would ccunteract the inhibitory effect of 2-5-3 component on MTP polymerization. Indeed it appears that in these patients the concentration of 2-5-9 compound is higher than in patients with low NCV (Fig. 2).

Moreover, we correlate the inhibitory effect of several 2-5 fractions with the 2-5-9 level present in this fraction. This correlation appears to be negative. $Y = A \log X + B$ ; $A = -11$ ; $B = 20$ ; $R = 0.96$.

Thus we compared 3 parameters in these patients : NCV, percentage of inhibition of MT formation by 2-5 fraction and ratio of concentration |2-5-9|/|2-5-3| (Table 3). As a test probe, we also report results of a

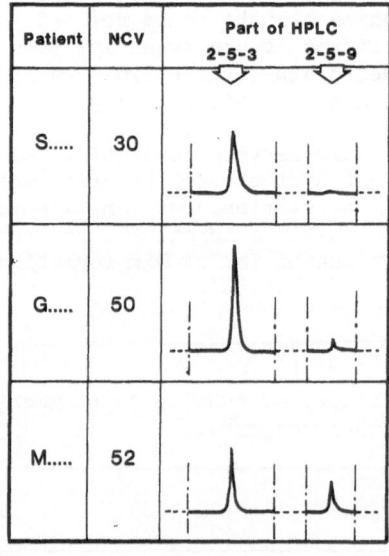

Figure 2. Part of high performance liquid chromatograms for 3 uremic patients.

patient with a low NCV. We note that the higher the |2-5-9|/|2-5-3| ratio, the lower the inhibitory effect, even if the level of 2-5-3 is the same.

So our results seem to support the hypothesis that 2-5-9 compound would counteract the inhibitory effect of 2-5-3 compound.

If this effect also occurs in vivo, it is possible to think that these two compounds might be a regulatory system for tubulin polymerization.

REFERENCES

1. R.E. Stephens and K.T. Edds, Microtubules structure, chemistry and and function, Physiol. Rev., 56 : 709 (1976).
2. R.E. Ahonen, Peripheral neuropathy in uremic patients and in renal transplant recipients, Acta Neuropathol., 54 : 43 (1981).
3. P. Gallice, J.P. Monti, A. Crevat, C. Durand and A. Murisasco, A compound from Uremic Plasma and from Normal Urine Isolated by liquid chromatography and identified by NMR, Clin. Chem., 31 : 1 (1985)
4. M.L. Shelanski, F. Gaskin and C.R. Cantor, Microtubules assembly in the absence of added nucleotides, Proc. Natl. Acad. Sci. USA 70 : 765 (1973).
5. M.D. Weingarten, A.M. Lockwood, S. Hwo and N.W. Kirschner, Protein factor essential for microtubule assembly, Proc. Natl. Acad. Sci. USA, 72 : 1858 (1975).
6. D. Braguer, P. Gallice, J.P. Monti, A. Murisasco and A. Crevat, Inhibition of microtubule formation by uremic toxins : action mechanism and hypothesis about the active component, Clin. Nephrol., 25 : 4 (1986).
7. T. Niwa, H. Asada, K. Maeda, K. Yamaha, T. Ohki and A. Soeito, Profiling of organic acids and polyols in nerve of uraemic and nonuraemic patients, J. Chrom., 377 : 15 (1986).

# ENCEPHALOPATHIC TOXICITY:  AN EXPERIMENTAL MODEL

# OF UREMIA AND SOLUTE-SPECIFIC DIALYSIS

Paul E. Teschan, J.J. Lipman, P. Lawrence, and D. BeBoer

Division of Nephrology
Vanderbilt University
Nashville, Tennessee

## INTRODUCTION

The questions before us are:  (1) How does renal failure make patients sick? and (2) How does dialysis make them better?  Put another way:  What is the connection, if any, between abnormal extracellular fluid (ECF) chemistry and patients' symptomatic illness, since both are produced bys severe renal failure and both are improved by dialysis?

With George Schreiner[1] we perceived the "mental and personality changes" of the uremic sickness as encephalopathic--as a neurobehavioral syndrome.  Logically, therefore, we chose quantitative phychometric and electroencephalographic measures to index patients' clinical symptomatic illness, and found that they did so in normal, azotemic, dialyzed and transplanted patients[2].  Among dialyzed patients,these measures also reflected changes in the amount of dialysis[3].

## METHODS

Proceeding directly from that clinical experience, and to examine the relationship between ECF chemistry and "clinical uremic toxicity" beyond the limitations of clinical studies, we have developed a model in conscious, ambulatory, bilaterally-nephrectomized rats[4].  The independent variable--ECF chemistry--is varied by chronic peritoneal dialysis. Dialysate is introduced through a catheter that enters the peritoneum via a chronically implanted catheter guide.  The catheter is partially withdrawn between exchanges, then replaced by an occluding plastic obturator tube between daily dialysis sessions (see figure 1).  To avoid blood sampling, equilibrated peritoneal dialysate (EPD) samples are obtained after a 60 minute dwell time before and after dialysis sessions.  An equilibration study revealed variable dialysate/blood concentration ratios that approximated 1.0 more often in the nephrectomized than in the sham-nephrectomized animals.  Intermittent peritoneal dialysis was performed in daily sessions at the rate of 8 exchanges/day, with 30 minute dwell times, using dialysate values approximating 10% of body weight.

In two replicate experiments, commercial therapeutic dialysate with 1.5% dextrose was used in 8 concurrent renoprival controls.  The mock "uremic dialysate" used in 7 renoprival experimental animals was devised to match the average plasma concentration in pre-terminal rats of 10 constituents that are conventionally measured in patients with renal failure. The quantified EEG, as the dependent/outcome variable, was recorded from chronically implanted epidural electrodes. Cables connected each rat through a computer-driven switching device, an EEG recorder and an analog-digital converter to an IBM/PC XT computer for power spectrum analyses[4] and their display in compressed spectral arrays. One-hundred-second EEG epochs were recorded and computed automatically from each rat every two hours between 1500 hours and 0700 hours each night.

RESULTS

As in man, our initial experiments revealed that EEG power (amplitude) in uremic rats declined progressively and shifted to lower frequencies with time. THus the computed ratio of slow-to-fast-wave-associated power, e.g. the theta/alpha ratio (TAR), the average of the nine computations each night from each rat, became more abnormal on successive days in the control group as did the equilibrated peritoneal dialysate urea nitrogen (EPDUN). Both TAR and EPDUN values returned toward control levels in the concurrently studied experimental animals that were dialyzed with standard commercial therapeutic dialysate.

In the present experiments, the effects of mock "uremic dialysate" (UD) were compared with those of standard commercial therapeutic dialysate (TD) in the model. Our prediction was that there would be no difference. We were wrong. While the average survival of 5.5 days was longer in the UD-treated than in undialyzed animals, TD-treated animals survived more than twice as long (average 13 days).

Dialysis with UD and TD modified the composition of equilibrated peritoneal dialysate (independent variable), as expected. Accordingly EPD urea nitrogen, potassium, and phosphate concentrations in UD-treated animals exceeded those of TD-treated animals while concentration curves for creatinine, calcium and magnesium did not diverge.

The dependent variable--the EEG/TAR ratio--was abnormal in UD-treated rats, but remained at control levels in the TD-treated controls.

We then examined all concurrent values of averaged TAR and averaged post-presession EPD chemical values in both UD and TD-treated animals. Three statistically significant low-level correlations were found with very low coefficients of determination: TAR vs. EPD.Urea Nitrogen; TAR vs. EPD.potassium; and TAR vs. EPD.Phosphate. TAR showed no correlation with EPD.calcium. See Table 1.

CONCLUSION

1. As in man, renal failure with "chemical uremia" in this model produces a decrease in EEG power and a shift in the EEG power spectrum toward lower frequencies.

2. Both the ECF chemical changes and the EEG abnormalities of renal failure are mitigated by peritoneal dialysis with TD and exacerbated by UD along with increased mortality.

3. One or more constituents of UD are directly or indirectly responsible for the differences in TAR and survivorship in this model, although (subject to further statistical analysis) the low correlations for urea, potassium and phosphate and the non-divergent creatinine, calcium and

magnesium data fail to provide convincing evidence as yet for any one of them.

   4.  Using this model, further solute-specific modifications of dialysate composition are expected to help identify which one or more of these solutes is (are) parameter(s) of the toxic effect.

## TABLE 1

### CORRELATIONS OF EEG THETA/ALPHA POWER RATIO VS:

| EPD Concentration of: | r | $r^2$ | n | p |
|---|---|---|---|---|
| Urea N | 0.47 | 0.22 | 137 | <.001 |
| K | 0.24 | 0.06 | 108 | <.01 |
| Pi | 0.28 | 0.08 | 110 | <.01 |
| Ca | 0.07 | 0.005 | 112 | ns. |

Cr, Mg not different $\pm$Nx.

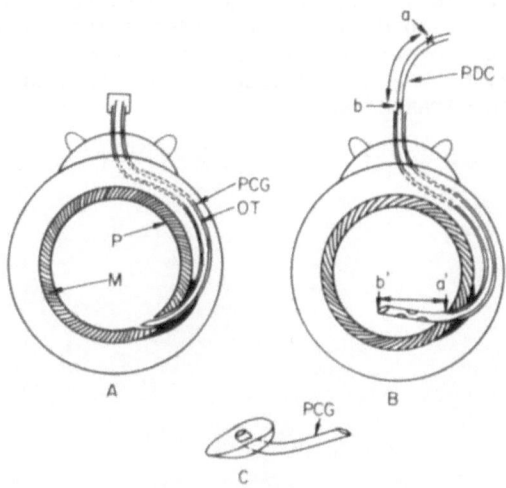

Figure 1.  A Peritoneal Dialysis in the Laboratory Rats.

   A:  Transverse section of peritoneal cavity showing peritoneal membrane (P) abdominal musculature (M),PCG peritoneal catheter guide (PCG) also shown at C; and occluding obturator tube (OT) extending to the peritoneal membrane.

   B:  The peritoneal dialysis catheter (PDC) moves from the peritoneal ingress point (a) into the peritoneal cavity ($b^1$) as the tube is advanced from position (a) to position (b).

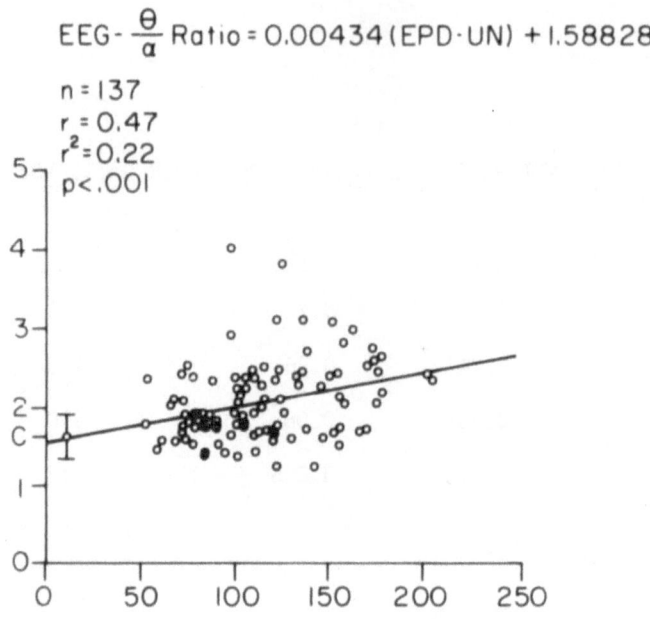

Figure 2. EEG-TAR ratio vs. equilibrated peritoneal dialysate urea nitrogen. Each data pair represents the average of 9 TAR values sampled overnight (y axis) and the average of the corresponding (bracketing) EPD urea nitrogen concentrations (X axis) for both UD--andTD--treated animals.

REFERENCES

1. G. E. Schreiner, Mental and personality changes in the uremic syndrome, <u>Medical Annals of the District of Columbia</u>, 28: 316-324 (1959).
2. P. E. Teschan, H. E. Ginn, J. R. Bourne, J. W. Ward, B. Hamel, J. C. Nunnally, M. Musso, and W. K. Vaughn, Quantitative indices of clinical uremia, <u>Kidney International</u>, 15: 676-697 (1979).
3. P. E. Teschan, H. E. Ginn, J. R. Bourne, J. W. Ward, J. D. Schaffer, A prospective study of reduced dialysis, <u>ASAIO Journal</u>, 6: 108-122 (1983).
4. D. DeBoer, J. J. Lipman, P. Lawrence, P. E. Teschan, Automated EEG monitoring system for quantification of uremic encephalopathy in a rodent model, <u>Proceedings of the American Association of Medical Instrumentation</u>, (1986).

# INTELLECTUAL IMPAIRMENT IN CHRONIC RENAL FAILURE

Takehide Takuma, Tsutomu Sanaka and Nobuhiro Sugino

Department of Medicine, Kidney Center
Tokyo Women's Medical College
Tokyo, Japan

## INTRODUCTION

As one of the various neurological disturbances in the patients with chronic renal failure (CRF), poor abilities for concentration, calculation, memory and abstract thinking were reported in the early stage of renal failure. However, the individual variations of these intellectual impairments and their changes under the influence of therapies have not been described. To evaluate individual intellectual impairment of patients with CRF, the Uchida-Kraepelin psychodiagnostic test was clinically applied. This paper reports the results mainly with regards to patient's prognosis and the effects of the induction of hemodialysis treatment and successful kidney transplantation.

## METHODS

The Uchida-Kraepelin psychodiagnostic test was originated by Emil Kraepelin, German psychiatrist, and modified practically by Yusaburo Uchida, Japanese psychologist. This test, abbreviated Kraepelin test, consists of continuous simple addition for 30 minutes. The number of total correct answers are correlated with intelligence quotient (I.Q.) with $r = 0.706$. Its advantage is in giving an objective, convenient mental evaluation in a short time without the examiner's bias. On the other hand, this test may not be applied for the patients with visual disturbance, paresis and tremor.

## RESULTS

(1) Kraepelin test for healthy controls

This test was applied to 23 healthy young adults, all 22 years of age. The mean S.D. of correct answers calculated were $70.7 \pm 11.3$ per minutes and errors were made in $0.43 \pm 0.35\%$ of answers. They were tested again 4 weeks later. The results for the second time were as follows; correct answers $77.1 \pm 11.4$ and errors $0.35 \pm 0.22\%$. The increase of 10% in calculations performed was probably due to practice.

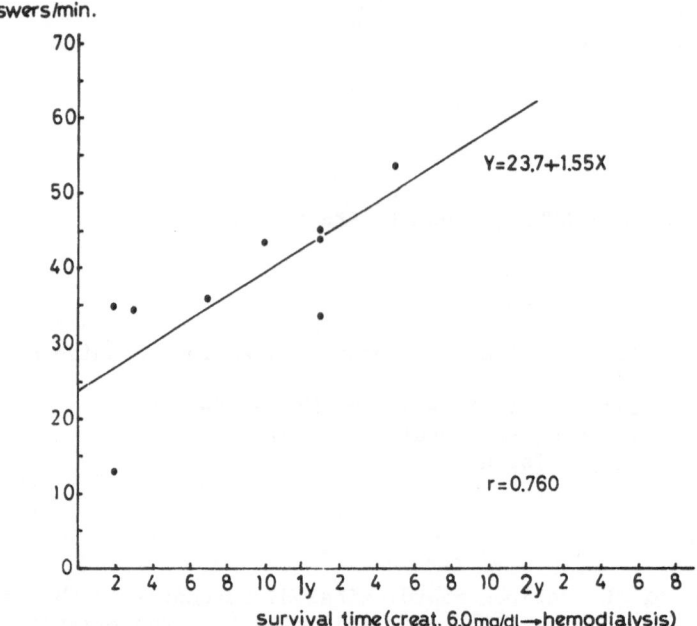

Fig. 1. In the 9 female patients with CRF, the number
of correct answers and survival time were
correlated closely with r=0.760 (0.01<p<0.02).


(2) Intellectual impairment of patients with CRF

The 19 patients were evaluated by the Kraepelin test at the time
when serum creatinine levels were 5-6mg/dl, and they were followed in our
ambulatory clinic until the induction of dialysis treatment. Among these
patients, some yielded speedy calculation, others showed sluggish calculation and numerous errors. Intellectual ability among individual patients
varied greatly.

The correlation between number of correct answers and survival time
was investigated. The survival time means in this study the timespan
from the timepoint when serum create was 6.0mg/dl until the initial hemodialysis treatment. In the 10 male patients these variables were correlated poorly. In contrast to this, they were closely correlated in a
group of 9 female patients (r=0.760). The mentally impaired female
patients probably have difficulties to prepare the regulated diets.
Results of the Kraepelin test for female patients with CRF may have some
prognostic significance. (Fig. 1)

Effect of some dietary therapies including low protein, high energy
in addition to essential amino acid and keto-acid were tested. These
dietary management did not bring about significant intellectual improvement in results of the Kraepelin test, although azotemia was reduced to
some extent.

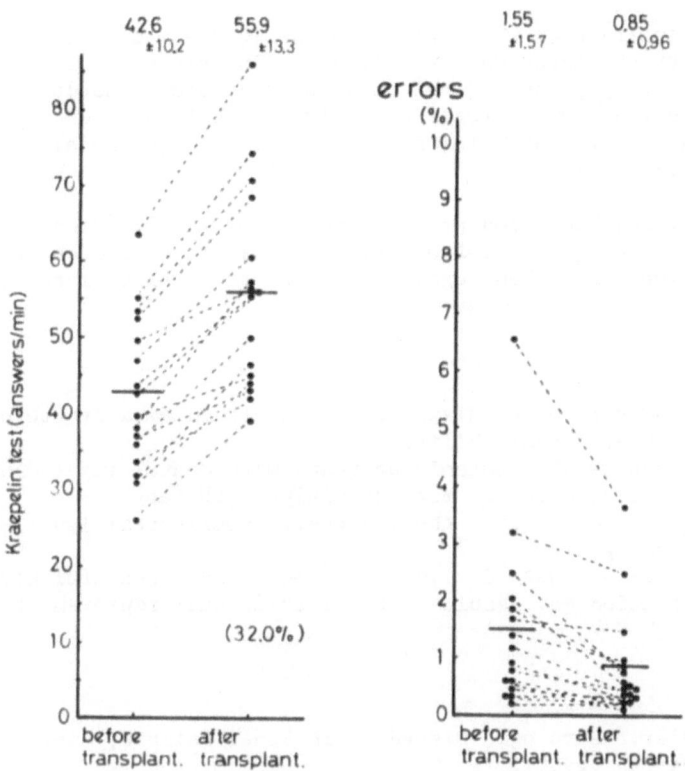

Fig. 2. After the successful kidney transplantation,
increase of 32.0% in calculations and decrease
in errors were observed in all 16 patients in
the results of Kraepelin test.

(3) Intellectual improvement after the induction
    of hemodialysis treatment

The Kraepelin test was applied to 23 patients with CRF a few weeks
before the initial dialysis.  They calculated $40.9 \pm 13.2$ answers per
minute and showed $3.0 \pm 3.3\%$ errors in average.  They received the same
test for the second time 2-3 weeks after the induction of dialysis treatment.  Individual variations were remarkable under this condition.  They
calculated more quickly, $51.3 \pm 11.9$ answers (30.3% increase), and made
$1.36 \pm 1.2\%$ errors.

Adequate dialysis treatment has resulted in intellectual improvement.  The responsible, partially dialysable substances that induce
mental deterioration are under investigation.

A few dialysis patients were tested periodically once a year.  Their
calculating abilities stayed almost constant during 3 years of dialysis
treatment.

(4) Success in kidney transplantation and
    intellectual improvement

The Kraepelin test was applied to the longterm dialysis patients for whom kidney transplantation were planned. In this study 16 patients were collected and they performed this test once before transplantation and twice when they no longer required dialysis treatment. On this trial too, remarkably intellectual improvement was observed in all the cases we studied. (Fig. 2)

Kidney transplantation provides normal or nearly normal renal function. Following change from chronic dialysis to successful kidney transplantation, they then regain their proper level of intelligence.


CONCLUSIONS

1) Uchida-Kraepelin psychodiagnostic test are clinically useful to evaluate intellectual abilities.
2) Non-dialysed female azotemic patients with mental impairment may predict earlier need for regular dialysis therapy.
3) During conservative diet therapy their intellectual level stays almost unchanged.
4) Induction of adequate dialysis treatment and successful kidney transplantation may result in the intellectual improvement.


REFERENCES

1. H. R. Tyler, Neurologic Disorders in Renal Failure, Amer. J. Med. 44:734-748 (1968).
2. C. F. Bolton, W. J. Johnson, and P. J. Dyck, Neurologic Manifestations of Renal Failure, in Strauss and Welt's Diseases of the Kidney, Third Edition, L. E. Earley and C. W. Gottschalk, ed., Little, Brown and Company, Boston (1979).
3. A. English, R. D. Savage, P. G. Britton, M. K. Ward, and D. N. S. Kerr, Intellectual Impairment in Chronic Renal Failure, Brit. Med. J. 1:888-890 (1978).

THE FRACTION b 4-2: ISOLATION, CHARACTERIZATION AND BIOLOGICAL ACTIVITIES

WITH REFERENCE TO UREMIC POLYNEUROPATHY

N.K. Man*, G. Cueille**, P. Faguer*, J. Boudet*, D.Pierrat*
J. Zingraff*, A. Sausse*, and J.L. Funck-Brentano*

*Department of Nephrology, Necker Hospital, Paris, France
**Rhône-Poulenc Research Center, Paris, France

161 Rue de Sèvres, Paris, France

INTRODUCTION

The toxic role of numerous compounds accumulated in the body fluids of uremic patients is not well established. Instead of comparing the over-all chromatographic plasma pattern of uremic patients to that of healthy subjects (1), one uremic symptom, namely active recent polyneuritis, was chosen and the correlation between the evolution of symptoms and biochemical patterns as modified by hemodialysis was evaluated. This correlation was complemented by a bioassay in which the fraction isolated previously exhibits a specific in vitro activity similar to that observed in vivo. These clinical, biochemical and bioassay techniques each contribute to form a correlative approach (2). Uremic polyneuritis was chosen to test the "middle molecule hypothesis" because it was found to respond favorably to any method which increases middle molecule transfer, i.e., more frequent or more prolonged dialysis or the use of a more porous membrane (3). In six patients with progressive and complete motor nerve paraplegia due to inadequate dialysis, the rapid improvement of nervous symptoms when using a dialysis membrane highly permeable to middle molecules brought us the first clinical evidence for the validity of the hypothesis (4).

I. Analytical methods

Plasma and urine from healthy subjects and plasma, urine, hemodialysate, hemofiltrate and peritoneal dialysate from uremic patients with and without neuropathy were submitted to high performance gel chromatography (HPGC) on Sephadex G-15 followed by gradient ion exchange chromatography (IEC) on DEAE Sephadex A-25 (5). In order to remove high molecular weight plasmatic components (>20 kD) which could interfere in gel chromatography with the fractions of middle molecular weight solute, plasma and peritoneal dialysate are ultrafiltered through semi-permeable membranes (AN-69 Rhône-Poulenc, Amicon Pm 10 or Millipore PTGC) before analysis, whereas urine, hemodialysate and hemofiltrate are treated without ultrafiltration. By HPGC on Sephadex G-15 equipped with U.V. detection at 254 nm, nine peaks designated "a" to "i" were detected in urine of healthy subjects. These peaks are very small or undetected in plasma of healthy subjects. All these peaks are detected in plasma, hemodialysate, hemofiltrate and peritoneal dialysate of uremic patients, indicating that all these peaks contain dialyzable solutes. Since only peak b was directly correlated with uremic polyneuropathy, further analysis of this peak was performed by IEC

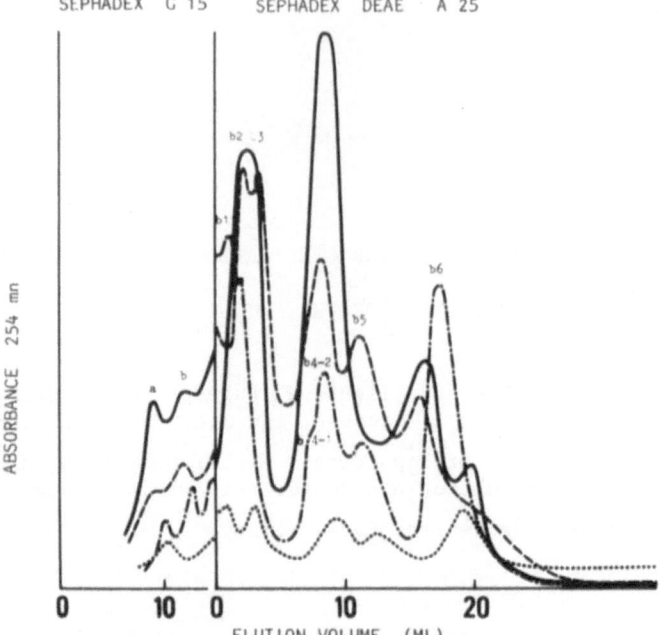

Fig. 1 Gel-permeation (Sephadex G-15) chromatography
followed by anion exchange (Sephadex DEAE A-25)
chromatography. Chromatograms of plasma (---)
and urine (-·-) from a healthy subject, or plasma
from a uremic patient with neuropathy (——) and a
uremic patient without neuropathy (– –).

on Sephadex DEAE A-25 with U.V. detection at 254 and 206 nm. Seven sub-
peaks designated b 1, b 2, b 3, b 4-1, b 4-2, b 5 and b 6 were detected
(Fig. 1). The sub-peak b 4-2 is the only one which correlates with uremic
polyneuropathy. This sub-peak b 4-2 is detected in urine and to a lesser
extent in plasma of healthy subjects. Molecular weight determination of
peak b 4-2 by ultrafiltration through cellulose acetate membrane gives a
rejection ratio compatible with an apparent MW of 537 D.

## II. Isolation procedure

Peak b and sub-peak b 4-2 collected from analytical columns are not
suitable for immediate analysis. Peak b contains sulphate and phosphate
and sub-peak b 4-2, numerous solutes which should be removed. The following preparative procedure was used to achieve these technical goals: 1)
preparative Sephadex G-15 chromatography, 2) removal of sulphate and
phosphate by Sephadex QAE-25, 3) isolation of b 4-2 fraction by Sephadex
DEAE Q-25, 4) removal of sodium by Sephadex SP C-25 chromatography and
finally 5) purification by kiselguhr and cellulose chromatography (6).
At each step the purity of the product was monitored by thin layer chromatography on silica gel. Dying reagents (periodic acid/phenyl hydrzaine,
p. anisidine/periodic acid) indicate that b 4-2 is an acidic polyol derivative with a 254 nm U.V. absorbing part. The frog sural nerve test (cf.
infra) reacts positively to the fraction b 4-2 isolated with this procedure. This purification procedure allowed the preparation of 53 mg b 4-2
solute from 60 L of urine of healthy subjects and 29 mg from .100 L of
uremic hemofiltrate which were used for calibration of chromatographic
columns and for determination of structure.

### III. Clinico-chemical correlation in uremic polyneuritis

Six patients with severe (paresthesia and paraplegia) and recent (3 to 6 month duration) uremic polyneuropathy due to inadequate dialysis were submitted to a dialysis strategy with a high removal rate of middle molecular weight range using a highly permeable membrane (Polyacrylonitrile AN 69, Rhône-Poulenc, France). Among the middle molecule fractions obtained by HPGC on Sephadex G-15 followed by IEC on Sephadex DEAE A-25 from plasma ultrafiltrate of six polyneuropathic patients, only peak b 4-2 was at a significantly higher concentration ($15 \pm 0.9$ mg/L) than that obtained from uremic patients without neuropathy ($4.6 \pm 1$ mg/L). When the neurological status of these patients improved after 3 months of adequate dialysis, the plasma concentration of peak b 4-2 returned to within the range ($5.3 \pm 0.6$ mg/L) observed in uremic patients without neuropathy (Fig. 2).

CLINICOCHEMICAL CORRELATIONS IN UREMIC POLYNEUROPATHY

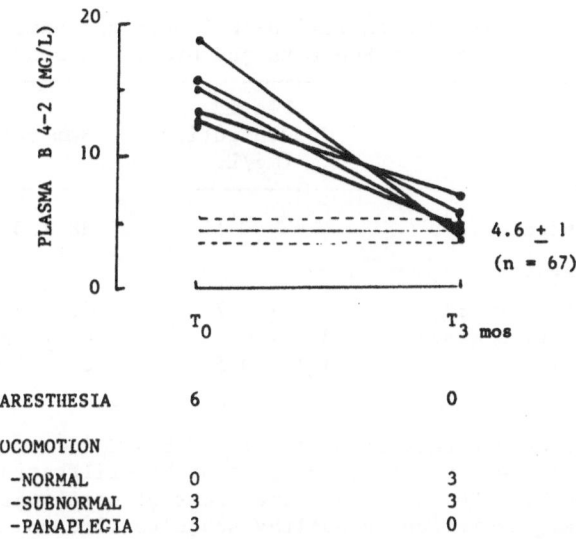

| | | |
|---|---|---|
| PARESTHESIA | 6 | 0 |
| LOCOMOTION | | |
| -NORMAL | 0 | 3 |
| -SUBNORMAL | 3 | 3 |
| -PARAPLEGIA | 3 | 0 |

Fig. 2 Correlation between plasma b 4-2 concentrations after 3 months of hemodialysis with a high flux membrane and the evolution of neurologic conditions in 6 uremic patients with recent and severe poly-neuropathy. Plasma b 4-2 concentrations (mean $\pm$ SEM) of uremic patients without neuropathy are shown: $4.6 \pm 1$ mg/L.

Numerous arguments suggest that inadequate hemodialysis favors the development of uremic neuropathy which might be due to the plasma retention of middle molecular weight solutes. Uremic polyneuritis was found to respond favorably to any method which increases middle molecule transfer. The rapid improvement that we have obtained by using a dialysis membrane which has high mass transfer for middle molecules seems to confirm this hypothesis (7). In the search for uremic toxins, high liquid performance chromatography was used to resolve specific U.V. absorbing constituants in body fluids (1). Generally this biochemical approach failed to show any positive correlation between the chromatographic profile and symptoms of the uremic syndrome. Conversely, investigations on clinico-chemical correlations in uremic polyneuritis led us to the isolation of a specific neurotoxin. The b 4-2 fraction thus appeared to be the sole plasma fraction which correlaated with active recent uremic polyneuropathy. This does not exclude, however, nutritional or depletion factors in the genesis of this uremic complication. During the course of polyneuropathy in the six patients studied, the improvement of neurological conditions was not found to be correlated either with plasma urea or with motor nerve conduction velocity. These data suggest that urea per se plays no role in the development of uremic polyneuritis and that motor nerve conduction velocity is not a sensitive criteria for clinical monitoring of adequate dialysis.

## IV. Plasma concentration and removal rate of fraction b 4-2 in healthy subjects and uremic patients

Quantitative determinations were made possible by calibration of analytic columns with standard solutions prepared from purified solute. Table 1 shows the b 4-2 plasma concentration of uremic patients on regular dialysis treatment (hemodialysis, hemofiltration and CAPD) without neuropathy where the concentration is about four times higher than that of healthy subjects (8, 9, 10).

Table 1.  B 4-2 Concentration in Plasma and Removal Rate in Healthy Subjects and Uremic Patients

|  | Plasma Concentration (mg/L) | Removal Rate |
|---|---|---|
| Healthy subjects (n=9 | < 1 | 38 $\pm$ 3 mg/24 h |
| Uremic patients |  |  |
| Hemodialysis (n=6) | 4.0 $\pm$ 0.7 | 57 $\pm$ 9 mg/session |
| Hemofiltration (n=7) | 3.5 $\pm$ 0.6 | 61 $\pm$ 12 mg/session |
| CAPD (n=8) | 4.1 $\pm$ 0.5 | 25 $\pm$ 4 mg/24 h |

The calculated weekly removal rate of b 4-2 in uremic patients treated 3 times a week by hemodialysis or by hemofiltration and in uremic patients treated by CAPD is of the same order of magnitude as the calculated weekly urinary excretion in healthy subjects (180 vs. 260 mg/week). In a healthy subject, the urinary excretion rate of b 4-2 was found rather constant (40 $\pm$ 4.8 mg/day) during 3 years, measured on 37 occasions, despite the fact that he received an uncontrolled free diet. When the subject received only glucose during 3 days (500 g/day), the urinary excretion rate decreases significantly, reaching its lowest level (24 mg/day) on the fourth day, and returns to within the previous excretion range when the subject regains a free diet (Fig. 3). Owing to the protein sparing effect of glucose diet, these data strongly suggest that b 4-2 has an endogenous protein origin.

## V. Characterization

Gas-liquid chromatography coupled with mass spectrometry (chemical and electronic impact ionization) on the permethylated components, which were previously isolated from fraction b 4-2 by thin layer chromatography, indicate clearly that glucuronococonjugates are the main component of fraction b 4-2, in addition to various small peptides. But acidic and enzymatic hydrolysis did not give exploitable results. Further mass spectrometry experiments were made on the trimethylsilyl derivative of the b 4-2 methly ester. Mass spectrometry data were compatible with the fact that b 4-2 is a glucuronide of MW 568 with an aglycon of MW 392. This weight may include 3 methyl ester groups; thus, the native MW of b 4-2

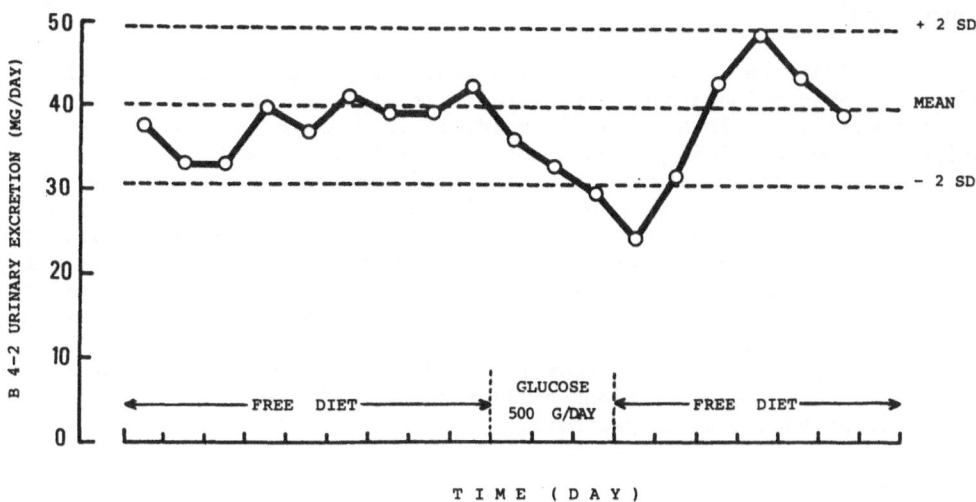

Fig. 3 Changes of b 4-2 urinary excretion in a healthy subject on a
glucose diet during 3 days compared to free diet periods.

could be 526. Moreover, mass spectrometry confirms that b 4-2 isolated
from urine of healthy subjects and from hemofiltrate of uremic patients
are identical (6).

## VI. Bioassay: in vitro frog sural nerve test

In order to evaluate the in vitro neurotoxicity of fractions obtained
by chromatographic analytical methods from plasma of uremic and polyneuropathic patients and from urine of healthy subjects, we performed an
in vitro test on the isolated frog sural nerve (11). The nerve was incubated in various media and stimulated by a rectangular shock (0.05 ms, 1
Hz, intensity supramaximal). Action potential was recorded. Plasma
ultrafiltrates of uremic polyneuropathic patients exhibit an inhibition
of action potential amplitude. No inhibition was observed when the nerve
was incubated with Ringer solution or plasma ultrafiltrates of either
healthy subjects or uremic patients without neuropathy. Among the fractions obtained by HPGC combined with gradient IEC, only fraction b and
fraction b 4-2 exhibit an inhibition of action potential amplitude. A
positive correlation was found between the inhibition index and plasma
ultrafiltrates or standard solutions at various b 4-2 concentrations.

## VII. Animal model

In order to study the renal handling of fraction b and sub-fraction
b 4-2, experimental chronic renal failure was induced in male Wistar rats
by electro-coagulation of renal cortex and controlateral nephrectomy.
Sham-operated rats stand for control (12). Integrated absorbance of peak
b correlates well with plasma creatinine concentration in the range from
50 to 1000 μmol/L, according to a linear relationship. Thus, plasma concentration fraction b increases in rats with the impairment of renal function. In uremic rats (mean plasma creatinine of $361 \pm 8$ μmol/L) sub-peak
b 4-2 was found at very high concentration whereas it was undetectable,
as in man, in plasma of sham-operated control rats (mean plasma creatinine
of $52 \pm 1$ μmol/L). Nerve conduction velocity, measured on the caudal
nerve, remains in the control range (35.5 - 43.5 m/sec) until renal function, assessed by endogenous creatinine clearance, decreases below 10% of
the initial value ($49 \pm 5$ vs. $539 \pm 12$ μL/min.100g).

## CONCLUSION

The fraction b 4-2 is a water-soluble compound with negative charge, high U.V. absorption and high resistance to storage and whose isotachophoretic mobility is close to citric acid. It is a glucuronic conjugate whose aglycon is not peptidic. It does not contain free NH2 and does not seem to have aromatic proton. It is a normal constituant of plasma and urine. It accumulates in the plasma of uremic polyneuropathic patients. It has a neurotoxic action in vitro and it is presumably of endogenous origin. The b 4-2 solute might be a good candidate for uremic toxin according to the criteria suggested by Bergström, for the b 4-2 is quantitatively measurable in biological fluid, the plasma concentration of the solute is higher in uremic than in nonuremic subjects, a high concentration is related to a specific symptom, namely polyneuritis, and the toxic effects of the solute were obtained at concentrations similar to those found in the body fluids of uremic polyneuropathic patients.

## REFERENCES

1. P. FÜRST, L. ZIMMERMAN, J. BERGSTROM. Determination of endogenous middle molecules in normal and uremic body fluids. Clin. Nephrol. 5:178, 1976.
2. J.L. FUNCK-BRENTANO, G. CUEILLE, N.K. MAN. A defense of the moddle molecule hypothesis. Kidney Int. 13 (Suppl. 8):S31, 1978.
3. A.L. BABB, P.C. FARRELL, D.A. UVELLI, B.H. SCRIBNER. Hemodialyzer evaluation by examination of solute molecular spectra. Trans. Am. Soc. Artif. Intern. Organs 18:98, 1972.
4. N.K. MAN, B. TERLAIN, J. PARIS, G. WERNER, A. SAUSSE, J.-L. FUNCK-BRENTANO. An approach to "middle molecules" identification in artificial kidney dialysate, with reference to neuropathy prevention. Trans. Am. Soc. Artif. Intern. Organs 19:320, 1973.
5. G. CUEILLE. Mise en évidence et évaluation des "Moyennes Molécules" de la taille de la Vitamine B12 présentes dans les liquides biologiques de sujets normaux et de patients urémiques. J. Chromatogr. 146:55,1978.
6. G.CUEILLE, N.K. MAN, A. SAUSSE, J.P. FARGES, J.L. FUNCK-BRENTANO. Further characterization of a neurotoxic uremic molecule, in "Proceedings 8th International Congress of Nephrology," W. ZURUKZOGLOU, M. PAPADIMITRIOU, M. PYRPASOPOULOS, M. SION, ed., S. Karger, Basel, 1981.
7. N.K. MAN, G. CUEILLE, J. ZINGRAFF, T. DRUEKE, P. JUNGERS, A. SAUSSE, J. BOUDET, J.L. FUNCK-BRENTANO. Evaluation of plasma neurotoxin concentration in uraemic polyneuropathic patients. Proc. Eur. Dial. Transplant Assoc. 15:164, 1978.
8. N.K.MAN, G. CUEILLE, J. ZINGRAFF, J. BOUDET, A. SAUSSE, J.L. FUNCK-BRENTANO. Uremic neurotoxin in the middle molecular weight range. Artificial Organs 4:116, 1980.
9. N.K.MAN, C. VERGER, P. FAGUER, J. VANTELON, J.L. FUNCK-BRENTANO. Middle molecule b 4-2 removal rate in patients on continuous ambulatory peritoneal dialysis, in: "Advances in Peritoneal Dialysis", G.M. GAHL, M. KESSEL, K.D. NOLPH, ed., Excerpta Medica, Amsterdam, 1981.
10. P. FAGUER, N.K.MAN, G. CUEILLE, S. DI GIULIO, J.L. FUNCK-BRENTANO. Improved separation and quantification of the "middle molecule" b 4-2 in uremia. Clin. Chem. 29:703, 1983.
11. J. BOUDET, G. CUEILLE, J.M. BENOIST, N.K. MAN, J.L. FUNCK-BRENTANO. In vitro frog sural nerve test: a monitor for detecting nerutoxin solutes. Artificial Organs 4 (Suppl): 94, 1980.
12. J. BOUDET, N.K.MAN, G. CUEILLE, Y. LEGRAIN, J.L. FUNCK-BRENTANO. Relationship between plasma concentration of middle molecular weight fraction b, motor nerve conduction velocity and plasma creatinine in experimental chronic renal failure in rats. Artificial Organs 4 (Suppl):115, 1980.

# INTRACELLULAR ACID-BASE AND ENERGY METABOLISM IN UREMIA: A PRELIMINARY STUDY ON FIVE PATIENTS BEFORE AND DURING DIALYSIS TREATMENT

Stefano Del Canale, Enrico Fiaccadori, Achille Guariglia,
Emilio Coffrini and Alberico Borghetti

Istituto di Clinica Medica e Nefrologia
Università di Parma
Parma, Italy

## INTRODUCTION

Failure of uremic patients to excrete end-products of endogenous metabolism necessarily results in a positive acid balance with consequent recruitment of both bone and soft tissue buffers.

Despite large evidence for strict interrelationships between cell buffering processes and metabolism during acidosis (1-2), little information on this subject is currently available on uremia.

This is likely to be due to technical difficulties inherent in the measurement of intracellular acid-base status.

The aim of the present study was thus to evaluate the main parameters of both intracellular acid-base and energy metabolism in a group of uremic patients before and during standard dialysis treatment.

## PATIENTS AND METHODS

Five uremic patients (2M-3F, mean age 56 years, range 44-66) suffering from end-stage renal failure (mean creatinine values $13.5 \pm 4$ mg%) and presenting severe metabolic acidosis ($HCO3_a = 15 \pm 4.5$ mEq/L) were admitted to the present study.

In all subjects (whose consent was previously obtained) muscle needle biopsies from the lateral portion of the quadriceps femoris were performed after at least two hours of bed rest.

Muscle samples were separated from the fascia and any visible fat or connective tissue, then divided into portions of suitable size (40-80 mg); some of them were than used for intracellular bicarbonate ($HCO3_i$) and pH ($pH_i$) determination by the use of $TCO_2$ method (3). The other samples were used for metabolite analyses according to enzymatic techniques (4). Energy charge potential (ECP) was calculated according to Atkinson (5).

The same parameters have been reevaluated while the patients were on standard hemodialysis treatment (4 hours 3 times weekly, dialysis bath acetate 40 mEq/L, duration 18-24 months).

Dietary recommendations during dialysis treatment included a 0.6 to 1.0 gr. protein/Kg BW, 30 KCal/BW diet, low in phosphate, satured fatty acids, ethanol and oligosaccharides.

At the time of the second biopsy all patients showed significant improvement of renal indexes and of metabolic acidosis ($HCO3_a$ = 22.1 ± 5.0 mEq/L). No patient had had serious illnesses during treatment; all of them were underweight.


## RESULTS

As shown in Figure 1, at the time of the first biopsy all patients were characterized by low levels of both $HCO3_i$ and $pH_i$ as compared to normal values.

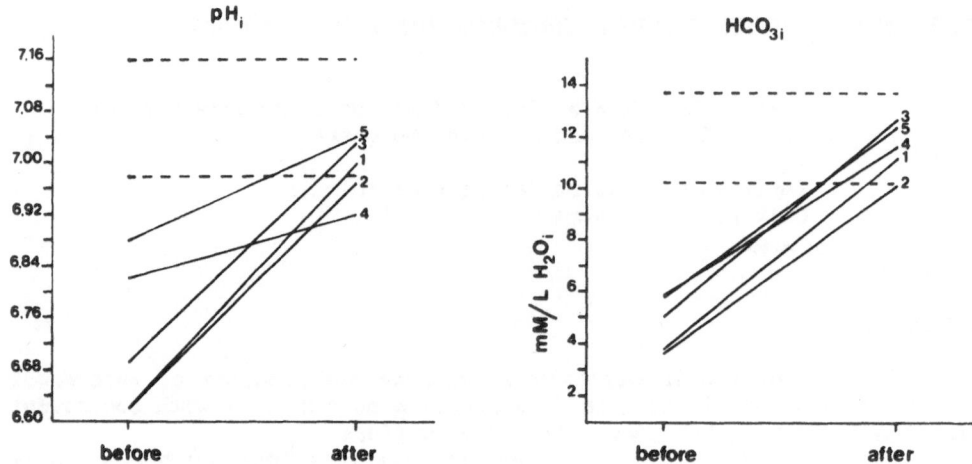

Figure 1: Intracellular pH and bicarbonate values of the five uremic patients before and during dialysis treatment are illustrated. Dashed lines indicate normal range values.

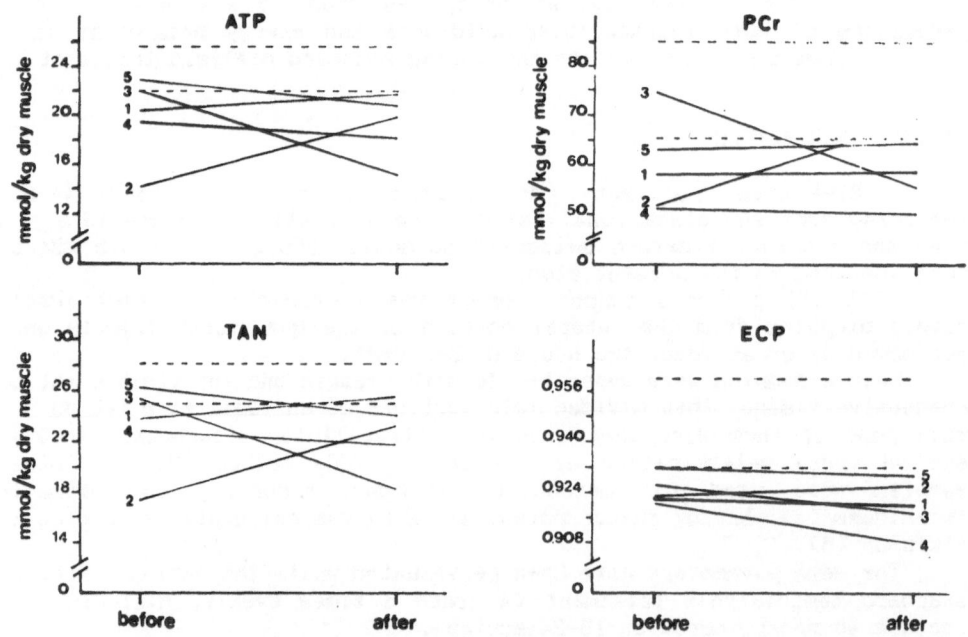

Figure 2: ATP, phosphocreatine (PCr), total adenine nucleotides (TAN) and energy charge potential (ECP) of the 5 patients as determined before and during dialysis treatment are illustrated. Dashed lines indicate normal range values.

A marked depression of ATP, phosphocreatine (PCr), total adenine nucleotide (TAN) and calculated BCP was also found and illustrated in Figure 2.

Reduced muscle lactate content as compared to control values (7.25 ± 2.6 vs. 9.96 ± 0.52 mmol/Kg FFS, p<0.05) was also detected (not reported in the figures).

At the time of the second biopsy, a full recovery of $HCO3_i$ levels with consequent nearly complete normalization of pHi is illustrated in Figure 1, while energy metabolism parameters were persistently low (cfr. Figure 2). Still significantly reduced muscle lactate levels were measured.

## DISCUSSION

A series of papers recently described the metabolic imbalance affecting the uremic cell. In particular defects in all steps of muscle cell energy production and utilization (6) as well as evidence for a glycolytic block and depressed high-energy phosphate content (7) have been reported both in experimental and human studies.

The results of the present study would confirm that end-stage uremic patients present profound alterations of cell energy metabolism. In fact, decreased ATP and PCr as well as lactate content would further support the hypothesis that impaired high-energy phosphate production together with disturbances of glycolytic sequences are characteristic of this stage of uremia.

As suggested elsewhere (7), in our patients too, intracellular acidosis may be viewed as the link between acid-base and energy metabolism derangements in advanced renal failure. In fact, increased cell acidity has been reported to inhibit main regulatory steps of glycolysis (8) as well as possibly interfere with mitochondrial phosphorilation (9): thus, the hypothesis to ascribe some of the metabolic disturbances encountered in our patients to intracellular acidosis has theoretical support.

However, when we reevaluated our patients after two years of hemodialysis treatment, a full replenishment of intracellular buffer stores with consequent pHi normalization was found. On the contrary the same abnormalities of muscle cell energy metabolism were still present.

It is then concluded that muscle energy metabolism disturbances of end-stage uremic patients are likely to be of multifactorial origin. As the effects of acidosis would be critical in the predialytic period, other factors not considered in this study, namely lack of substrate and catabolism may represent determining factors during dialysis treatment.

## REFERENCES

1. R.D. Cohen and R.A. Iles. Intracellular pH. Measurement, control, and metabolic interrelationships. CRC Crit. Rev. Clin. Lab. Sci. 6: 101-143 (1975).

2. E. Hultman and K. Sahlin. Acid-base balance during exercise. In: "Exercise and sport sciences reviews", Hutton RS, Miller DI eds. The Franklin Institute Press, Philadelphia: 4-128, (1980).

3. E. Flaccadori, S. Del Canale, U. Arduini, C. Antonucci, E. Coffrini, P. Vitali, R. Melej and A. Guariglia. Intracellular acid-base and electrolyte metabolism in skeletal muscle of patients with chronic obstructive lung disease and acute respiratory failure. Clin. Sci. 71: 703-712 (1986).

4. R.C. Harris, E. Hultman and L.O. Nordesjo. Glycogen, glycolytic intermediates and high energy phosphates determined in biopsy samples of musculus femoris of man at rest. Methods and variance of values. Scand. J. Clin. Lab. Invest. 33: 109-120 (1974).

5. D.E. Atkinson. The energy charge of the adenylate pool as a regulatory parameter. Interaction with feed-back modifiers. Biochemistry 7: 4030-4034 (1968).

6. N. Brautbar. Skeletal myopathy in uremia: abnormal energy metabolism. Kidney Int. 24 (Suppl.16): 81-86 (1983).

7. S. Del Canale, E. Fiaccadori, N. Ronda, K. Soderlund, C. Antonucci and A. Guariglia. Muscle energy metabolism in uremia. Metabolism 35 (11): 981-983 (1986).

8. A.S. Relman. Metabolic consequences of acid-base disorders. Kidney Int. 1: 347-359 (1972).

9. J.M. Lowenstein and B. Chance. The effect of hydrogen ions on the control of mitochondrial respiration. J. Biol. Chem. 243 (n.14): 3940-3946 (1968).

IS SELENIUM DEFICIENCY THE CAUSE OF URAEMIC CARDIOMYOPATHY?

E.R. Maher, B. Sampson, and J.R. Curtis

Departments of Medicine and Chemical Pathology
Charing Cross Hospital, London, England

SUMMARY

We studied the relationship between serum selenium (Se) and left ventricular performance in 33 patients on maintenance haemodialysis. Low serum Se was frequent. However, there were no significant differences in echocardiographic indices of left ventricular function between patients with serum Se <0.9 umol/l and those with serum Se >0.9 umol/l. We conclude that Se deficiency is not an important cause of cardiac failure in uraemia.

INTRODUCTION

Impaired cardiac performance is an important cause of morbidity and mortality in patients with severe renal failure. Multiple factors may contribute to the development of cardiac failure in patients on maintenance haemodialysis e.g. arterial hypertension, anaemia, fluid overload and arteriovenous fistulae. In addition it appears that the uraemic state per se may impair myocardial performance i.e. a uraemic cardiomyopathy (1). Selenium is an essential part of glutathione peroxidase and Se deficiency may cause a congestive cardiomyopathy (2-4). Serum Se is reduced in chronic renal failure (5,6). We therefore investigated whether there is a relationship between serum Se and left ventricular function in patients on maintenance haemodialysis.

METHODS

We studied 33 patients (20 males) with chronic renal failure on maintenance haemodialysis for between 6 months and 19 years (mean = 6.1 years). Mean patient age was 39 years (range 19 to 55 years) and none had clinical or electrocardiographic evidence of ischaemic heart disease.

Left ventricular performance was assessed after haemodialysis by m-mode echocardiography. Left ventricular end diastolic diameter (EDD) and end systolic diameter (ESD) were used to calculate fractional shortening (FS = EDD-ESD/EDD). The velocity of circumferential fibre shortening (Vcf) was derived from (EDD-ESD)/EDD.LVET, where LVET is the left ventricular ejection time which was calculated from the echocardiogram according to the method of Cooper et al (7). All measurements were averaged over at least 5 cycles.

Serum Se was measured by electrothermal atomisation atomic absorption spectroscopy. Patients were divided into two groups : those with serum Se >0.9 umol/l (Group A, n = 15) and those with serum Se <0.9 umol/l (Group B, n = 18). Results are expressed as mean +SD and statistical significance was taken at the 5% level.


RESULTS

The mean serum Se of Group A and of Group B were 1.2 + 0.2 umol and 0.6 + 0.1 umol/l respectively. The two groups did not differ in age, duration of haemodialysis, mean arterial blood pressure or packed cell volume (see Table 1).

There were no significant differences between the two groups in any of the measured indices of myocardial function (see Table 2).


Table 1 : Clinical and laboratory characteristics of the 2 groups of patients

|  | Group A | Group B |  |
|---|---|---|---|
| Number of patients | 15 | 18 |  |
| Age (years) | 39 +10 | 39 +11 | NS |
| Years on haemodialysis | 7.4 +5.8 | 5.1 +3.6 | NS |
| Mean blood pressure (mmHg) | 107 +14 | 112 +9 | NS |
| Packed cell volume | .23 +.05 | .23 +.05 | NS |


Table 2 : Cardiac indices in the two groups of patients

|  | Group A | Group B |  |
|---|---|---|---|
| Vcf (circum/sec) | 0.9 +0.3 | 0.9 +0.2 | NS |
| Fractional shortening | .29 +.07 | .29 +.07 | NS |
| End diastolic diameter (mm) | 45 +7.5 | 47 +7.2 | NS |
| End systolic diameter (mm) | 32 +6.4 | 34 +7.3 | NS |

# DISCUSSION

Most of our patients had serum Se levels below normal limits. Others groups have also found low serum Se (5,&) and reduced plasma and erythrocyte glutathione peroxidase (6) in patients with renal failure. This Se depletion may be caused by reduced dietary intake or loss through the dialysis membrane. The selenoenzyme glutathione peroxidase is an important part of the cellular antioxidant defence mechanisms, and Se deficiency may lead to tissue injury from lipid peroxides and free radicals. Selenium deficiency is responsible for Keshan disease, an endemic cardiomyopathy in North China (2), and cardiomyopathy in animals (8). Cardiac failure in a patient on total parenteral nutrition has been attibuted to Se deficiency, but blood levels were much lower than those found in haemodialysis patients (3). We could not demonstrate any relationship between serum Se and left ventricular function in patients on maintenance haemodialysis. Many factors may contribute to depressed myocardial performance in renal failure, but Se deficiency does not appear to be a major cause cardiac failure in uraemia.

# REFERENCES

1. Drueke T, Le Pailleur C, Zingraff J, Jungers P. Uremic cardiomyopathy and pericarditis. Advances in Nephrology 1980; 9: 33-70

2. Chen X, Yang G, Chen J, Chen X, Wen Z, Ge X. Studies on the relations of selenium and Keshan disease. Biol Trace Elements Res 1980; 2: 91-107

3. Fleming CR, Lie JT, McCall JT, O'Brien JF, Baille EE, Thistle JL. Seleniumlenium deficiency and fatal cardiomyopathy in a patient on home parenteral nutrition. Gastroenterology 1982; 83: 689-93

4. Johnson RA, Baker SS, Fallon JT, Maynard EP, Ruskin JN, Wen Z, Ge K, Cohen HJ. An occidental case of cardiomyopathy and selenium deficiency. N Engl J Med 1981; 304: 1210-12

5. Kallistratos G, Evangelou A, seleniumferiadas K, Vezyraki P, Barboutis K. Seleniumlenium and haemodialysis : seleniumrum selenium levels in healthy persons, non-cancer and cancer patients with chronic renal failure. Nephron 1985; 41: 217-22

6. Leung A, Henderson I, Fell G, Hall D, Kennedy AC. Seleniumlenium deficiency in chronic uraemia and dialysis. Proc EDTA 1985; :1134-8

7. Cooper RH, O'Rourke RA, Karliner JS, Petersen KL, Leopold GR. Comparison of ultrasound and cineangiographic measurements of the mean rate of circomferential fiber shortening in man. Circulation 1972; 46: 914-923

8. Van Fleet JF, Ferrans VJ, Ruth GR. Ultrastructural alterations in nutritional cardiomyopathy of selenium - vitamin E deficient swine. I Fiber lesions. Lab Invest 1977; 37 (2): 188-200

# DOES HYPERMAGNESEMIA SUPPRESS PARATHYROID ACTIVITY AND THEREFORE PLAY A ROLE IN AMELIORATION OF UREMIC TOXICITY ?

P.L. Oe, J. van der Meulen, P.T.A.M. Lips, P.M.J.M. de Vries,
H. van Bronswijk and A.J.M. Donker

Department of Medicine, Academic Hospital Free University
de Boelelaan 1117, 1081 HV  Amsterdam, The Netherlands

Key words: Magnesium hydroxide - Parathyroid activity - Uremic toxicity.


## INTRODUCTION

Parathyroid hormone satisfies all criteria of a uremic toxin, due to a cause and effect relationship between excess parathyroid hormone and disturbances of the nervous system in uremia[1]. Reports concerning the suppression of parathyroid activity by hypermagnesemia are few and contradictory[2,3]. Therefore, the long-term administration of magnesium hydroxide ($Mg(OH)_2$) as phosphate binder in patients on hemodialysis was an excellent opportunity for an own observation of the relationship between hypermagnesemia and parathyroid activity.


## PATIENTS AND METHODS

Fourteen patients, all on in-center dialysis twice a week (total dialysis time 10-12 h/week) for more than 6 months, participated after informed consent was obtained. Dietary intake was kept unchanged. Calcium carbonate (3g/day) orally was prescribed. It was not necessary to administer vitamin D in all patients. The dose of phosphate binders was adjusted in order to keep serum phosphate level <1.5 mmol/l. However, adjusting of the dose of $Mg(OH)_2$ was also done in order to keep serum magnesium (Mg) level <2.3 mmol/l and to avoid troublesome diarrhea. Therefore, magnesium hydroxide alone could not control hyperphosphatemia at the desired level and aluminium hydroxide ($Al(OH)_3$ had to be added.

The study was divided in three periods. In period I (=baseline) $Al(OH)_3$ was administered during the meals for 6 to 9 (mean 8.1) months. In period II $Mg(OH)_2$ was given for 2 to 6.5 (mean 3.7) months. In period III $Al(OH)_3$ was added to the established dose of $Mg(OH)_2$ for 4 to 13 (mean 10.4) months. The dialysate was devoid of Mg whereas in all three periods the calcium (Ca) content was 1.5 mmol/l.

Serum levels of Ca, phosphate, alkaline phosphatase (AF), Mg, parathyroid hormone (PTH) and aluminium (Al) were monitored regularly. Ca, phosphate, Af, Mg, K and Hb were determined according to routine laboratory methods, reference ranges being 2.20 - 2.60 mmol/l, 0.8 - 1.3 mmol/l,

25 - 100 U/1, 0.80 - 1.05 mmol/1, 3.9 - 5.1 mmol/1 and 8 - 10 mmol/1, respectively. Serum PTH was analysed with a two-step immunochemical method, which measures the intact PTH (1 - 84) molecule and does not detect fragments[4]. The assay involves a N-terminal immuno-extraction followed by a midregion and C-regional immunoassay. The upper reference limit is 15 pmol/1. The interassay coefficient of variation is 5.5 %. Serum Al level was determined by atomic absorption spectrophotometry[5]. The normal level is below 20 microg/1 in healthy persons. The interassay coefficient of variation is 5.2 %. For statistical analysis conventional methods were used. All values are given as mean ± SD. Comparison of mean values was done by the two-tailed-t-test for paired data.

RESULTS

The dose of $Mg(OH)_2$ could not be increased sufficiently in 7 patients as serum Mg exceeded 2.3 mmol/1. Troublesome diarrhea was the limiting factor in the remaining 7 patients.

Mean $Al(OH)_3$ dose was 6.6 ± 2.3 g/day in period I. Mean $Mg(OH)_2$ dose was 2.4 ± 0.6 g/day in period II. In period III 2.6 ± 1.2 g $Mg(OH)_2$ together with 3.3 ± 1.4 g $Al(OH)_3$ were administered daily.

Fig. 1 shows the decrease of serum PTH level from 26.6 ± 21.8 to 18.6 ± 20.6 pmol/1 (p<0.005) in period II, whereas a further decrease to 7.8 ± 6.0 pmol/1 (p<0.005) occurred in period III.

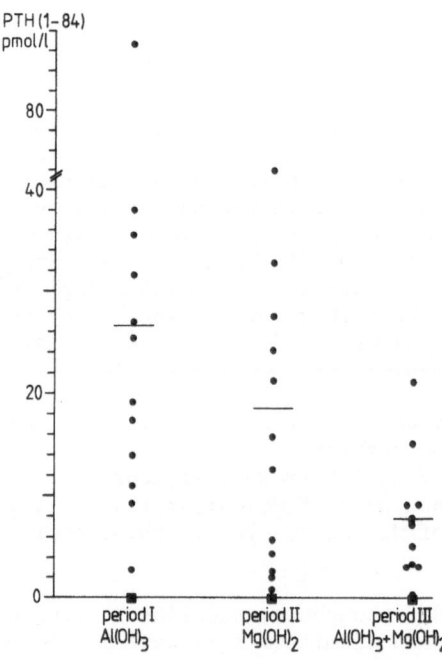

Fig. 1. Serum PTH levels. See text.

Table I. Relevant observed data during the study. See text.

|  | Period I | Period II | Period III |
|---|---|---|---|
| Magnesium (mmol/1) | 1.43 ± 0.19 | 1.72 ± 0.30 | 1.68 ± 0.26 |
| Phosphate (mmol/1) | 1.42 ± 0.24 | 2.01 ± 0.49 | 1.63 ± 0.21 |
| Calcium (mmol/1) | 2.29 ± 0.07 | 2.22 ± 0.07 | 2.23 ± 0.11 |
| Aluminium (microg/1) | 57.1 ± 27.3 | 30.7 ± 15.6 | 55.8 ± 19.1 |
| Ratio serum Al/daily consumption of Al(OH)$_3$ | 9.8 ± 4.6 |  | 17.8 ± 10.2 |
| Potassium (mmol/1) | 5.1 ± 0.4 | 5.7 ± 0.3 | 5.5 ± 0.4 |

Table I shows the significant (p<0.005) increase of serum Mg level in period II. No further increase was seen in period III. Substitution of Mg(OH)$_2$ for Al(OH)$_3$ resulted in a significant (p<0.005) rise of serum phosohate in period II, whereas a significant (p<0.005) decrease in serum phosphate level occurred after addition of Al(OH)$_3$ to Mg(OH)$_2$. The changes of serum Ca were the reverse. Serum Al levels decreased significantly (p<0.005) after substitution of Mg(OH)$_2$ for Al(OH)$_3$. However, levels of serum Al rose again significantly (p<0.005) and were comparable to the base-line values in period I, after the addition of Al(OH)$_3$ in period III. There was a significant (p<0.005) rise of the ratio between serum Al and the daily consumption of Al(OH)$_3$ in period III compared with the ratio in period I. Another phenomenon of the substitution of Mg(OH)$_2$ for Al(OH)$_3$ was the significant (p<0.005) increase of serum K, although in period III a significant (p<0.005) decrease of serum K could be noted. No significant changes were seen in serum AF and Hb levels.

An inverse relationship (R=-0.76, n=12, p<0.005) was noted between serum Mg and PTH levels.

DISCUSSION

No toxic manifestations such as central nervous system depression and electrocardiographic abnormalities[6] were observed during the long-term administration of Mg(OH)$_2$. However, the same criterion of the upper limit for serum Mg (2.3 mmol/1) was used as suggested in a short-term study[7]. The maintenance dose of Mg(OH)$_2$ could be increased following the suggestion of using a dialysate devoid of Mg[8].

Mg(OH)$_2$ alone as alternative phosphate binder could not decrease serum phosphate <1.50 mmol/1. Therefore, Al(OH)$_3$ had to be added. The dose of this agent could be halved by Mg(OH)$_2$ administration. However, serum Al levels rose to base-line values and were accompanied by an increased ratio between serum Al and daily consumption of Al(OH)$_3$ indicating an increased fractional Al absorption from the gut[9]. This might be explained by the dominant effect of Mg(OH)$_2$ as phosphate binder, by which Al(OH)$_3$ is only partially converted to Al phosphate and therefore absorbed.

Serum K increased after substitution of $Mg(OH)_2$ for $Al(OH)_3$, although this phenomenon was less in period III. This might be explained by displacement of intracellular K by Mg[10].

Suppressed parathyroid activity by hypermagnesemia was an important observation of this study and previously suggested by other investigators[2]. As PTH satisfies all criteria of a uremic toxin, due to a cause and effect relationship between excess PTH and disturbances of the nervous system in uremia[1], hypermagnesemia might play an important role in amelioration of uremic toxicity. However, further study i.e. histomorphometry of bone biopsies is necessary.

REFERENCES

1.  S. G. Massry, Neurotoxicity of parathyroid hormone in uremia, Kidn. Intern. 28:suppl. 17 S-5 (1985).
2.  S. G. Massry, J. W. Coburn, and C. R. Kleeman, Evidence for suppression of parathyroid gland activity by hypermagnesemia, J. Clin. Invest. 49:1619 (1979).
3.  M. Gonella, L. Moriconi, G. Betti, F. Bonaguidi, G. Buzzigoli, V. Bartolini, and G. Mariani, Serum levels of PTH, Mg, Ca, inorganic phosphate and alkaline phosphatase in uraemic patients on differentiated Mg dialysis, Proc. EDTA 17:362 (1980).
4.  A. W. Landall, J. Elting, J. Ells, and B. A. Roos, Estimation of biologically active intact parathyroid hormone in normal and hyperparathyroid sera, J. Clin. End. Metab. 57:1007 (1983).
5.  G. B. v.d. Voet, E. J. M. de Haas, and F. A. de Wolf, Monitoring of aluminum in whole blood, plasma, serum and water, J. Anal. Tox. 9:97 (1985).
6.  R. E. Randall, D. Cohen, C. C. Spray, and E. C. Rossmeisl, Hypermagnesemia in renal failure, Ann. Intern. Med. 61:73 (1964).
7.  A. P. Guillot, V. L. Hood, C. F. Runge, and F. J. Gennari, The use of magnesium-containing phosphate binders in patients with end-stage renal disease on maintenance hemodialysis, Nephron 30:114 (1982).
8.  F. P. Brunner, and G. Thiel, The use of magnesium-containing phosphate binders in patients with end-stage renal disease on maintenance haemodialysis, Nephron 32:266 (1982).
9.  J. v.d. Meulen, P. D. Bezemer, P. Lips, and P. L. Oe, Individual differences in gastrointestinal absorption of aluminum, N. Engl. J. Med. 311:1322 (1984).
10. R. O'Donovan, C. Monitz, D. Baldwin, M. Rogerson, and V. Parsons, Hyperfosfataemia is controlled by oral magnesium carbonate while on zero magnesium dialysate without aluminium binders, Proc. EDTA-ERA 22:1229 (1985).

# SYNTHESES AND STRUCTURE-ACTIVITY RELATIONSHIPS OF THYMOPOIETIN

Takashi Abiko

Kidney Research Laboratory
Kojinkai
Higashishichiban-cho 84
Sendai 980, Japan

Two peptides, thymopoietins I and II, which are T-cell differentiating hormones of thymus were purified and characterized by Goldstein,[1] using an in vitro neuromscular assay. Each peptide contains 49 amino acid residues and both peptides were shown to be related and to differ by only three amino acid residues.[2]

Fujino and coworkers[3] reported the first synthetic peptide to exhibit the full activity of thymopoietin II in 1977.

In 1975, a chemical synthesis by Schlesinger et al.[4] revealed that biological activity was exhibited by fragment (29-41) of thymopoietin II. Subsequent the pentapeptide corresponding to residues 32-36 of thymopoietin II was shown to retain the biological activity of thymopoietin II.[5]

Then we reported that the decapeptide (32-41)[6] of thymopoietin II and the pentapeptide (32-36)[7] induce some recovery of E-rosette formation in the uremic state.

In 1981, we also reported that the octadecapeptide (32-49)[8] of thymopoietin II induces some recovery of E-rosette formation in blood of patients with rheumatoid arthritis.

In 1981, the proposed structures of thymopoietins I and II were revised by Goldstein et al.[9] In 1982, we reported[10] the synthesis of the octadecapeptide (32-49), which corresponds to a part of revised structure of thymopoietin II and showed that the biological activity of revised thympoietin II fragment (32-49) on low E-rosette-forming cells of an aged patient with chronic renal failure was equal to that of unrevised thymopoietin II fragment (32-49).

In 1981, Goldstein et al.[9] elucidated the primary structure of thymopoietin III, which was isolated from bovine spleen. The three peptides, thymopoietins I, II and III, have identical sequences except for amino acid residues at positions 1, 2, 34 and 43.

In 1985, we also reported[11] the synthesis of the nonatracontapeptide corresonding to the entire amino acid sequence of thymopoietin I and showed that this peptide could restore the E-rosette-forming capacity in a uremic patient.

```
I   H-[Gly-Gln]-Phe-Leu-Glu-Asp-Pro-Ser-Val-Leu-Thr-Lys-Glu-Lys-Leu-
                    5                    10                    15

II  H-[Pro-Glu]-Phe-Leu-Glu-Asp-Pro-Ser-Val-Leu-Thr-Lys-Glu-Lys-Leu-

I     Lys-Ser-Glu-Leu-Val-Ala-Asn-Asn-Val-Thr-Leu-Pro-Ala-Gly-Glu-
                        20                    25                    30

II    Lys-Ser-Glu-Leu-Val-Ala-Asn-Asn-Val-Thr-Leu-Pro-Ala-Gly-Glu-

I     Gln-Arg-Lys-Asp-Val-Tyr-Val-Glu-Leu-Tyr-Leu-Gln-[His]-Leu-Thr-
                        35                    40                    45

II    Gln-Arg-Lys-Asp-Val-Tyr-Val-Glu-Leu-Tyr-Leu-Gln-[Ser]-Leu-Thr-

I     Ala-Leu-Lys-Arg-OH
                    49

II    Ala-Leu-Lys-Arg-OH
```

Fig. 1. Comparison of Amino Acid Sequences of Thymopoietin

I and Thymopoietin II

Differences in the sequences are enclosed in boxes.


In 1984, Goldstein et al.[12] reported that thymopentin corresponding to residues 32-36 of thymopoietin II needed the addition of an octapeptide corresponding to thymopoietin II fragment (38-45) for full competition with native thymopoietin in a radioreceptor assay derived from the human T-cell line CEM. The octapeptide (38-45) was not completely active biologically but addition of this peptide to thymopoietin enhanced its biological potency. Their results suggest that thymopoietin binds to its receptor on T-cells two distinct sites one created by amino acids 32-36 (active site) and other by amino acids 38-45 (binding site). We are interested in the peptide fragment corresponding to amino acids 32-45 of thymopoietin II, which has two important sites (active and binding) within its molecule. In the course of our synthetic study on thymopoietin II fragment 32-45 and its analogs, it was observed for the first time, that tha presence of amino acid having a bulkier side chain is needed at position 37 for high activity of the transformation of T-cells in common variable immunodeficiency into lymphoblasts with mitotic activity after PHA stimulation.[12]

Further, it was found that the presence of an amino acid having a hydrophobic and bulkier side chain is needed at postion 37 for high activity and conversely the presence of an amino acid having a bulkier and hydrophilic side chain at position 37 induces lower potency.[13]

Recently, we reported[14] the synthesis of the nonatetracontapeptide corresponding to the entire amino acid sequence of thymopoietin III and showed that this peptide found to have restoring activity on the T-cell transformation by PHA but the restoring activity of this peptide was lower than that of the synthetic thymopoietin I.

In our previous papers,[7] we also reported that the increased activity of E-rosette-forming cells from uremic patients induced by Arg-Lys-Glu-Val-Tyr, which corresponds to residues 32-36 of thymopoietin III, was lower than that of Arg-Lys-Asp-Val-Tyr, which corresponds to residues 32-36 of thymopoietin I. These results seem to suggest that replacing Asp[34] of thymopoietin I by Glu gave an analog with lower potency for restoring cell-mediated immunological activities.

```
        1                    5                      10                    15
 I   H-[Gly-Gln]-Phe-Leu-Glu-Asp-Pro-Ser-Val-Leu-Thr-Lys-Glu-Lys-Leu-

III  H-[Pro-Glu]-Phe-Leu-Glu-Asp-Pro-Ser-Val-Leu-Thr-Lys-Glu-Lys-Leu-
                        20                    25                    30
 I   Lys-Ser-Glu-Leu-Val-Ala-Asn-Asn-Val-Thr-Leu-Pro-Ala-Gly-Glu-

III  Lys-Ser-Glu-Leu-Val-Ala-Asn-Asn-Val-Thr-Leu-Pro-Ala-Gly-Glu-
                            35                    40                    45
 I   Gln-Arg-Lys-[Asp]-Val-Tyr-Val-Glu-Leu-Tyr-Leu-Gln-His-Leu-Thr-

III  Gln-Arg-Lys-[Glu]-Val-Tyr-Val-Glu-Leu-Tyr-Leu-Gln-His-Leu-Thr-

 I   Ala-Leu-Lys-Arg-OH

III  Ala-Leu-Lys-Arg-OH
```

Fig. 2. Comparison of Amino Acid Sequences of Thymopoietin I and Thymopoietin III; Differences in the Sequences are Enclosed in Boxes

REFERENCES

The amino acid residues mentioned in this paper are L-configuration except for glycine. The abbreviations used to denote amino acid derivatives and peptides are those recommended by the IUPAC-IUB Joint Commission on Biochemical Nomenclature: European J. Biochem., 138: 9 (1984); Int. J. Peptide Protein Res., 24, No. 1 (1984).

1. G. Goldstein, Nature (London), 247: 11 (1974).
2. D. H. Schlesinger and G. Goldstein, Cell, 5: 361 (1975).
3. M. Fujino, S. Shinagawa, T. Fukuda, M. Takaoka, H. Kawaji and Y. Sugino, Chem. Pharm. Bull., 25: 1486 (1977).
4. D. H. Schlesinger, G. Goldstein, M. P. Scheid and E. A. Boyse, Cell, 5: 367 (1975).
5. G. Goldstein, M. P. Scheid, E. A. Boyse, D. H. Schlesinger and J. Van Wauwe, Science, 204: 1309 (1972).
6. T. Abiko, M. Kumikawa and H. Sekino, Chem. Pharm. Bull., 27: 2233 (1979).
7. T. Abiko, I. Nonodera and H. Sekino, Chem. Pharm. Bull., 28, 2507 (1980).
8. T. Abiko, I. Onodera and H. Sekino, Chem. Pharm. Bull., 29, 2322 (1981).
9. T. Audha, D. H. Schlesinger and G. Goldstein, Biochemistry, 20: 6195 (1981).
10. T. Abiko and H. Sekino, Chem. Pharm. Bull., 30: 3271 (1982).
11. T. Abiko and H. Sekino, Chem. Pharm. Bull., 33: 1583 (1985)
12. T. Abiko, H. Shishido and H. Sekino, J. Appl. Biochem., 7: 408 (1985).
13. T. Abiko, H. Shishido and H. Sekino, Biotechnol. Appl. Biochem., 8: 408 (1985).
13. T. Abiko, H. Shishido and H. Sekino, Biotechnol. Appl. Biochem., 34, 2133 (1986).

THE IMMUNOSUPPRESSIVE FRACTION ISOLATED FROM UREMIC ULTRAFILTRATES :

ATTEMPTED CHARACTERIZATION OF THE MECHANISM OF ACTION

J. Navarro, M.C. Grossetête, J.L. Touraine, and J. Traeger

Inserm U.80, CNRS UA 1177, UCB

Lyon, France

Since many years our attention has been focused on the isolation of an immunosuppressive fraction from the serum of uremic patients. This fraction which was obtained after a single chromatography of uremic fluids on a Sephadex G25 column, was shown to inhibit markedly the lymphocyte proliferation induced by several mitogens and by allogeneic cells[1]. A continuous infusion of this material in rats allowed a significant prolongation of skin allograft survival[1]. By contrast the antibody production to sheep red blood cells was not affected by such a treatment. Furthermore, this crude fraction reduced significantly the graft-versus-host reaction induced in lethally irradiated mice injected with allogeneic spleen lymphocytes previously incubated in the presence of the uremic factor[1]. Inhibition of cell proliferation appears to be the common denominator of all these in vitro and in vivo immunosuppressive effects.

We have further analysed the mechanism of action of the uremic fraction concurrently with fractionation assays which are detailed in the following paper.


KINETIC EXPERIMENTS

Lymphocyte cultures were performed as previously described[1]. In kinetic experiments, the uremic fraction was added at different times of the culture stimulated either with phytohemagglutinin (PHA) or with allogeneic cells. Lymphocyte proliferation was measured by adding 3H-thymidine during the last 18H period. Cells were then harvested and processed for liquid scintillation counting.

The inhibitory effect of the uremic factor on lymphocyte proliferation was not mediated by a mere cytotoxic effect. Cells were alive at the end of the culture as shown by the trypan blue exclusion test. Furthermore, the uremic fraction must be added at the beginning ot the culture to obtain a significant inhibitory effect : within the first 24 hours in cultures stimulated with PHA and within the first 3 days in a mixed lymphocyte culture. These results suggest that the inhibitory factor isolated from uremic sera interferes with an early event in lymphocyte stimulation.

It is now generally accepted that T cell proliferative responses to mitogens and antigens result from the production of interleukin 2 (IL2) and from the interaction of this peptide with specific membrane receptors which appear on activated T cells. We have thus analysed the effect of the uremic solute on both IL2 production and acquisition of its receptors. These experiments have been performed with fraction 2 recovered after the last purification step using a C18 Resolve column.

EFFECT ON IL2 PRODUCTION

Peripheral blood lymphocytes from healthy donors have been stimulated with PHA for 18 hours, in the presence or not of the immunosuppressive fraction. Cells were then discarded by centrifugation and IL2 activity in the supernatant was determined using the growth of the IL2 dependent CTLL2 cell-line. When the cells had been cultured with the mitogen and medium alone, the supernatant contained IL2 which allowed the growth of the CTLL2 cell line (Table I). However, this growth was not supported by the supernatant recovered from cells incubated with the uremic fraction, indicating a defect in IL2 production. These results could not be explained by a direct inhibitory effect of the active solute on the CTLL2 cell line. When cultured in IL2 containing medium, these cells were not affected by the presence of the uremic factor.

EFFECT ON IL2 RECEPTOR ACQUISITION

IL2 receptor induction on lymphocyte membrane was obtained by culturing the cells in the same conditions as described above. IL2 receptor expression was analysed using an indirect immunofluorescence technique with a murine monoclonal antibody directed against these receptors. Fluorescence intensity of the cells was then analysed by flow cytometry (Fig 1).

When the cells had been incubated in the presence of the uremic fraction a significant decrease in fluorescent cell number was observed indicating a decrease in IL2 receptor expression.

Table I. Inhibitory Effect of Fraction 2 from C18 Resolve on IL2 Production

| Supernatant dilutions | CTLL2 proliferation in the presence of cell cultures supernatants from : | |
| :---: | :---: | :---: |
| | Control medium + PHA | Fraction 2 + PHA |
| 1/2 | 10 726 $\pm$ 560 * | 1 718 $\pm$ 135 |
| 1/4 | 4 896 $\pm$ 119 | 1 030 $\pm$ 293 |
| 1/8 | 2 506 $\pm$ 24 | 667 $\pm$ 137 |
| 1/16 | 1 201 $\pm$ 45 | 612 $\pm$ 287 |

* Results are expressed as mean dpm $\pm$ sd

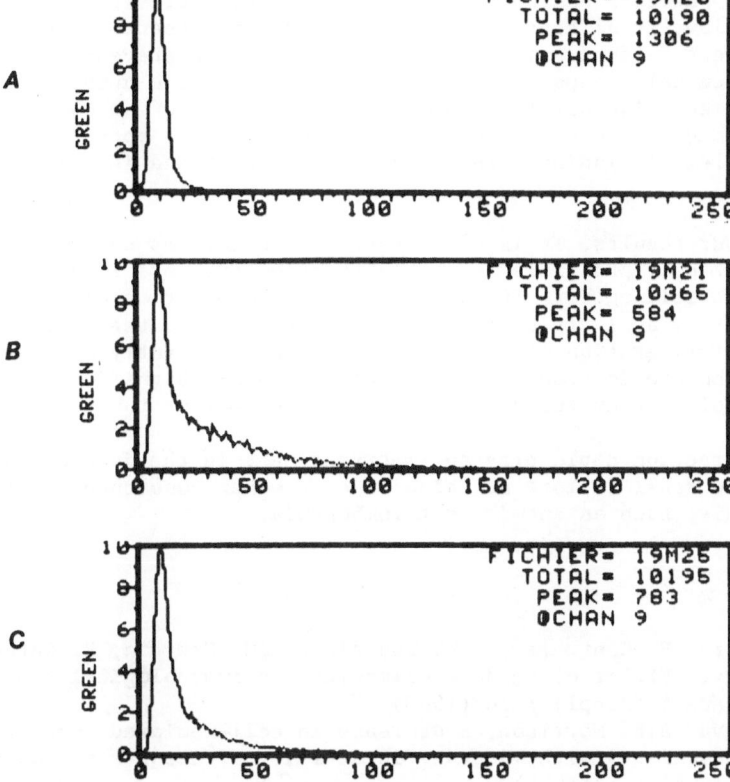

Fig 1. Analysis by flow cytometry of the IL2 receptor expression by :
A) non-stimulated cells,
B) PHA-stimulated cells,
C) PHA-stimulated cells in the presence of fraction 4-3-36-36-2
X-axis : fluorescence intensity
Y-axis : cell number

EFFECT ON B LYMPHOCYTE PROLIFERATION

We have also investigated the effect of the uremic solute on B lymphocyte proliferation. These experiments were carried out using murine B
lymphocytes which are specifically activated by the lipopolysaccharide
from Escherichia Coli (LPS). Results showed that B cell proliferation was
significantly decreased by the fraction isolated from uremic sera.

DISCUSSION

There is a general agreement that patients with chronic renal failure
have impaired immune responses. This immunodeficient state is attested by
prolongation of skin allograft survival, infections, and delayed skin
reactions. The mechanism of this pathological suppression of immune responses has not yet been elucidated. The in vitro reactivity of lymphocytes

from uremic patients has been shown either to be decreased [2,3] or quite normal [4]. Alevy et al have recently reported the presence in uremic rats of an adherent Ia suppressive cell population which prevents the proliferative spleen cell responses [5]. In addition there is a universal agreement that uremic sera inhibit the proliferative capacity of normal lymphocytes. Thus the immunodeficiency secondary to renal failure appears to be a composite probleme including both cellular impairment and inhibitory serum factors.

From our results, it is clear that uremic samples contain a potent inhibitor of cell proliferation. This fraction prevents T lymphocyte activation by inhibiting both IL2 production and acquisition of its receptors. This inhibitory activity however is not restricted to the T cell lineage, since the proliferation of B cells are also significantly depressed. Furthermore, the uremic fraction was shown to inhibit significantly the proliferation of some cultured cell lines from normal or tumorous origin.

This fraction could play an important part in the immunodeficiency secondary to renal failure and also in some other consequences of impaired DNA synthesis, such as anemia or thrombopenia.

REFERENCES

1. J. Navarro, P. Contreras, J.L. Touraine, A.M. Freyria, R. Later, J. Traeger, Effect of "middle molecules" on immunological fonctions, Art. Organs 4 (suppl) : 76 (1980)
2. J. Raskova, A.B. Morrison, A decrease in cell-mediated immunity in uremia associated with increase in activity of suppressor cells, Am. J. Pathol. 84 : 1 (1976)
3. P. Kurz, H. Kohler, S. Meuer, T. Hütteroth, K.H. Meyer zum Büschenfelde Impaired cellular immune response in chronic renal failure : evidence for a T cell defect, Kidney Int. 29 : 1209 (1986)
4. J.L. Touraine, F. Touraine, J.P. Revillard, J. Brochier, J. Traeger T-lymphocytes and serum inhibitors of cell-mediated immunity in renal insufficiency, Nephron 14 : 195 (1975)
5. Y.G. Alevy, K.R. Mueller, J.R. Anderson, P.S. Hutcheson, R.G. Slavin The role of monocytes and responder cells in suppression of antigen specific T cell proliferation in chronic renal failure, Int. Arch. Allergy appl. Immunol. 73 : 97 (1984)

# ISOLATION OF AN IMMUNOSUPPRESSIVE FRACTION FROM UREMIC ULTRAFILTRATES

J.Navarro, M.C.Grossetête, J.L.Touraine, and J. Traeger

Inserm U.80, CNRS UA 1177, UCB

Lyon, France

Cell mediated immune responses were decreased in patients with chronic renal failure (CRF) and this deficiency was shown to be mainly dependent on toxic or inhibitory serum factors. In previous study we have reported the isolation of a crude fraction from uremic sera with immunosuppressive properties[1,2]. This crude fraction which has a molecular weight (MW) below 5.000 daltons, inhibits markedly the lymphocyte proliferation induced by several mitogens or by allogeneic cells[1]. Further attempts to purify and characterize this immunosuppressive factor are presented. The inhibitory effect of each eluate fraction on in vitro lymphocyte proliferation has been assessed at each step of purification.

## MATERIAL AND METHODS

### Samples

Serum ultrafiltrates (1-21) were obtained from 9 patients with CRF. As controls we used the sera from 7 healthy donors. These sera were filtrated on a PMIO membrane, the cut-off of which was about the same as Cuprophane membrane (ca 10.000).

### Purification procedure

It was summarized in Table I. At each step of purification each eluate fraction was lyophilized and its inhibitory activity was checked on lymphocyte proliferation induced by allogeneic cells.

Preparative gel filtration. 100 ml of filtrates from uremic or normal sera were applied on a Sephadex G15 column (K5 x 100 cm). Elution was performed with ammonium bicarbonate (0.03M, pH 7.2), and the absorbance of the eluate was detected at 254 nm.

High performance liquid chromatography (HPLC). Inhibitory fractions were successively separated on the following prepacked columns (Waters Associates) : µ-Bondapak C18 (7,8 x 300 mm),2 I60 columns (7,8 x 300 mm) attached in serie and C18 Resolve (3.9 x 150 mm). Elution was performed with a volatile buffer (triethyl-ammonium-acetate 0.02 M , pH 6) and eluate was monitored at 254 nm.

Table I.  Purification Procedure

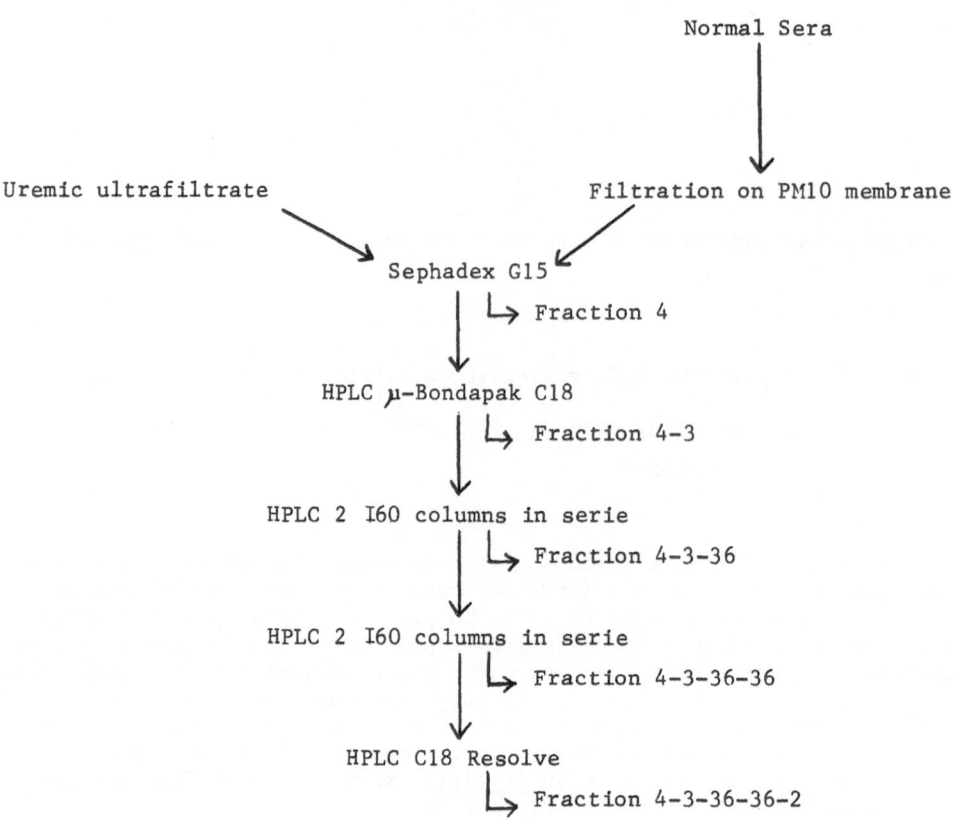

Normal Sera

Uremic ultrafiltrate → Sephadex G15 ← Filtration on PM10 membrane

└→ Fraction 4

HPLC μ-Bondapak C18

└→ Fraction 4-3

HPLC 2 I60 columns in serie

└→ Fraction 4-3-36

HPLC 2 I60 columns in serie

└→ Fraction 4-3-36-36

HPLC C18 Resolve

└→ Fraction 4-3-36-36-2

## Enzyme assays

Inhibitory fractions were incubated for 18 hours at 30°C in the presence of insoluble enzymes : pronase, papaïne or trypsin (Sigma). Enzymes were discarded by centrifugation and the supernatants were tested in the lymphocyte culture assay.

## Lymphocyte culture

Peripheral blood lymphocytes were obtained from 10 healthy donors, pooled and stocked at -70°C until used. At the time of experiment a cell aliquot was defrozen and lymphocyte suspension was prepared as previously described[3]. Cultures were performed in microplate wells by adding 0.1 ml of cell suspension (3 x $10^6$ cells/ml) to 0.1 ml of the eluate fractions. Before testing, each fraction was lyophilized and adjusted to the initial volume of ultrafiltrate. Lymphocyte proliferation was measured by 3H-thymidine incorporation into the cells from day 5 to day 6.

## RESULTS

## Gel filtration

After gel filtration of normal and uremic ultrafiltrates about 12 fractions of 100 ml each were collected, fraction 1 corresponding to the exclusion peak[3]. A major inhibition of lymphocyte proliferation (c.a 90 %) was noticed in fraction 8 obtained after filtration of both uremic and normal samples. This fraction was however discarded since it contained many electrolytes interfering with our biological test[4]. A significant inhibitory activity (c.a 60 %) was also noticed in fraction 4 obtained after the filtration of 8 under 9 uremic samples and of 2 of the 7 normal sera.

Trypsin or papain treatment of fraction 4 led to a slight decrease of the inhibitory activity. A more significant decrease (from 68 to 26 %) was found after pronase treatment, suggesting that at least some substance (s) responsible for the inhibitory effect was (were) of peptidic nature[3].

HPLC

μ-Bondapak C18 : Fraction 4 was separated into several peaks after chromatography on the μ-Bondapak C18 column[3]. Inhibitory activity was only recovered in the first eluted fraction (4-3). A significant decrease of the inhibitory activity was found after pronase treatment of fraction 4-3 substantiating the above results obtained with fraction 4 (Table II).

2 I60 columns in serie : According to the manufacturer these columns allowed separation of molecules ranging from 1.000 to 20.000. The calibration of the column with solutes of known MW did not permit, however, to determine the precise MW of the active fraction 4.3.36.

Recycling of this fraction on the 2 I60 columns led to a single peak 4.3.36.36 showing a significant inhibitory activity.

C18 Resolve : Inhibitory activity was recovered in the first eluted fraction (4.3.36.36.2). This last fraction contained a peptidic component as shown by total acid hydrolysis. We could not ascertain, however, that this peptidic factor was responsible for the observed effect since inhibitory activity was not affected by proteolytic enzyme treatment.

DISCUSSION

These results showed improved purification of the immunosuppressive fraction found in uremic ultrafiltrates. This fraction with a MW below 1500 daltons did not contain any salts which could interfere with our biological test. This fraction was present in some normal sera, suggesting

Table II. Inhibitory Effect of Fraction 4-3 from $C_{18}$ Column Before and After Pronase Treatment

| Experiments | Before | After |
|---|---|---|
| 1 | 73 % | + 4 % |
| 2 | 47 % | 17 % |
| 3 | 23 % | + 2 % |
| 4 | 70 % | 29 % |
| 5 | 75 % | 61 % |
| 6 | 51 % | +14 % |

the retention or the hyperproduction of a physiological compound in the uremic sera.

Enzyme assays performed on the active fractions recovered from G25 and μ-Bondapak C18 columns, suggested that at least some substance (s) with biological properties is (are) of peptidic nature. This fact, however, has been infirmed by consecutive experiments carried out on the last isolated fraction, since activity of fraction 4.3.36.36.2 was not abrogated after pronase treatment. Based on these results at least two hypothesis could be presented :

1 - The last recovered active compound was of peptidic nature, but the loss of one or several amino acids during fractionation prevented enzyme proteolysis.

2 - The inhibitory factor obtained after separation on G15 or μ-Bondapak C18 column contained two active components. The first, of peptidic nature, might be altered during the consecutive steps of fractionation. The second recovered after C18 Resolve chromatography was not a peptide and complementary analysis of this fraction are in progress.

REFERENCES

1. J. Navarro, J.L. Touraine, C. Corre, J. Traeger, Effet in vitro des "moyennes molécules" sur la prolifération lymphocytaire. Path. Biol. 24 : 189 (1976)

2. J. Navarro, J.L. Touraine, C. Corre, J. Traeger, Prolongation of skin allograft survival and inhibition of graft versus host reaction in rodents treated with "middle molecules". Cell. Immunol. 31 : 349 (1977)

3. J. Navarro, M.C. Grossetête, A. Defrasne, J.L. Touraine, J. Traeger, Isolation of an immunosuppressive fraction in ultrafiltrate from uremic sera, Nephron 40 : 396 (1985)

4. P. Contreras, R. Later, J. Navarro, J.L. Touraine, A.M. Freyria, J. Traeger, Molecules in the middle molecular weight range. Critical review of methods of separation from fluid of uremic patients, Nephron 32 (3) : 193 (1982)

# BINDING OF ORGANIC ACIDS TO SURFACE RECEPTORS OF

# LYMPHOCYTES AS AN IMMUNOSUPPRESSIVE MECHANISM IN UREMIA

Tsutomu Sanaka, Yutaro Hayasaka, Yoichiro Kawashima,
Takehide Takuma, Nobuhiro Sugino, Kazuo Ota and
Paul F. Gulyassy

Kidney Center, Tokyo Women's Medical College, Tokyo, Japan

## INTRODUCTION

Some organic acids which inhibit binding of small ligands to plasma, so called protein binding inhibitors (PB-Ix), accumulate in body fluids of patients with renal failure. PB-Ix are believed to bind to sites on the surface of albumin and inhibit its detoxification, transport and reservoir functions[1][2][3]. They may play a wider toxic role, such as immunosuppression in uremic patients. PB-Ix have been shown to inhibit the mitogenic activity of lymphocytes[4].

The present study was aimed to clarify the mechanism of the immunosuppressive effects of PB-Ix to lymphocytes.

## MATERIALS AND METHODS

### Lymphocyte isolation and culture

Lymphocytes were isolated from one normal volunteer as previously described[4]. Subsequently they were incubated at 37°C in 95% $O_2$/5% $CO_2$ and 100% humidity, and stimulated with phytohemagglutinin (PHA).

### Extraction, fractionation and assay of immunosuppressive PB-Ix

A large amount of plasma was collected from one hemodialyzed patient, with severe chronic renal failure, who was undergoing therapeutic plasmapheresis. Five hundred ml of azotemic plasma was acidified to pH 3.0 with 3 N HCl and adsorbed with 90 ml of the non-ionic resin XAD-2 (Bio-beads, Bio Rad, USA)[1][2][3].

After the mixture was stirred for two hours, plasma was removed, the XAD-2 was washed with distilled water and PB-IX were eluted with 200 ml of methanol. The methanol was evaporated to dryness. The residue was dissolved in 50 nM formic acid and passed through an ion-exchange column, SP-Sephadex (Pharmacia, Sweden). The elute from this column was lyophilyzed and dissolved in 5 ml of distilled water. This sample was further fractionated with a 30 × 1.5 cm DEAE-Sephadex and subsequent followed with 100 × 0.9 cm Sephadex G15 column.

## Effect PB-Ixl on lymphocyte blastformation

The possible effect of PB-Ix as immunosuppressive factors was determined with fraction No. 4 of the Sephadex G15 column.  Fraction No. 4
tentatively named PB-Ixl was diluted and tested at 5, 10, and 20 Δ%/ml
of lymphocyte suspension.

## Response of lymphocytes pretreated with PB-Ixl to PHA

For studies in the time of onset and reversibility of PB-Ix effects
lymphocytes were pretreated with 20 Δ%/ml of PB-Ixl for 0.5, 1, 2 and
24 hours respectively.  Then they were washed three times with physiological saline and incubated without PB-Ix in PHA containing fresh
culture medium for 72 hrs.

## Effect of PB-Ixl on lymphocytes determined with fluorescence labeled PHA

(1) Normal lymphocytes were pretreated for 120 minutes with different
concentrations of PB-Ix (10, 5 and 2.5 Δ%/ml).  Then 1.5 μg/100 ul of a
fluorescence labeled PHA (FLPHA, E-Y Laboratories, USA) was added.

After 60 min the lymphocyte being washed three times with physiological saline, their fluorescence activity was measured by fluorospec-
trometer using 450 nM excitation and 520 nM emission (Hitachi, Japan).

(2) The FLPHA (50 or 100 μg/ul) and quinolinic acid (10 or 50 μg/ul)
were simultaneously added to normal lymphocytes.  Quinolinic acid was identified as one compartment of the crude fraction PB-Ixl.  One hour after
the addition, the lymphocytes were washed three times and the histogram
of the lymphocytes were analyzed with a fluorescence activated cell sorter
(FACS 440).

## Reactivity of normal lymphocytes pretreated with PB-Ix to anti Leu4 and anti HLA-DR monoclonal antibody

Monoclonal antibodies anti Leu4 and anti HLA-DR were added to normal
lymphocytes preincubated with PB-Ixl (10 Δ%/ml), quinolinic acid (10 μg/ul),
hippuric acid (10 μg/ul) or m-hydroxyphenylhydracrylic acid (HPHA) (10
μg/ul) for 12 hours.  Scatter gate sets on the lymphocyte fraction were
analyzed using FACS 440.


## RESULT

## Effect of PB-Ixl on lymphocyte blastformation

Blastogenesis in the lymphocyte added the other gelfiltration fraction without PB-Ix activity showed approximately the same count as the
control.  The lymphocytes incubated with 10 Δ%/ml of PB-Ixl showed the
remarkably low mean blastogenic index of 30.6%.

## Response of lymphocytes pretreated with PB-Ixl to PHA

Blastogenic response of lymphocytes pretreated with PB-Ixl were consistently less than that of the control without any PB-Ix.  The longer
the pretreatment time, the lower was the blastogenesis.

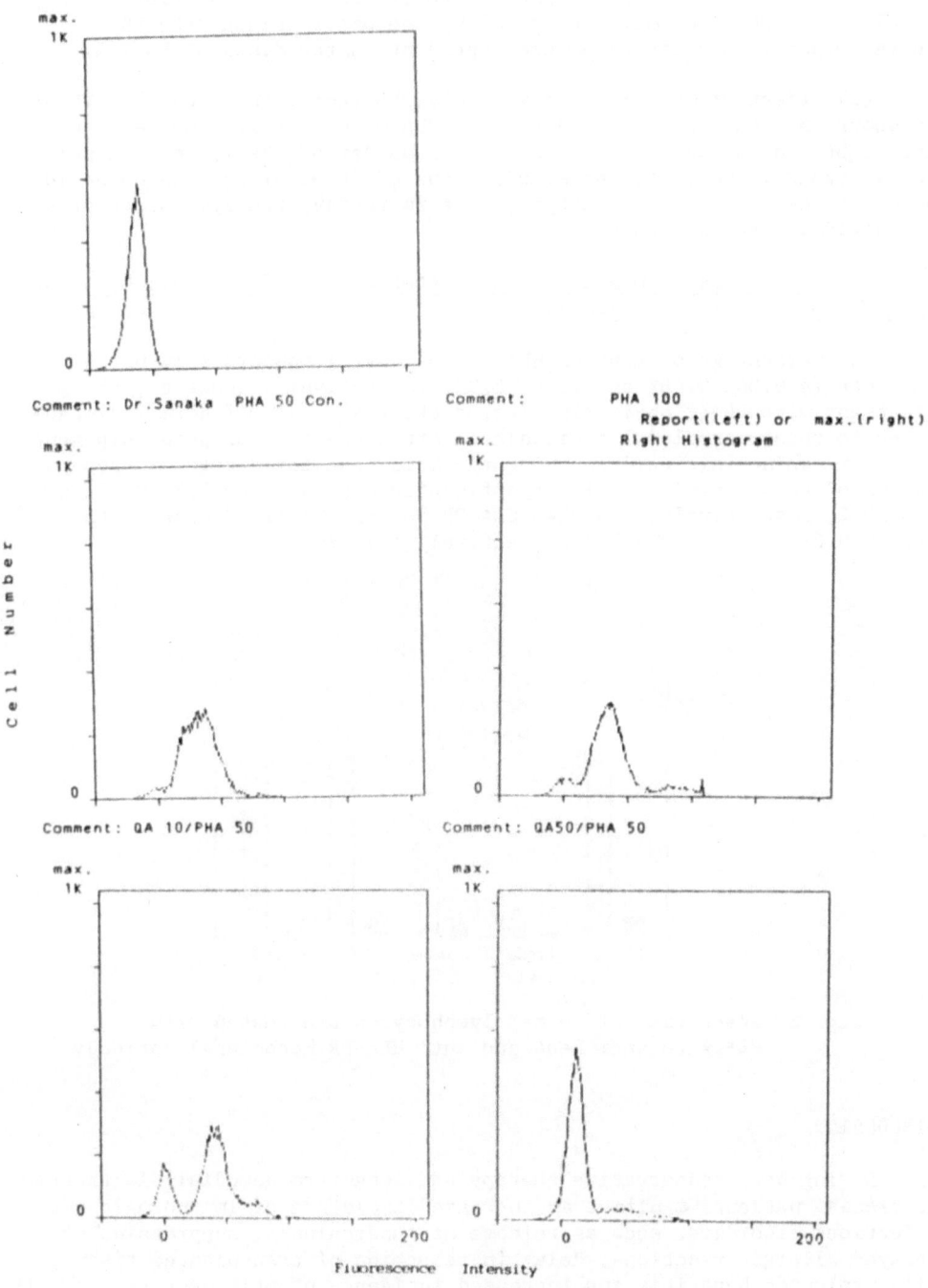

Fig. 1. Effect of Quinolinic Acid on
lymphocytes determined with FLPHA

Effect of PB-Ixl on lymphocytes determined with FLPHA

(1) With regard to the effect of PB-Ixl on FLPHA, the fluorescence levels of lymphocytes were decreased by the pretreatment with PB-Ixl. Those tendencies to decrease were dependent on the dosis of PB-Ixl.

(2) Effect of Quinolinic Acid on lymphocytes determined with FLPHA is shown in Figure 1. The addition of PHA moved the lymphocyte peak to the right side by one scale. The left shoulder of the lymphocyte peak was irregular. However, the simultaneous addition of PHA and quinolinic acid produced a different pattern, that is to say, the lymphocyte peak was divided into two fractions.

Reactivity of normal lymphocytes pretreated with PB-Ix to anti Leu4 and anti HLA-DR monoclonal antibody

The percentage of Leu4 or HLA-DR positive lymphocytes in normal subjects is $9.36 \pm 0.46\%$ and $75.6 \pm 5.9\%$, respectively. However, the pretreatment with PB-Ix obviously altered the reactivity of normal lymphocytes to these monoclonal antibodies. After addition of pure components of PB-Ix, quinolinic acid and hippuric acid, the percentage of subpopulation of Leu4 positive cells significantly decreased to $3.2 \pm 0.27$ and $2.8 \pm 0.2$, respectively, and also HLA-DR positive cells decreased to $46.3 \pm 10.6\%$ and $67.66 \pm 4.7\%$, respectively (Figure 2).

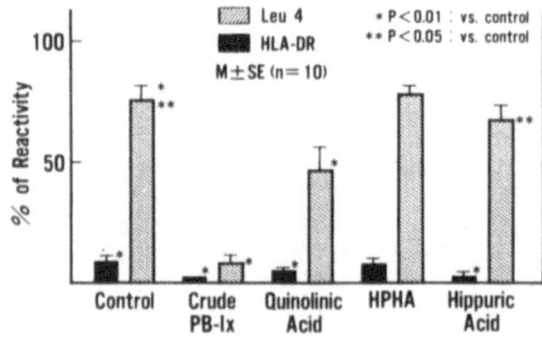

Fig. 2. Reactivity of normal lymphocytes pretreated with PB-Ix to anti Leu4 and anti HLA-DR monoclonal antibody

DISCUSSION

During both conservative therapy and long-term hemodialysis treatment of uremia, patients manifest an increased incidence of immunologic and infectious disorders, such as relapse of tuberculosis, suppression of delayed allergic reactions, delay in rejection of transplanted tissue, mild prolonged hepatitis and increased incidence of malignant tumors. The main reasons for these abnormalities have been considered to be immunocyte abnormalities, existence of an immunosuppressive factor, and a disorder in humoral antibody production.

PB-Ixl was shown to have a suppressive effect on lymphocyte blast formation and the degree of its effect was dependent upon its concentration. We found that lymphocytes pretreated with PB-Ixl apparently had a reduced response to PHA. It became clear that PB-Ix(s) not only bind to sites on the surface of albumin but also inhibit the mitogenic activity of lymphocytes.

On the other hand, PB-Ix has been found to be a mixture of several chemical substances, including quinolinic and hippuric acids, and possibly furan dicarboxylic acid[5)6)]. Quinolinic acid was recognized to have powerful suppressive effect to PHA induced lymphocyte blastogenesis[7)].

Normal lymphocytes preincubated with PHA and quinolinic acid were divided into two fractions by means of fluorescence activated cell sorter (FACS 440). The right hand side peak in Figure 1 might be PHA bound lymphocytes and the left hand side might be lymphocytes lacking PHA. Moreover, as the concentration of quinolinic acid was increased, the lymphocyte peak presumed to be PHA-free lymphocytes became a single, tall peak similar to the control, and in the same position. Thus quinolinic acid appears to have an inhibitory effect at the PHA receptor on the surface of lymphocytes.

The following study was aimed to further evaluate whether PB-Ix has a high affinity to the lymphocyte surface and thereby influences on the reactivity of lymphocytes to anti Leu4 and/or anti HLA-DR monoclonal antibody or not. Although results were not demonstrated in a figure, lymphocytes pretreated with crude PB-Ix, quinolinic acid, and hippuric acid were scattered in area No. 2, while control and HPHA treated lymphocytes were not in this area.

Therefore, these results suggest to us that PB-Ix bind to the surface of lymphocytes and exert possible inhibitory effects on mitogenic activity and PHA, anti Leu4, and anti HLA-DR receptors of lymphocytes. These in vitro finding may reflect, at least partly, an important aspect of the mechanism of the immunosuppressive state in uremia[8)].

REFERENCES

1) T. A. Depner and P. F. Gulyassy: Plasma protein binding in uremia-- Excretion and characterization of an inhibitor. Kidney International 18:86-94, 1980.
2) T. A. Depner, L. A. Stanfel, E. A. Jarrard and P. F. Gulyassy: Impaired plasma phenytoin binding in uremia--Effect of in vitro acidification and anion exchange resin. Nephron 25:231-237, 1980.
3) T. Sanaka, H. Kasai, Y. Kawashima, F. Takahashi, Y. Hayasaka, K. Ota and N. Sugino: Chemical identification of protein binding inhibitor in uremic patients. Japanese Journal of Kidney and Dialysis 14:75-81, 1983.
4) T. Sanaka, Y. Kawashima, F. Takahashi, et al: Suppressive effect of uremic protein binding inhibitor on cultivated lymphocytes. Japanese J Nephrology 15:1063-1069, 1983.
5) T. Sanaka: Protein binding inhibitor in uremia. Japanese J Circulation 17:400-409, 1985.
6) K. Koide, J. Touyama, N. Inoue, et al: Oral Adsorbent. Japanese J Clinician (Nihon Rinsyou in Japanese) 43:422-440, 1985.
7) Y. Kawashima, T. Sanaka, N. Sugino: Suppressive effect of quinolinic acid on bone marrow erythroid growth in uremia. Abstract of Symposium Uremic Toxin, Gent, 1986.
8) T. Sanaka and K. Ota: Lymphocyte blastformation in uremic patients. Japanese J Clin Immunol 8:881-887, 1976.

# UREMIC DYSMETABOLISM AND ITS EFFECT ON IMMUNOCOMPETENT AND ERYTHROID CELL FUNCTION

Tor-Erik Widerøe

Department of Nephrology
University hospital of Trondheim
Norway

## INTRODUCTION

From numerous of publications it is evident that an immunodeficiency exists in end-stage renal failure (ESRF) patients. To give a strict definition of these defects is, however, difficult.

The methods used for investigation during the last decade have changed from detailed but in part unspecific measurements of functions in vitro of different parts of the immune responding system, to the current studies of cells employing monoclonal antibodies. These studies give partly conflicting results concerning the immunological implications in uremia. Also patients affected by ESRF show frequently multifactorial disorders influencing the immune function. In summary it can, however, be stated that:

a) Nutritional deficiency is the main contributor giving an immunodeficiency partly indistinguishable from uremic toxenemia (1).

b) In diseases resulting in ESRF (2) a high frequency deviation in immunoglobulin allotypes, phenotypes and haplotypes has been observed.

c) An abnormally high prevalence of autoantibody in dialysis patients has been noted. This can be multifactorial but might also reflect an imbalance in immunological control mechanisms (3).

d) Blood transfusions affect the immunological status and have not been interpreted adequately in several discussions of the observed results (4).

e) The recent years interpretation of the immunological nature of bioincompatibility reactions achieved during hemodialysis may explain some of the immunological behaviour of these patients (5) and has mainly not been taken into consideration.

f) The different drugs and supplements needed during ESRF have been suggested to interfere with the results obtained (6,7).

To compare findings in experimental animal models and in patients account for probably many of the difficulties in evaluating the results since patients are more inhomogenous. Moreover, highly diversified results have been harvested when testing normal cells in uremic plasma compared to uremic cells in normal plasma or growth media. The culture media used are indeed essential for the obtainable results as normal plasma, calf serum and bovine serum albumin contain growth factors not available in artificial media (8). Extensive work has, however, been done and the methodological pitfalls increasingly discussed.

GENERAL IMMUNOLOGICAL FINDINGS

It is clearly established that uremia <u>per se</u> must be regarded as the main cause of an acquired impairment of immunological functions. These defects include both the humoral and the cellular immune system. Particularly the cellular immune system has been defined. Both intrinsic defects within the cells (8-27) and extrinsic factors of inhibitory nature in the uremic body fluid (28-34) have been found. Figure 1 shows differentiation and interactions of mononuclear cells. It must

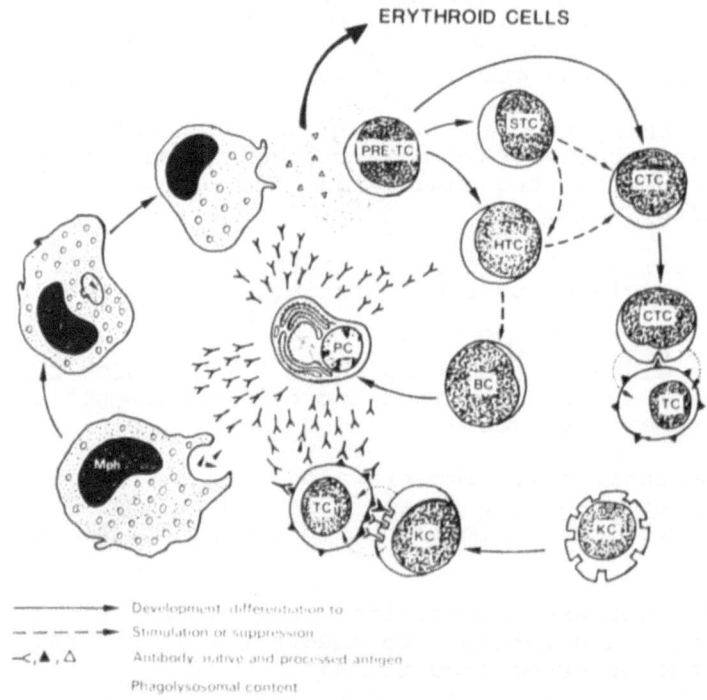

Fig. 1. Interaction of mononuclear cells in response to antigen. PRE-TC: passing T-cell, S-TC: suppressor T-cell, H-TC: helper T-cell, C-TC: cytotoxic T-cell, TC: target-cell, BC: B-cell, KC: killer-cell, PC: plasma cell, Mph: macrophage.

be pointed out that the macrophages have a significant direct and indirect effect in the cellular and in the humoral immune system. Not only are these interactions between the mononuclear cells important, but it is also well accepted that these cells play a major role for the development of early erythroid progenitor cells (35) and for the erythropoietin-like production responsible for the hypoproliferative anemia during ESRF (36). One will briefly try to depict and define aspects of this imbalance in uremia.

IMMUNOCOMPETENT CELLS

## Uremic cells (intrinsic defects)

Unfortunately, interpretation of leucocyte function in cells from uremic patients is difficult because cell cultures in vitro usually require supplementation with blood sera from either normal controls or non-human species. In addition to providing growth factors for normal lymphocytes (8), these sera may interfere with the investigation of abnormal lymphocyte responses. This complication may be particularly important in uremia, because the mononuclear cells from such patients circulate in blood plasma that contains inhibitors of lymphocyte function. It is then needed to monitore the proliferative response in serum-free media. Using this method Kleinman et al (8) recently showed that there was a considerable day to day variation in the response to mitogen (Concavalin-A) in both controls and uremic patients, but in general significantly depressed in patients before dialysis as compared to controls. It was also demonstrated that this cell response was increased immediately postdialysis in most patients investigated. Using thymidine incorporation in lymphoblasts having increased mitotic activity after antigen stimulation, depressed DNA-synthesis was seen. This is in agreement with earlier investigations (9,19,20,24,26,27), though results to the contrary have also been published (16,23,37). Heterogenicity concerning nutrition is suggested to be an important factor. Donati et al (24) found in what they called class-1 EDTA rehabilitated patients on hemodialysis that the dose dependent lymphocyte response to polyclonal mitogens, i.e. Concavalin-A (Con-A), pokeweed mitogen (PWM) and phytohemoagglutinine (PHA), showed large variability but in general lower than in the control group.

The cytotoxic response of lymphocytes to alloantigens, as typified by the mixed lymphocyte culture (MLC) reaction, has been found to be depressed in a majority of uremic patients (9,20,23,26). Using different target cells, several investigators have not found any change in natural killer (NK) and killer (K) cell activity of uremic patients (12,21,22). Suppressed activity is, however, also suggested (38).

B-B-lymphocyte function has not been so extensively investiated as compared to the T-lymphocytes. B-lymphocyte differentiation in response to mitogen stimulation is known mainly to be T-cell dependent. Since a body of data suggests a defect in the T-cell function of the uremic patients, Kurz et al (26) have examined T-helper function in a T-B-cell collaboration to determine IgG production of unfractioned lymphocytes in a mitogen driven system. Uremic B-cells secreted significantly less IgG than the controls. Such findings (2, 27) shed considerable light on the observation that a high proportion of uremic non-responders exist following exposure to hepatitis B (39,40) and other vaccines (41,42,43).

If, however, corrective therapy is to be approached, an understanding of the cell to cell mediated defects is essential. Monocyte mediated suppression of immune response is essential. The mononuclear macrophage function can be quantified directly by phagocytosis and indirectly via processing functions on accessory cells like lymphocytes and erythroid cells.

Limited studies have been carried out in man with respect to uremic effects on the mononuclear phagocytose system and the data available are conflicting. Monocytes and macrophages from uremic patients have been found to have decreased (44) or normal (45) phagocytic function. In experimental uremia Nelson et al (13) recorded normal phagocytic function by such cells but depressed catabolism of radiolabelled microaggregated human serum albumine. These differences could be explained by the fact that attachment of particle to a cell membrane is energy independent while cell ingestion and catabolism is energy dependent (46).

Although monocytes are required for optimal T-cell proliferative responsiveness (47,48) or mitogen induced immunoglobuline production (49,50), they also show suppressive activity (48,50) in these responses. Furthermore, Alevy et al (11,17) have shown that adherent suppressor cells decrease lymphocyte response in experimental uremia. Several studies have later been conducted to clarify in detail these interactions in cells from uremic patients (8,14,23,26,27). The most extensive work has recently been published by Tsakalos et al (27) showing that:

1) uremic monocytes suppressed the mitogenic response in T-and B-cells. This effect was not seen by normal monocytes,

2) uremic monocytes, but not normal monocytes depressed the T-and B-cell coloniforming properties.

3) T-cells from uremic patients were significantly less effective than control cells in supporting B-cell colony growth. It is thus suggested that monocyte-mediated suppressive activity is an important mechanism contributing to impaired lymphocyte responsiveness in ESRF patients. Additional or relaxed abnormalities in monocyte "accessory" function may also exist in such patients.

Recently Kurz et al (26) found reduced Interleucin-2 (IL-2) production in stimulated T-lymphocytes from uremic patients. This result call for further investigations on different monokines that promote T-cell activation , B-cell colony stimulation or growth factors that could give therapeutic benefits. Work on recombined produced stimulating factors within the white blood cell system is fortunately in progress (51).

To further clarify the immunological defects of uremic patients several investigators have used cytoflowmetric counting of the subpopulations of the white blood cell lines but with different results (16,21,14,52). These divergencies reflect probably more the heterogenity of the patients than the method used. Donati et al (24) showed in their class-1 EDTA rehabilitated patients mainly normal numbers of total white blood cells. Lymphopenia was found in all patients, and the mean value of the T-cell subsets appeared to be lower than in healthy controls. The ratio T4/T8 was normal in most patients. NK-cells and B-cells were considered normal. This is in agreement with the findings in our laboratory showing also increased number of monocytes in uremic patients (52).

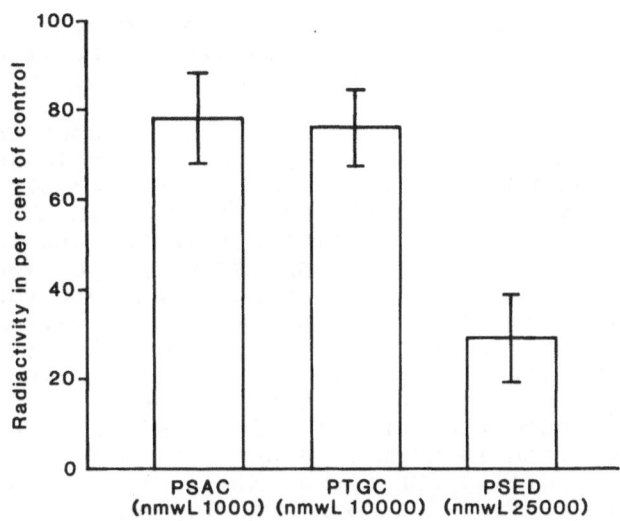

Fig. 2. Effect on phagocytosis of radiolabelled Candida Albicans by human mononuclear phagocytes cultured in different permeates obtained by filtration of uremic plasma through Pellicon filters (PSAC, PTGC, PSED). Total radioactivity reflects cell detachment during the culture period expressed as percentage of controls (59).

## Inhibitors in uremic plasma (extrinsic defects)

Uremic blood plasma influences normal cell functions. Several basic nucleic functions are affected resulting in depressed DNA-synthesis (53,54). Inhibition of membrane-bound Na-K-ATP-ase (55) takes place, and the pentose-phosphate shunt is altered giving increased lipid peroxydation as demonstrated for the plasma membrane of red blood cells (56,57).

Mononuclear phagocytosis in uremic milieu has mainly been investigated by Jørstad et al (58,59,60) showing inhibition by untreated uremic and postdialysis plasma, by ultrafiltrate of uremic plasma and in plasmafractions corresponding to high molecular weight area. Figure 2 shows macrophage function when cultured in different permeates obtained by filtration of uremic plasma through Pellicon filters (59).

The general findings when using lymphocyte function as target are: 1)MLC-blocking factors (28) in uremic sera and in ultrafiltrate ("cut-off" point 50.000 daltons), 2)E-rosette-blocking factors decreasing postdialysis (34), and 3)suppression of mitogen stimulated lymphocytes (31). In most cases the inhibitory factors are suggested to be in the middle molecular area. Measuring thymidine incorporation in Hela cells and fibroblasts, Delaporte et al (30) found that ultrafiltrable uremic plasma-fractions in the middle molecular area were not cytotoxic by themselves but needed normal plasma - possible macromolecules - to excert toxic effect in the model used.

This result can explain the findings by Jørstad of inhibited monocytic phagocytosis in vitro mainly by the uremic ultra-filtrate and plasma fractions in the macromolecular area (59, 60).

Several humoral factors affect the immunoregulatory cells. Raskova and Raska (29) have shown that a uremic immunosuppresive factor belongs to a family of immunoregulatory lipoproteins. The most active fractions are found in the VLDLs (29). Uremic VLDLs, perhaps containing distinctly abnormal apoproteins, possess strong immunosuppressive properties and could account for the observed defects in lymphocyte response in MLC (29). Other lipoproteins in human serum have been shown to have potent T-cell immunoregulatory properties (61). Whether these abnormal lipoproteins account for the immune distubance in uremic patients remain to be determined.

It is imposed that uremic serum possibly inhibits thymidine-incorporation in normal lymphocytes because of an abnormal cyclic GMP/cyclic AMP intracellular ratio restored by Levamisole (62). Likewise both zinc (7) and pyridoxin (6) therapy enhances previously depressed reactivity of normal white blood cells in uremic plasma. High $\beta_2$-microglobulin concentration is observed in blood of ESRF patients (63). This protein has been shown to enhance the expression of surface receptors of IgG on human lymphocytes (64). Lymphocytes from uremic patients failed to respond to either this heptapeptide or its des-His-fragment as opposed to normal lymphocytes (33). It is suggested that continuous interaction between lymphocytes and high concentration of $\beta_2$-microglobulin or its fragment(s) in uremia may have altered sensitivity to the immuno-modulatory effects of the peptides.

## Effects of blood transfusion

Watson et al (4) concluded that the low cellular reactivity in uremic patients was associated with blood transfusion and influenced the outcome of renal transplantation. Blood transfusion given to a previous non-transfused patient produced an increase in T-suppressor cell activity (16) so also macrophage induced suppression (65). Bender et al (66) compared the subpopulations of blood leucocytes in nontransfused and transfused patients in hemodialysis. Changes relative to normal controls occurred almost exclusively in the group who had received erythrocyte transfusion. The most obvious change took place in the T4-cells which resulted in a significant decrease in T4/T8 ratio. Inhibition of uremic plasma on normal lymphocyte function is also correlated to anti-HLA antibodies (28) and to prior blood transfusion (28,32). These data may also explain some of the earlier mentioned controversial findings in uremic patients.

## Effect of bioincompatibility during hemodialysis

Cumulative effect of long-term intermittent exposure to hemodialysis membranes may also contribute to changes in the immune system by the ability to activate complement via the alternative pathway or through direct shearing forces on the immunocompetent cells (5,67,68). In parallell with activation of complement during one dialysis session using cuprophane membrane equipment, we have found an increase of body clearence for the anaphylatoxins C3a and C5a. As shown in Figure 3

# Complement Activation during Hemodialysis

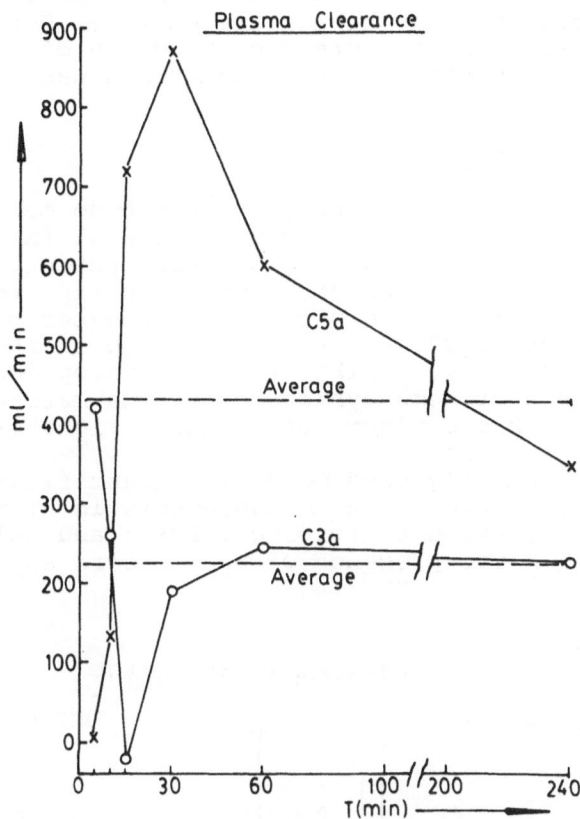

Fig. 3. Calculated body clearance for C3a and C5a during
4 hours of hemodialysis with Cuprophan membrane equipment.
Total body clearance of C3a and C5a from circulating blood
plasma were based on dialyzer generation rate, patient plasma
accumulation rate, and arterial plasma concentrations (69).
Average values during 4 hours of dialysis are indicated by
the broken lines.

this increase was particularly pronounced for C5a reaching a
maximum of 900 ml/min and with an average of 30 ml/min during
4 hours of treatment (69). Human polymorphnuclear leucocytes
and monocytes possess specific receptors which after interaction with C5a lead to stimulation of several cellular responses (70). The obtained results demonstrate decreased C5a
receptor availability of these cells in dialyzed patients (71).
Factors contributing to the heterogenity of these patients
may also be the individual ability to respond on the interaction with these artificial membranes as measured by complement activation (72). Using a method in vitro for complement
activation we have found that individual donor plasmas show
significantly different reactivity when incubated with different menbranes (in press).

Although different data are reported, changes in DNA
content and surface antigen of T-cell subsets have been shown

during exposure to dialysis membranes (36,52,73,74). Decreased
response to mitogen stimulation of lymphocytes accompanied by
a decrease in IL-2 production after single passage of blood
through the dialyzer has been measured (37). Recently, Bingel
et al published data on increased induction of human IL-1
during dialysis in vitro when an endotoxin was assed to the
dialysate (75).

## Erythroid progenitor cells

The hypoproliferative state of the bone marrow in chronic
renal failure patients can be due to several factors. First;
it is suggested that the kidney in these patients fails to produce significant amount of erythropoietin (Ep) to meet the increased demands of new red blood cells created by increased
hemolysis (56,57,76,77,78), blood loss and inhibitors of erythropoiesis. Secondly; inhibitors of erythropoiesis which include inhibitors of heme synthesis in bone marrow have been
demonstrated in blood plasma of uremic patients (79,80,81,82,
83).

The most commonly used model for quantification of erythropoiesis has, however, been measurements in vitro of erythroid
progenitor cells by burst-forming (BFU-E) and colony-forming
(CFU-E) units in the presence of uremic sera as compared to

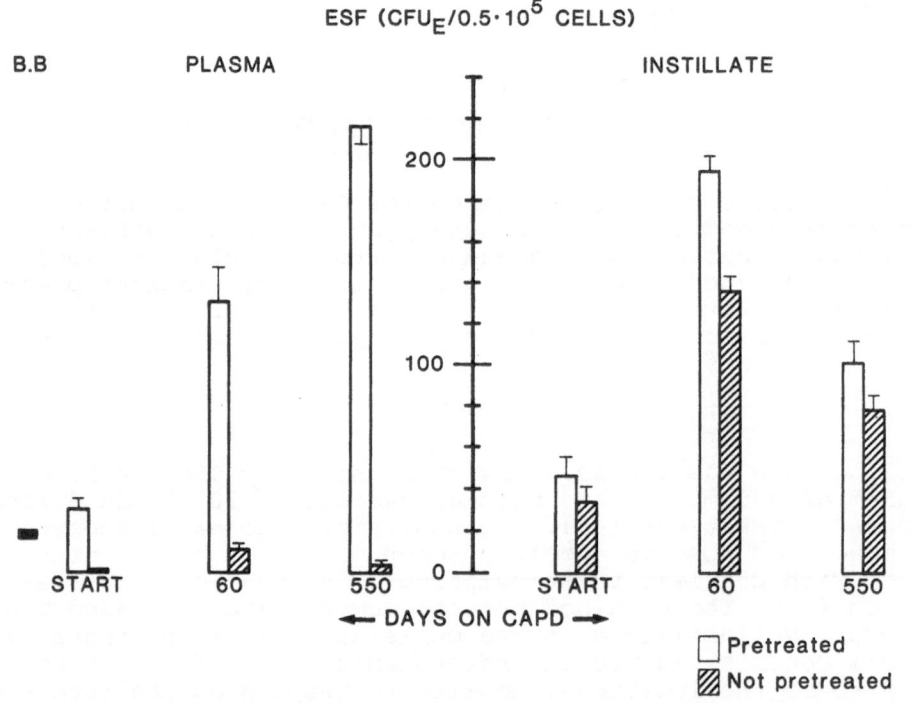

Fig. 4. Measurement of erythropoiesis stimulating factor(s)
(ESF) by colony-forming units (CFU-E) of neonatal mouse liver
cells (83). Comparison of pretreated (dialyzed in vitro)
and not pretreated samples from blood plasma and from 4 hours
instilled peritoneal dialysate during CAPD treatment in one
anephric patient is depicted.

normal sera. Consequently, the findings with uremic sera could reflect a deficiency of stimulators or the presence of inhibitors. A true quantification of EP is thus difficult to achieve. Furthermore, several pitfalls can be found in the different procedures used for pre-treatment of the samples to be tested. The cell culture system used is often affected by inhibitors or toxic substances present in the crude samples (83) and not fully inactivated by heating up to 56°C (80). As shown in Figure 4 it is found in our laboratory that dialysis in vitro of blood samples using a membrane with a "cut-off" at 12.000 daltons removes substances with inhibitory effect on the assay used and revealed changes in plasma erythropoiesis stimulating factors during CAPD treatment. Such changes are not found in 4 hours instilled peritoneal dialysate (83). It is indicated that uremic sera contain dialyzable inhibitors of erythropoiesis in vitro, but these inhibitors lack spesificity as also is the situation for granulocyte-macrophage and megakariocytic colonies (84). The phagocytic function of macrophages is inhibited (83). Figure 5 shows the effect of uremic plasma when taken by start of CAPD and after 60 days of treatment when respectively 12.5 and 50 mU/ml of standard sheep Ep were added to the same culture of erythroid cells (83). The plasma had a significant inhibitory effect of the added Ep, but normal values were achieved during CAPD treatment.

Normally there is a negative correlation between hematocrit and erythropoiesis measured in vitro (82,84). In uremic patients, however, a positive correlation is found (82,83,84)

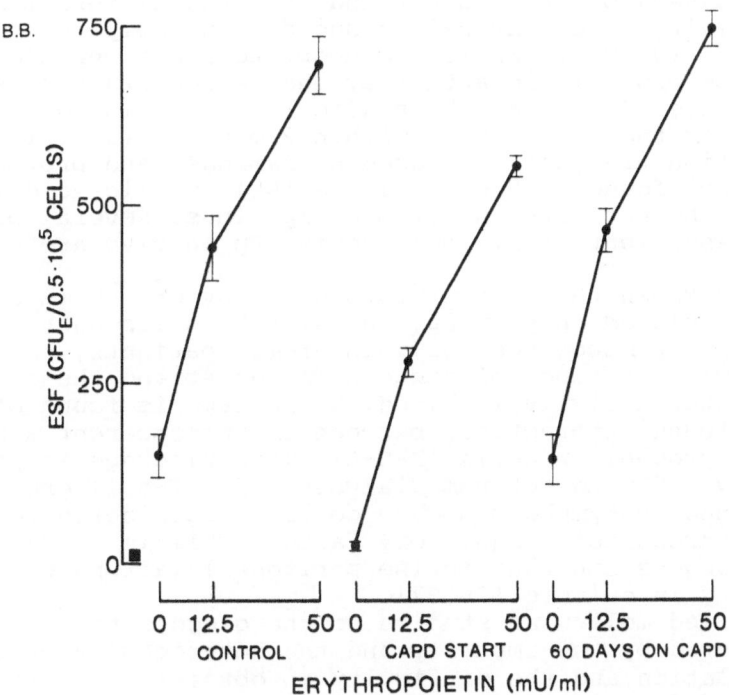

Fig. 5. Measurement of ESF in vitro. Effect of plasma from an anephric patient on the CFU-E properties of 12.5 and 50 mU/ml standard Ep by CAPD start and after 60 days of treatment. Control represents plasma from healthy adults (83).

indicating the existence of inhibitory factors. It could also indicate, however, that patients with minimal residual kidney function are in need of extrarenal Ep production and it is suggested that this Ep production is less sensitive to changes in tissue oxygenation than Ep produced by the kidney (85,86).

Radke et al have reported that the polyamine spermine might contribute to the anemia in ESRF (81). While this material appears to be inhibitory in vitro others have provided preliminary evidence that spermine, as well, is not specific for erythropoiesis (87). Furthermore, Spragg et al have contested reports of elevated plasma spermine levels in these patients (88). Earlier reports that the parathyreoidea hormone might be responsible for the inhibition of red cell production (89) could not be confirmed when the pure hormone or a synthetic peptide of a partial sequence of the hormone were tested in culture (90). When partially purified parathyreoid extract preparations were tested, inhibition of erythropoiesis in vitro was not specific and the extract also inhibited granulopoiesis in vitro as well (90). Part of the problem in defining the rose of inhibitors in the anemia of ESRF is the lack of appropriate control data that would support the possible specificity of the observed inhibitors.

Improved radioimmunological methods measuring Ep have given further information on Ep production and concentration in uremic fluids (82). Low, normal and high values are found but without the normal correlation whether to hematocrit or to erythropoiesis in vitro (82). These findings need further evaluation since Sherwood and Goldwasser (91) found that immunoactive fragments of Ep were detectable in uremic serum. The immunoassay of Ep does not indicate the different biological activity of extrarenal produced Ep as compared to renal produced Ep (91,92). Ep is a glucoprotein hormone and is protected from proteolytic attack by the carbohydrate present in its structure. If 10% of the sialic acid is removed Ep is cleared from the circulation within 2 minutes (93). Increased desialination (94) and increased neuramidase and protease activity are found in uremic plasma (95). Sialic acid is necessary for the activity in vitro of Ep. Thus, several pitfalls in the measurement on biologic active Ep in vivo are still with us.

It is known that the erythroid colony growth is not necessarily regulated only by Ep, but also by cells such as T-lymphocytes or monocytes (35). In uremic patients, particularly those showing presence of reduced T4 subset and the T-cell growth factor activity in blood, a blockage is found of the normally found interactions between immunocompetent and early erythroid progenitor cells (BFU-E). This blockage is partly removed by CAPD and not hemodialysis (35). T-cell growth factor produced by lympho-monocyte cells is considered to be an early erythropoiesis regulatory factor. This is in agreement with the suggestion that murine peritoneal macrophages release Ep activity in culture (96,97).

Detailed molecular studies of the glucoprotein Ep have been hampered by the impurity and the heterogenity of target cell population and the difficulty in obtaining significant quantities of purified Ep. Recently Jacobs et al (98) have been successful in characterization and production of recombinant Ep with biologic activity. This preparation is now ready for clinical trials. Published data by Eschbach and Mladenovic using sheep Ep in an experimental model (99,100)

and unpublished data (personal communication) using recombinant Ep on uremic patients are promising. They confirm the
suggestion of a relative Ep deficiency in these patients.

The studies with Ep should, however, stimulate further
research with different stimulatory factors within the white
and the red blood cell lines to normalize the hampered cyc-
ling and interactions between these cells in ESRF patients.
The measurements should be less based on phenomena, but on
basic molecular biologic functions and on specific transcellular membrane transport properties of substances.

ACKNOWLEDGEMENT

The author would like to thank Dr Jørstad, Dr Smeby and
Dr Eik-Nes for all kind help in supplying scientific data and
in preparing the manuscript.

REFERENCES

1.  R.J. Glassock, Nutrition, immunology, and renal disease,
    Kidney Int. 26:Suppl. 16:194 (1985).
2.  G.J. Laundy, and B.A. Bradley, Immunoglobulin allotypes
    in patients in end-stage renal failure, J Immunogenetics
    12:181 (1985).
3.  R.F. Gagnon, J. Shuster, M. Kaye, Auto-immunity in patients
    with end-stage renal disease maintained on hemodialysis
    and continuous ambulatory peritoneal dialysis, J Clin
    Lab Immunol. 11:155 (1983).
4.  M.A. Watson, J.D. Briggs, A.A. Diamandopoulous, D.N.H.
    Hamilton and H.M. Dick, Endogenous cell-mediated immunity,
    blood transfusion and outcome of renal transplantation,
    Lancet. 1:1323 (1979).
5.  A.K. Cheung, and L.W. Henderson, Effects of complement
    activation by hemodialysis membranes. Am J Nephrol. 6:
    81 (1986).
6.  D.A. Casciato, L.P. McAdam, J.D. Kopple, R. Bluestone,
    L.S. Goldberg, P.J. Clements, and B.W. Knutson, Immuno-
    logic abnormalities in hemodialysis patients; Improve-
    ment after Pyridoxin therapy, Nephron. 38:9 (1984).
7.  L.D. Antoniou, and R.J. Shalhoub, Zinc-induced enhance-
    ment of lymphocyte function and viability in chronic
    uremia, Nephron. 40:13 (1985).
8.  K.S. Kleinman, and D.C. Zoschke, Suppression of human lym-
    phocyte responses in chronic renal failure mediated by
    adherent cells: Analysis in serum-free media. J Lab
    Clin Med. 106:262 (1985).
9.  T. Kunori, I. Feherman, O. Ringdén, and E. Möller, In
    vitro characterization of immunological responseiveness,
    Nephron. 26:234 (1980).
10. J.E. Lortan, P. Kiepiela, H.M. Coovadia, and Y.K. Seedat,
    Suppressor cell assayed by numerical and functional
    tests in chronic renal failure, Kidney Int. 22:192 (1982).
11. Y.G. Alevy, K.R. Mueller, and R.G. Slavin, Immune response
    in experimentally induced uremia. VI. Uremic macrophages
    are defective in their ability to present antigen to
    T cells. Clin Immunol and Immunopath. 29:433 (1983).

12. B. Charpentier, P.H. Lang, B. Martin, J. Noury, D. Marthieu, and D. Fries, Depressed polymorphnuclear leucocyte functions associated with normal cytotoxic functions of T and nature Killer cells during chronic hemodialysis, Clin Nephrology. 19:288 (1983).

13. J. Nelson, D.J. Ormrod, and T.E. Miller, Host immune status in uremia. IV. Phagocytosis and inflammatory response in vivo. Kidney Int. 23:312 (1983).

14. K. Osaki, H. Otsuka, K. Uomizu, R. Harada, Y. Otsuji, and S. Hashimoto, Monocyte-mediated mitogen responses of lymphocytes in uremic patients. Nephron. 34:87 (1983).

15. J. Rubin, L.M. Lin, R. Lewis, J. Cruse, and J.D. Bower, Host defence mechanisms in contnuous ambulatory peritoneal dialysis. Clin Nephrology. 20:140 (1983).

16. M.D. Smith, G. Hardy, J.D. Williams, G.A. Coles, Suppressor cell numbers and activity in non-transfused renal dialysis patients. Clin Nephrology. 20:130 (1983).

17. Y.G. Alevy, R. Mueller, J.R. Anderson, P.S. Hutcheson, and R.G. Slavin, The role of monocytes and responder cells in suppression of antigen-specific T cell proliferation in chronic renal failure. Int Archs Allergy Appl Immun. 73:97 (1984).

18. I. Jatoi, R.G. Slavin, and Y.G. Alevy, Immune response in experimental uremia. VII Uremic thymocytes amplify the response of control lymph node cells to alloantigens. Clin Immunol and Immunopath. 30:80 (1984).

19. S. Mezzano, A.J. Pesce, V.E. Pollak, and J.G. Michael, Analysis of humoral and cellular factors that contribute to impaired immune responsiveness in experimental uremia. Nephron. 36:15 (1984).

20. J. Raskova, J. Ghobrial, S. Shea, R. Eisinger, and K. Raska jr., Suppressor cells in end-stage renal disease. Functional assays and monoclonal antibody analysis. Am J Med. 76:847 (1984).

21. W.C. Waltzer, R.J. Bachvaroff, A.P. Raisbeck, B. Egelandsdal, C. Pullis, L. Shen, and F.T. Rapaport, Immunological monitoring in patients with end-stage renal disease. J Clin Immunol. 4:364 (1984).

22. E. Langhoff, and J. Ladefoged, In vitro natural killer and killer cell functions in uremia. Int Archs Allergy Appl Immun. 78:218 (1985).

23. E. Mirapeix, J. Montolin, C. Suárez, J. Lopez Pedret, and L. Revert, Defective radioresistant suppressor cell activity in hemodialysis patients. Nephron. 41:184 (1985).

24. D. Donati, P. Cervini, O. Amatruda, L. Baratelli, D. Cassani, A. De Maio, G. Frattini, M. Martegani, and L. Gastaldi, Nutritional state and immune function in uremic patients on maintenance hemodialysis. Proc Int Symp on Immune and Metabolic Aspects of Therapeutic Blood Purification Systems, Trondheim 1985, ed. Smeby LL, Jørstad S, Widerøe T-E, Karger Basel, pp 303-309. (1986).

25. R.F. Gagnon, Delayed-type hypersensitivity skin reaction in the chronically uremic mouse: Influence of severity and duration of uremia on the development of response. Nephron. 43:16 (1986).

26. P. Kurz, H. Köhler, S. Meuer, T. Hütteroth, and K-H Meyer zum Büschenfelde, Impaired cellular immune responses in chronic renal failure: Evidence of T cell defect, Kidney Int. 29:1209 (1986).

27. W.D. Tsakolos, Th.C. Theoharides, E.D. Hendler, J. Goffi-
    net, J.M. Dwyer, R.L. Whisler, and P.W. Askenase, Immune
    defects in chronic renal impairment: evidence for defec-
    tive regulation of lymphocyte response by macrophages
    from patients with chronic renal impairment on hemodia-
    lysis, Clin Exp Immunol. 63:218 (1986).
28. I. Fehrman, O. Ringdén, and J. Bergström, MLC-blocking
    factors in uremic sera, Clin Nephrol. 14:183 (1980).
29. J. Raskova, and K. Raska, Humoral inhibitors of the immune
    response in uremia. IV. Effects of serum and of isolated
    serum very low density lipoprotein from uremic rats on
    cellular immune reactions in vitro, Lab Invest. 45:410
    (1981).
30. G. Delaporte, F. Gros, C. Jonsson, and J. Bergström, In
    vitro cytotoxic properties of plasma samples from uremic
    patients, Clin Nephrol. 17:247 (1982).
31. K. Kamata, M. OKubo, and M. Sada, Immuno-suppressive fac-
    tors in uremic sera are composed of both dialyzable and
    non-dialyzable components, Clin Exp Immunol. 54:277 (1983).
32. P.K. Donnelly, B.K. Shenton, A.M. Alomran, D.M.A. Francis,
    G. Proud, and R.M.R. Taylor, N new mechanism of humoral
    immunosuppression in chronic renal failure and its impor-
    tance to dialysis and transplantation, Proc Europ Dial
    Transpl Assoc. 20:297 (1983).
33. M. Rola-Pleszczynski, D. Bolduc, S. Forand, G.E. Plante,
    and S. St-Pierre, Cellular immune function in uremia:
    Altered cytotoxic and suppressor cell responses to an
    immunomodulating heptapeptide, Clin Immunol and Immuno-
    path. 28:177 (1984).
34. P. Youing, J. Cledes, J.P. Herve, P. Miossec, and W.J.W.
    Morrow, Low-affinity E-rosette-blocking factor in hemo-
    dialysis-treated patients in chronic renal failure,
    Clin Immunol and Immunopath. 26:423 (1983).
35. J.F. Mangan, C. Chikkappa, L. Bieler, W.B. Scharfman, and
    D.R. Parkinson, Regulation of human blood erythroid
    Burst-Forming-Unit (BFU-E) proliferation by FC receptors
    and monoclonal antibodies, Blood. 59:990 (1982).
36. S. Lampari, and S. Carozzi, T lymphocytes, monocytes and
    erythropoiesis disorders in chronic renal failure,
    Nephron. 39:211 (1985).
37. K.G. Chandy, M. Pahl, N.D. Vaziri, and S. Gupa, Acute
    effects of dialysis on T lymphocytes in patients with
    end-stage renal disease, J Clin Lab Immunol. 17:119
    (1985).
38. A.M. Badger, D.B. Bernard, B.A. Idelson, and S.R. Cooper-
    band, Depressed spontaneous cellular cytotoxicity asso-
    ciated with normal or enhanced antibody-dependent cel-
    lular cytotoxicity in patients with chronic hemodialysis,
    Clin Exp Immunol. 45:563 (1981).
39. P. Jungers, J.F. DeLagneaw, P. Prunet, and J. Crosnier,
    Vaccination against hepatitis B in hemodialysis centers,
    Adv Nephrol. 11:303 (1982).
40. H. Köhler, W. Arnold, G. Renschin, H.H. Dormeyer, and
    K.H. Meyer zum Buschenfelde, Active hepatitis B vacci-
    nation of dialysis patients and medical staff, Kidney
    Int. 25:124 (1984).
41. F.G. Casio, G.S. Giebink, C. Le Than, and G. Schiffman,
    Pneumococcal vaccination in patients with chronic renal
    disease and renal allograft recipiens, Kidney Int. 20:
    254 (1981).

42.  R. Cappel, D. Beers, C. Liesnard, and M. Sratwa, Impaired humoral and cell-mediated immune responses in dialyzed patients after influenza vaccination, <u>Nephron.</u> 33:21 (1983).

43.  J. Nikoskelainen, M. Koskela, J. Forsström, A. Kasanen, and M. Leinonen, Persistence of antibodies to pneumo-coccal vaccine in patients with chronic renal failure, <u>Kidney Int.</u> 28:672 (1985).

44.  D. Urbanitz, and H.G. Sieberth, Impaired phagocytic acti-vity of human monocytes in respect to reduced antibacte-rial resistance in uremia, <u>Clin Nephrol.</u> 4:13 (1975).

45.  G. Lahnborg, L. Berghem, T. Ahlgren, L-G Groth, G. Lund-gren, and A. Tillegard, Reticuloendothelial function in human renal allograft recipients, <u>Transplantation.</u> 28: 111 (1979).

46.  T.M. Saba, Physiology and physiopathology of the reticuli-endothelial system. <u>Arch Intern Med.</u> 126:1031 (1970).

47.  K.L. Rosenstreich, J.J. Farrar, and S. Dougherty, Absolute macrophage dependency of T lymphocyte activation by mi-togens. <u>J Immunol.</u> 116:131 (1976).

48.  J. Knop, Influence of various macrophage population on Con-A induced T cell proliferation. <u>Immunology</u> 41:379 (1980).

49.  S.A. Rosenberg, and P.E. Lipsky, The role of monocyte fac-tors in the differentiation of immunglobulin secreting cells from human peripheral blood B cells. <u>J Immunol.</u> 125:232 (1980).

50.  F.G. Meyling, and T.A. Waldmann, Human B cell activation in vitro: augmentation and suppression by monocytes of the immunglobulin production induced by various B cell stimulants. <u>J Immunol.</u> 126:529 (1981).

51.  D. Metcalf, C.G. Begley, G.R. Johnson, N.A. Nicola, M.A. Vadas, A.F. Lopez, D.J. Williamson, G.G. Wong, S.C. Clark, and E.A. Wang, Biologic properties in vitro of a recombinant human granulocyte-macrophage colony- stimu-lating factor. <u>Blood</u> 67:37 (1986).

52.  O. Bakke, T. Balstad, L.C. Smeby, S. Jørstad, and T-E. Widerøe, Mononuclear cells (monocytes (OK 141+), $T_4$ and $T_8$+ cells) during hemodialysis. Proc Int Symp on Immune and Metabolic Aspects of Therapeutic Blood Purification Systems, Trondheim 1985, ed. Smeby L.C, Jørstad S, Widerøe T-E, Karger Basel, pp 105-111 (1986).

53.  R. Korz, U. Loe Guitz, H. Brunner, and R. Heinz, Lympho-cyte enzymes of DNA synthesis in chronic renal failure. <u>Proc Eur Dial Transpl Assoc.</u> 13:528 (1976).

54.  R.A. Gutman, and A.T. Auang, Inhibitor of marrow thymidine incorporation from sera of patients with uremia. <u>Kidney Int.</u> 18:715 (1980).

55.  E.N. Wardle, A study of the possible toxic metabolites of uremia in red cell metabolism. <u>Acta Haematol.</u> 43:129 (1970).

56.  M. Kurada, T. Asaka, Y.M. Tofuku, and R. Takeda, Serum antioxidant activity in uremic patients. <u>Nephron.</u> 41: 293 (1985).

57.  M. Taccone-Galluci, O. Giardini, R. Lubrano, U. Mazzarella, D. Bardino, S. Khashan, O. Mannarino, M. Elli, M. Cozzari, U. Bounteristiani, and C.U. Casciani, Red blood cell membrane lipid peroxidation in continuous ambulatory peritonal dialysis patients. <u>Am J Nephrol.</u> 6:92 (1986).

58.  S. Jørstad and K.E. Viken, Inhibitory effects of plasma from uraemic patients on human mononuclear phagocytes cultured in vitro. <u>Acta Path Microbiol Scand Sect C</u> <u>859</u>:169 (1977).

59.  S. Jørstad, and S. Kværnes, Uraemic toxins of high molecular weight inhibiting human mononuclear phagocytes cultured in vitro. <u>Acta Path Microbiol Scand Sect C 86</u>: 221 (1978).

60.  S. Jørstad, L.C. Smeby, T-E. Widerøe, and K.J. Berg, Transport of uremic toxius through conventional hemodialysis membranes. <u>Clin Nephrol.</u> 12:168 (1979).

61.  L.K. Curtiss, and T.S. Edgington, Differential sensitivity of lymphocyte subpopulation to suppression by low density lipoprotein inhibitor, an immunoregulatory human serum low density lipoprotein. <u>J Clin Invest.</u> 63:193 (1979).

62.  D. Modai, J. Weissgarten, U. Shaked, S. Segal, A. Pik, and R. Fuchs, Levamisole circumvents inhibition of lymphocyte activation imposed by uremic serum. <u>Nephron.</u> 40:436 (1985).

63.  P.A. Peterson, P.E. Ervin, and J. Berggard, Differentiation of glomerular, tubular, and normal proteinuria: determinations of urinary excretion of $\beta_2$-microglobulin, albumin and total protein, <u>J Clin Invest.</u> 48:1189 (1969).

64.  R.E. Birch, M.W. Fanger, and G.M. Bernier, $\beta_2$-microglobulin enhances lymphocyte surface receptor expression for IgG. <u>J Immunol.</u> 122:997 (1979).

65.  P.A. Keown, and B. Descamps, Improved renal graft survival after blood transfusion: a nonspesific erythrocyte mediated immunoregulatory process. <u>Lancet</u> 1:20 (1979).

66.  B.S. Bender, J.L. Curtis, J.E. Nagel, F.J. Chrest, E.S. Kraus, G.R. Briefel, and W.H. Adler, Analysis of immune status of hemodialyzed adults: association with prior transfusions. <u>Kidney Int.</u> 26:436 (1984).

67.  R.M. Hakim, D.T. Fearon, M. Lazarus, and C.S. Perzanowski, Biocompatibility of dialysis membranes: effects of chronic complement activation. <u>Kidney Int.</u> 26:194 (1984).

68.  A.K. Cheung, D.E. Chenoweth, D. Otsuka, and L.W. Henderson, Compartmental distribution of complement activation products in artificial kidneys. <u>Kidney Int.</u> 30:74 (1986).

69.  L.C. Smeby, S. Jørstad, T. Balstad, and T-E. Widerøe, Dialyzer generation and plasma clearance of activated complement during hemodialysis. Proc Int Symp on Immune and Metabolic Aspects of Therapeutic Blood Purification Systems. Trondheim 1985, ed. Smeby LC, Jørstad S, Widerøe T.E, Karger Basel, pp 303 (1986).

70.  D.E. Chenoweth, and T.E. Hugli, Demonstration of specific $C_{5a}$ receptor on intact human polymorphonuclear leucocytes. <u>Proc Nath Acad Sci USA 75</u> (8), 3943 (1978).

71.  S.L. Lewis, D.E. Van Epps, and D.E. Chenoweth, $C_{5a}$ receptor modulation on neutrophils and monocytes from chronic hemodialysis and peritoneal dialysis patients. <u>Clin Nephrol.</u> 26:37 (1986).

72.  R.M. Hakim, J. Breillatt, J.M. Lazarus, and F.K. Port, Complement activation and hypersensitivity reactions to dialysis membranes. <u>Engl J Med.</u> 311:878 (1984).

73. S. Stefoni, A.N. Costa, M.P. Scolar, L. Coli, G. Mosconi, A. Buscaroli, P. Boni, S. Iannelli, T. D´Atena, G. Cianciolo, L. Bruni, and U. Bonomini, DNA content and surface antigen characteristics of T-cell subsets in chronic dialysis patients. Proc Int Symp on Immune and Metabolic Aspects of Therapeutic Blood Purifcation Systems, Trondheim 1985, ed. Smeby LC, Jørstad S, Widerøe T-E, Karger Basel pp 168 (1986).

74. Y. Chida, S, Sakurai, and N. Yoshiyama, The effect of hemodialysis on lymphocyte subsets during dialysis. Clin Nephrol. 25:159 (1986).

75. M. Bingel, G. Lonnemann, S. Shaldon, K.M. Koch, and C.A. Dinarello, Human interleucin-1 production during hemodialysis. Nephron. 43:161 (1986).

76. J.E. Hefti, A. Blumberg, and H.R. Marti, Red cell survival and red cell enzymes in patients on continuous peritoneal dialysis (CAPD). Clin Nephrol. 19:232 (1983).

77. R. Lerner, B. Werner, H. Asaba, B. Ternstedt, and E. Elmqvist, Assessment of hemodialysis in regular hemodialysis patients by measuring carbon monoxide production rate. Clin Nephrol. 20:239 (1983).

78. C.R. Angle, M.S. Swanson, S.J. Stohs, and R.S. Markin, Abnormal erythrocyte pyrimidin nucleotides in uremic subjects. Nephron. 39:169 (1985).

79. G. Goubeaud, H.W. Leber, H.H. Schott, and G. Schütterle, Middle molecules and haemoglobulin synthesis. Proc Europ Dial Transpl Assoc. 13:371 (1976).

80. R.A. Gutman, A.T. Huang, and N.S. Bouknight, Inhibitor of marrow thymidin incorporation from sera of patients with uremia. Kidney Int. 18:715 (1980).

81. H.W. Radtke, A.B. Rege, M.B. LaMarche, and D. Bartos, Identification of spermine as an inhibitor of Erythropoiesis in patients with chronic renal failure. Clin Invest. 67:1623 (1981).

82. R.J.S. McGonigle, J.D. Wallin, R.K. Shadduck, and J.W. Fischer, Erythropoietin deficiency and inhibition of erythropoiesis in renal insufficiency. Kidney Int. 25:437 (1984).

83. T.E. Widerøe, T. Sanengen, and S. Halvorsen, Erythropoietin and uremic toxicity during continuous ambulatory dialysis. Kidney Int. 24, Suppl 16, 208 (1983).

84. F. Delwiche, G.M. Segal, J.W. Eschbach, and J.W. Adamson, Hematopoietic inhibitors in chronic renal failure: lack of in vitro specificity. Kidney Int. 29:641 (1986).

85. J.W. Fisher, Mechanism of the anemia of chronic renal failure. Nephron. 25:106 (1980).

86. W. Fried, and A. Anagnoston, Extrarenal erythropoietin production. In: Kidney Hormones, ed. Fischer J.W. vol. II:pp 231 (Academic Press, London 1977).

87. J. Caro, and A.J. Erslev, Uremic inhibitors of erythropoiesis. Semin Nephrol. 5:128 (1985).

88. B.P. Spragg, D.P. Bentley, and G.A. Coles, Anaemia of chronic renal failure. Polyamines are not raised in uremic serum. Nephron. 38:65 (1984).

89. D. Meytes, E. Bogin, A. Ma, P.P. Dukes, and S.G. Massry, Effect of parathyroid hormone on erythropoiesis. J Clin Invest. 67:1263 (1981).

90. F. Delwiche, M.J. Garrity, J.S. Powell, R.P. Robertson, and J.W. Adamson, High level of the circulating form of parathyreoid hormone do not inhibit in vitro erythropoiesis. J Lab Clin Med. 102:613 (1983).

91.  J.B. Sherwood, and E. Goldwasser, A radioimmunoassay for erythropoietin. Blood 54:885 (1979).

92.  P.M. Cotes, Immunoreactive erythropoietin in serum. I. Evidence of the validity of the assay method and the physiological relevance of estimates. Br J Hemat. 50:427 (1983).

93.  M.C. Bordas, N.S. Serbource-Goguel, J.M. Feger, J.M. Maccario, J.M. Agneray, and G.M. Durand, Evaluation of the degree of desialynation of serum $\alpha_1$-acid glucoprotein and $\alpha_1$-antitrypsin. Clin Chim Acta. 125:311 (1982).

94.  H. Lewinsky, U. Gafter, J. Levi, and D. Allalouf, Neuramidase-like activity in sera of uremic anemic patients. Nephron. 37:35 (1984).

95.  J.S. Shannon, T.R.J. Lappin, G.E. Elder, G.M. Roberts, M.G. McGeown, and J.M. Bridges, Increased plasma glycosidase and protease activity in uraemia: possible role in the aetiology of the anaemia of chronic renal failure. Clin Chim Acta. 153:203 (1985).

96.  J.I. Kurland, A. Meyors, and M.A.S. Moore, Synthesis and release of erythroid colony-and burst-potentiating activities by purified populations of murine peritoneal macrophages. J Exp Med. 151:839 (1980).

97.  J.N. Rich, W. Heit, and B. Kubanek, An erythropoietic stimulating factor similar to erythropoietin released by macrophages after treatment with silica. Blut 4: 297 (1980).

98.  K. Jacobs, C. Shoemaker, R. Rudersdorf, S.D. Neill, R.J. Kaufman, A. Mufson, J. Seehra, S.S. Jones, R. Hewich, E.F. Fritsch, M. Kawakita, T. Shimizu, and T. Miyake, Isolation and characterization of genomic and cDNA clones of human erythropoietin. Nature 313:806 (1985).

99.  J.W. Eschbach, J. Mladenovic, J.F. Garcia, P.W. Wahl, and J.W. Adamson, The anemia of chronic renal failure sheep. Response to erythropoietin-rich plasma in vivo. J Clin Invest. 74:434 (1984).

100. J. Mladenovic, J.W. Eschbach, J.R. Koup, J.F. Garcia, and J.W. Adamson, Erythropoietin kinetics in normal and uremic sheep. J Lab Clin Med. 195:659 (1985).

CONVERSION OF AMMONIA TO GLUTAMATE BY L-GLUTAMIC DEHYDROGENASE, ALCOHOL

DEHYDROGENASE AND NAD$^+$ IMMOBILIZED WITHIN LIPID-POLYAMIDE

POLYETHYLENEIMINE MICROCAPSULES

Ehud Ilan* and Thomas Ming Swi Chang

Artificial Cells and Organs Research Centre, Faculty of
Medicine, McGill University, 3655 Drummond St.
Montreal, Canada H3G 1Y6

INTRODUCTION

The use of Albumin-Collodion coated Activated Charcoal (ACAC) hemoperfusion systems for treatment of end-stage renal failure has several advantages over the hemodialysis procedure. With coated charcoal ACAC hemoperfusion, the removal of uremic toxins, such as creatinine, uric acid, and middle molecules, is much more effective than with standard hemodialysis[1]. However, the ACAC hemoperfusion system does not remove water, electrolytes or urea. As the removal of excess water can be achieved with the help of a small ultrafiltrator[1,2], the remaining major problems are electrolytes and urea removal.

This report is concerned with the problem of urea removal. As the conversion of urea to ammonia can easily be carried out using microencapsulated urease (see Fig. 1 and references 3,4), this problem is therefore reduced to the removal of the ammonia resulting from the enzymatic reaction. In the present study ammonia was removed through conversion into glutamate, within lipid-polyamide microcapsules, by a multienzyme system (Fig. 1). L-Glutamic dehydrogenase, alcohol dehydrogenase, NAD$^+$ and α-ketoglutarate were all retained within the microcapsules without leakage by the lipid-polyamide membrane. Thus, the free cofactor could be recycled with the help of the two enzymes, in the presence of ethanol, and convert the permeating external lipophilic ammonia into glutamate. The possibility of using such systems for the conversion of urea and ammonia into amino acids has already been demonstrated in earlier studies[5,6]. In these studies, hemoglobin alone or hemoglobin together with polyethyleneimine was used as filler for the microcapsules. It is known, however, that hemoglobin preparations contain enzymes which deactivate NAD$^+$[7]. This factor might be the reason for the slow conversion rate of ammonia into glutamate, and the short activity times described in these communications. In the present study, we were successful in solving these problems by using pure polyethyleneimine as a microcapsule filler.

---

*Present address
Dept. of Biomedical Engineering, Technion - I.I.T., Haifa, 32000, Israel.

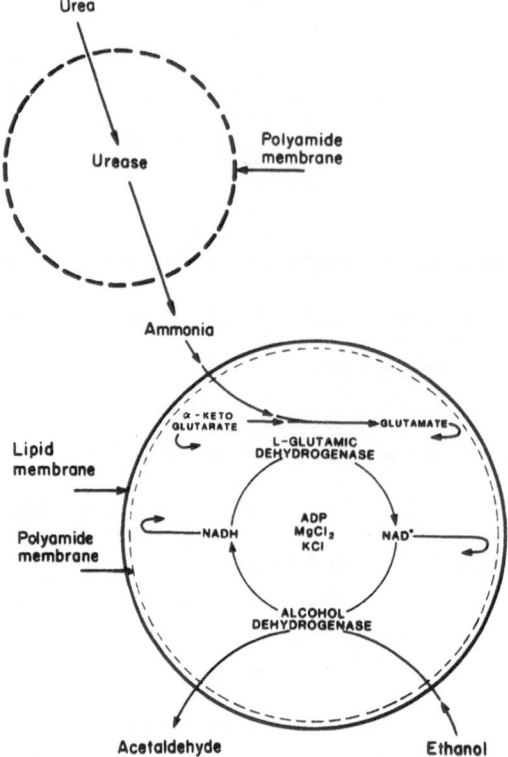

Fig. 1 Schematic representation of microcapsules system for the conversion of urea into glutamate

EXPERIMENTAL

Preparation of Microcapsules

These were prepared according to the procedure by Yu and Chang[6] with some modifications. The major modification was the omission of hemoglobin as microcapsule filler. A typical batch of microcapsules was 5.0 ml in volume and 4.8 g in weight.

Assay of Ammonia Conversion into Glutamate

In each experiment 2.5 ml of lipid-polyamide membrane microcapsules was added to 5.0 ml of ammonia-ethanol substrate solution in 10 ml glass-stoppered flask. The substrate solution was 20 mM in ammonium acetate, 200 mM in ethanol and 5 mM in Tris, and its final pH was adjusted to 8.8 with HCl. The reaction was carried out in a rotary shaker at 30°C and 140 rpm. The rate of ammonia conversion into glutamate was studied by measuring colorimetrically the rate of decrease of ammonia concentrations in the ammonia-ethanol substrate solution and also by direct measurements of the glutamate formed inside the microcapsules. The latter determination was carried out using a reverse phase HPLC technique based on Hill et al.[8].

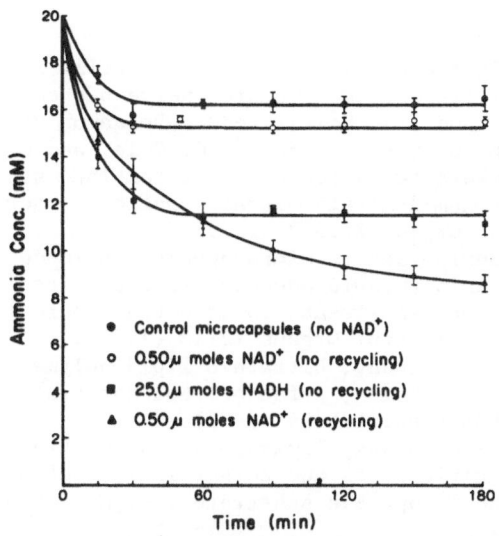

Fig. 2   Changes in ammonia concentrations after the addition of 2.5 ml batches of different types of lipid-polyamide membrane microcapsules to 5.0 ml batches of ammonia-ethanol substrate solution. The microcapsules marked 'no recycling' did not contain alcohol dehydrogenase.

RESULTS AND DISCUSSION

Assay of Ammonia Conversion into Glutamate

When using 'low NAD$^+$ (0.50 μmoles) microcapsules with recycling' it was seen from Fig. 2 and from the HPLC measurements, that the amounts of ammonia converted into glutamate are similar to the actual amounts of glutamate formed inside the microcapsules at different points of time during the reaction. We can calculate from Fig. 2 that about 40 μmoles of ammonia were converted into glutamate within the first 3 hours of the reaction. Thus, the 0.50 μmoles NAD$^+$ retained in the 2.5 ml batch of microcapsules could be recycled 80 times during this period of time. A much larger amount of NADH had to be encapsulated to convert ammonia into glutamate without cofactor recycling (Fig. 2).

Storage Stability

In reusing 'low NAD$^+$ microcapsules with recycling' stored at 4°C, 81% of the original activity was retained after 7 days and 56% was still retained after 16 days of storage. One can see from these results that by using pure polyethyleneimine as microcapsule filler it is possible to drastically increase the storage stability of the multienzyme system.

ACKNOWLEDGEMENTS

The support of the Medical Research Council of Canada, Natural Sciences and Engineering Research Council of Canada and the Quebec "Virage" Centre of Excellence Grant in biotechnology to TMSC is gratefully acknowledged. We thank Mr. Colin Lister for his technical assistance with the HPLC determinations.

REFERENCES

1. T.M.S. Chang, Biotechnology of artificial cells including application to artificial organs, in: "Comprehensive Biotechnology", Vol. 4, C.W. Robinson and J.A. Howell, eds., Pergamon Press, New York (1985).
2. T.M.S. Chang, E. Chirito, P. Barré, C. Cole, and M. Hewish, Clinical performance characteristics of a new combined system for simultaneous hemoperfusion–hemodialysis–ultrafiltration in series, Trans. Am. Soc. Artif. Intern. Organs 21:502 (1975).
3. T.M.S. Chang, Semipermeable microcapsules, Science 146:524 (1964).
4. T.M.S. Chang, Semipermeable aqueous microcapsules ("Artificial Cells"): with emphasis on experiments in an extracorporeal shunt system, Trans. Am. Soc. Artif. Intern. Organs 12:13 (1966).
5. Y.T. Yu, and T.M.S. Chang, Ultrathin lipid–polymer membrane microcapsules containing multienzymes, cofactors and substrates for multistep enzyme reactions, FEBS Lett. 125:94 (1981).
6. Y.T. Yu, and T.M.S. Chang, Lipid–polyamide membrane microcapsules immobilized multienzymes and cofactors for sequential conversion of lipophilic and lipophobic substrates, J. Microbial Enzyme Technol. 4:327 (1982).
7. J. Grunwald, and T.M.S. Chang, Immobilization of alchol dehydrogenase, malic dehydrogenase and dextran-NAD$^+$ within nylon–polyethyleneimine microcapsules: preparation and cofactor recycling, J. Molec. Catalysis 11:83 (1981).
8. D.W. Hill, F.H. Walters, T.D. Wilson, and J.D. Stuart, High performance liquid chromatographic determination of amino acids in the picomole range, Anal. Chem. 51:1338 (1979).

# A PURIFICATION METHOD BY UNCOATED CHARCOAL

Nicola Cerulli*, Laura Politi, Anna Rita D'Angelo* and
Roberto Scandurra

Clinica Urologica* and Dipartimento di Scienze Biochimiche
Universita' "La Sapienza", Roma, Italy

## INTRODUCTION

The persistent state of sickness of patients in periodical dialysis, has been attributed, during the last years, to the retention of
"uremic toxins" never identified with certainty. Parathyroid hormone,
guanidinic compounds, aromatic substances, indacan, middle molecules,
etc.have been proposed at various times the specific molecule responsible of the uremic state. Middle molecules which had in the past a very
large following,have been recently defined elusive by many authors[1-3].

We have long been convinced that the uremic state, which is persistent also with substitutive treatments, is supported not by intangible
uremic toxins, but by the engorgement of all the metabolic pathways
caused by the low and progressive accumulation of various substances
during years of renal failure. This accumulation cannot be reduced to levels compatible with a state of well-being even by the most efficient
dialytic treatments, since the hours/week of dialysis are insufficient to
remove all the substances accumulated during sickness and those daily
produced even because most of accumulants are hardly dialysable. The engorgement of all the metabolic pathways continues then to support the
state of sickness.

As a confirm of this hypothesis many authors, mostly Japanese,have
shown by chromatographic proceedures, the presence in biological fluids
of uremic patients, the most varied compounds, many of them ordinary me-
tabolytes reaching such high concentrations as to become toxic[4-6].

Other substances found in uremic fluids, absent or present only in
traces in healthy subjects, are in our opinion the result of the satura-
tion of some metabolic pathways followed by activation of others, the
products of which are not further metabolised and moreover are scarcely
removed by dialysis. Most of these accumulants have been considered markers of uremia.

It follows that further purification devices need to be added to
dialysis. To this purpose we have investigated the efficiency of charcoal in the purification of blood, by using a modified method employed
in the past by some authors[7-9] To monitor the efficiency of the
blood purification a chromatographic method recently proposed to evaluate uremia has been used[10,11]

## MATERIALS AND METHODS

The blood ultrafiltrate obtained by a PAN ASAHI 200 and 250 at a flow of 120-180 ml/min was made to flow in a charcoal cartridge containing about 500 g of uncoated granular charcoal(Dideco,Mirandola,Italy). Fines were removed by a 0.2 $\mu$m nylon filter(Pall DSLK-INFZP).The ultrafiltrate was then added to the particulate fraction of the blood and reinjected to the patient. An appropriate volume of ultrafiltrate was discharged to reach dry weight.

Perfusion on charcoal ( 3 hrs ) was followed by 1 hr dialysis to remove urea and electrolytes not adsorbed by charcoal.

The ultrafiltrate ( 2 ml ) was analysed on a chromatographic column (1.6 x 100 cm, K 16, Pharmacia, Sweden) filled with Biogel P2 (Bio Rad, USA)particles less than 40 $\mu$m to give a bed height of 83 cm. Elution was performed with 50 mM ammonium bicarbonate pH 8 at 0.5 ml/min obtained with a peristaltic pump and monitored at 254 nm with a Uvicord S detector( LKB, Sweden ).

HPLC experiments were performed as reported elsewere[10] .

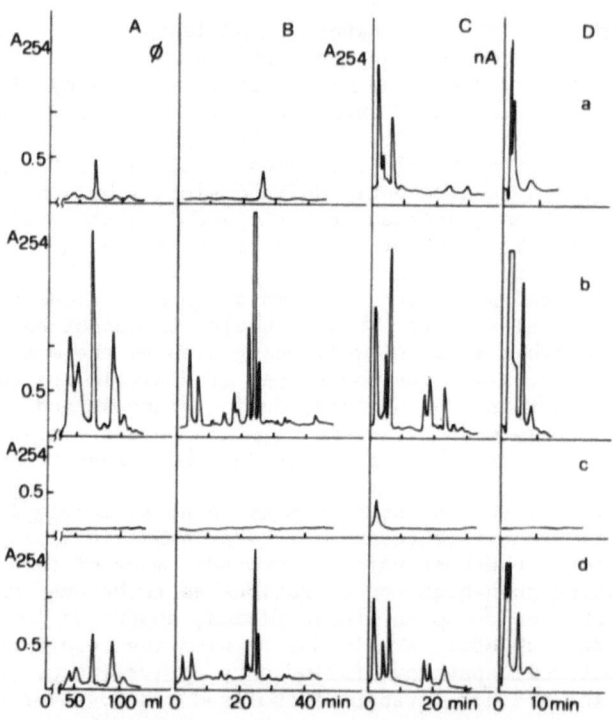

fig.1.Chromatographic patterns of ultrafiltrates of a
healthy subject ( line a ) and of a patient with
a severe uremic state, in the standard dialysis
program ( line b ). A, Biogel P2 chromatography;
B-D, HPLC:B, RP 18 column monitored in fluorescence ( exc.295 nm; emiss.405 nm ); C, monitored at
245 nm; D,monitored with electrochemical detector[10]
In line c the profiles of the ultrafiltrate downstream the charcoal cartridge after 1 and 3 hrs of
perfusion; in line d upstream the charcoal cartridge after 3 hrs of perfusion.

## RESULTS AND DISCUSSION

As reported in fig.1 line a, the ultrafiltrate of healthy subject has a low level of 254 nm absorbing material either detected with gel chromatography or with HPLC whilst that of the uremic subject has a large number of strongly absorbing peaks ( fig.1, line b ) which disappear after a percolation through the charcoal cartridge ( fig.1, line c).

After 3 hrs of charcoal treatment the chromatographic patterns of the ultrafiltrate taken upstream the charcoal cartridge are clearly reduced and behaves as reported in fig. 1, line d. The chromatograms of the ultrafiltrate taken downstream the charcoal cartridge also after 3 hrs are quite flat and identical as those reported in fig.1,line c.

Using this chromatographic approach in addition to the routine hematochemical analyses, a group of three patients in periodical hemodialysis from 7 to 12 years, selected for the severity of their clinical state ( hypertension, hyperPTH, creatinine > 12 mg ), were studied.

The patients were daily perfused on charcoal for three hrs then dialysed for one hr to remove urea and electrolytes not retained by charcoal and to normalize the acid-base balance.

Every day chromatographic checks performed before perfusion, resulted higher than those performed at the end of the previous perfusion but progressively lower than those performed before the first treatment.

After fifteen consecutive perfusions the gel chromatographic pattern of the ultrafiltrate ( fig. 2, B ) was dramatically decreased as compared to the initial one (fig. 2, A ) and was very similar to that of the healthy subject ( fig.1, line a ). Whilst the hematochemical parameters were not found to vary significantly, however, an improvement in the clinical state of each patient was invariably found: an increase of subjective well-being, attentiveness and resistance to fatigue.

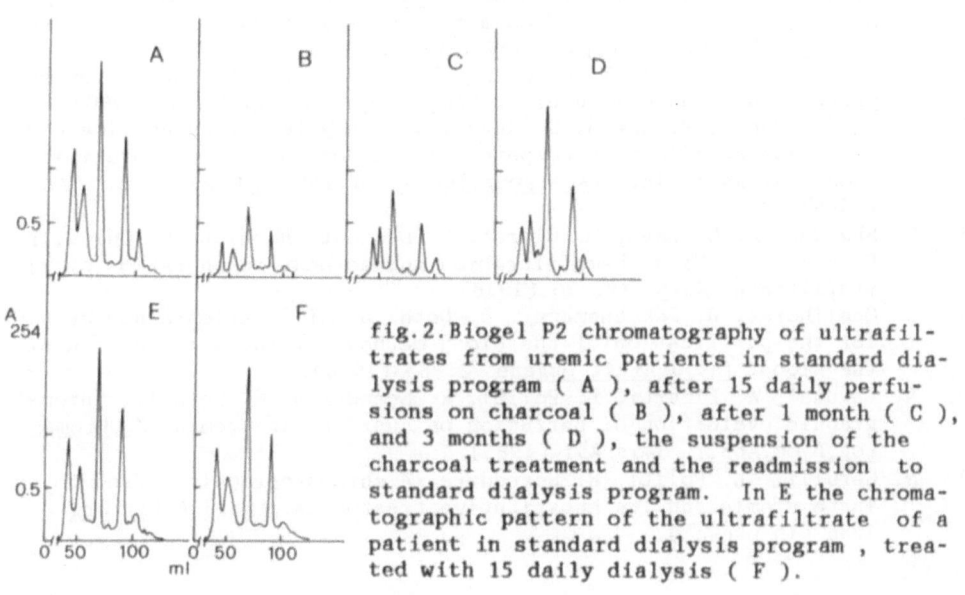

fig.2.Biogel P2 chromatography of ultrafiltrates from uremic patients in standard dialysis program ( A ), after 15 daily perfusions on charcoal ( B ), after 1 month ( C ), and 3 months ( D ), the suspension of the charcoal treatment and the readmission to standard dialysis program. In E the chromatographic pattern of the ultrafiltrate of a patient in standard dialysis program , treated with 15 daily dialysis ( F ).

Furthermore a marked reduction of the typical greyish color of the skin was invariably found and a sufficient decrease of the arterial pressure, to allow discontinuation of all hypotensive therapy .

Patients in clinical and chromatographic conditions ( fig.2, E ) similar to those perfused on charcoal were treated with a daily dialysis,

At the end of the intensive charcoal treatment,one patient was read-readmitted in the standard dialysis program ( 4 hrs thrice a week ) and his ultrafiltrates analysed with gel chromatography. Checks performed 1 and 3 months after the stoppage of the charcoal treatment ( fig. 2 C and D ) showed a slow and progressive increase of the chromatographic patterns, accompanied by a worsening of the clinical state. Both these results could be explained by the accumulation of almost undialyzable metabolites .

The other two patients were treated weekly with one charcoal substituting one of the three dialysis. This protocol maintained those clinical improvements and chromatographic results obtained by the intensive charcoal treatment.

Should this last result be confirmed by a large clinical trial, the proposed depurative treatment would not be much more expensive than dialysis.

The chromatographic approach ( overall the simplest chromatography on Biogel P2 ) brings out clinically relevant biochemical parameters not monitored by standard hematochemical analyses.

REFERENCES

1.  S. Giovannetti, G. Barsotti.Uremic intoxication. Nephron, 14:123 (1975).
2.  S. Contreras, R. Later, J. Navarro, J. L. Touraine ,A. M. Freyra, J. Traeger. Molecules in the middle molecular weight range. Nephron 32: 183 (1982).
3.  A. Schoots, F. Mikkers, C. Cramers, R. De Smet, S. Ringoir. Uremic toxins and elusive middle molecules. Nephron, 38: 1 (1984).
4.  B. H. Mamdani, M . Mashouf Shaykh ,M. A. Evenson, G. Dunea. Chromatographic studies of uremic plasma. Int.J.Art.Org. 2: 187 (1979)
5.  T. M .S. Chang, C. Lister. Middle molecules in hepatic coma and uremia. Artif.Org. 4(suppl) :169 (1980).
6.  A. Saito, I. Kanazawa, T. G. Chung ,K. Maeda. Analytic study for separation of middle molecules. Artif.Org. 4(suppl) 13: (1980).
7.  T. M. S. Chang, P. Barre, S. Kuruvilla. Long-term reduced time hemoperfusion-hemodialysis compared to standard dialysis:a preliminary crossover analysis. Trans.Amer.Soc.Artif.Intern.Organs 31: 572 (1985).
8.  S. Shaldon, M. C. Beau, G. Claret, G. Deshodt, H. Miot, R. Oules, P. Raperez, C. Mion. Hemofiltration and sorbent regeneration of ultrafiltrate. EDTA XV: 20 (1978).
9.  E. Quellhorst, B. Schuenemann, B. Doth. Hemofiltration:a new method for the purification of the blood method for the purification of the blood. Int.J.Artif.Organs 2: 83 (1978).
10. A. Lagana', A. Liberti, L. Politi, R. Scandurra, N. Cerulli. Chromatographic evaluation of perfusion on charcoal in uremia. J.Chromatog.Biom.Appl. 345: 251 (1985).
11. N. Cerulli, L. Politi, R. Scandurra. A chromatographic method to evaluate uremia and its substitutive treatments. Int.J.Artif.Org . 1986 (in press ).

AMINO ACID COMPOSITION OF UREMIC MIDDLE AND LOW MOLECULAR WEIGHT

RETENTION PRODUCTS

Nadine Bazilinski[1,2], Mashouf Shaykh[1], Sarosh Ahmed[1],
Theodore Musiala[3], Robert H. Williams[3], Ann Poulos[3],
Alvin Dubin[3], and George Dunea[1,2]

[1]Division of Nephrology, Cook County Hospital and Hektoen
  Institute for Medical Research, Chicago, IL 60612 (USA)

[2]University of Illinois at Chicago, College of Medicine,
  Chicago, IL 60612 (USA)

[3]Department of Biochemistry, Rush University and Hektoen
  Institute for Medical Research, Chicago, IL 60612 (USA)

ABSTRACT

    In order to characterize the spectrum of small peptides retained in
chronic renal failure, we carried out high pressure liquid chromatography
(HPLC) of serum ultrafiltrates from patients with chronic renal failure
(CRF), acute renal failure (ARF), and normal subjects.  HPLC patterns in
CRF resolved into more than twenty peaks; those in ARF contained fewer
peaks and resembled that of normals.  We carried out amino acid analysis
of HPLC fractions after hydrolysis with 6N HCl of four patients with CRF,
one patient with ARF, and one normal subject.  Following hydrolysis each
HPLC fraction yielded several amino acids.  Glycine, leucine, serine,
phosphoserine, glutamic acid, and phenylalanine were found in greatest
frequency in the four CRF patients.

INTRODUCTION

    It is known that small peptides are retained in the serum of patients
with renal failure.  These peptides, which may play a role in producing
uremic symptoms, have been usually identified by the ninhydrin reaction
or by chemical derivatization.  Only a few peptides have been isolated
and fully characterized[1].  In earlier studies we have analyzed the amino
acid composition of middle molecular weight fractions obtained by gel
filtration chromatography and further separated by thin layer chromatography[2].  Here, we extend our studies with HPLC analysis of serum ultrafiltrates from patients with chronic renal failure (CRF) and acute renal
failure (ARF), followed by amino acid analysis of acid hydrolysates of
each HPLC fraction.

PATIENTS

    We obtained 50 blood samples from patients with CRF, ARF, and from
normal controls for HPLC patterns.  For subsequent amino acid analysis

samples were obtained from patients F, J, and M with chronic renal failure on hemodialysis, and patient B, who was receiving intermittent peritoneal dialysis.  Mean BUN and serum creatinine were 97 mg% and 18 mg% respectively.  We also obtained blood from patient JJ with ARF in whom BUN was 110 mg% and creatinine 13.7 mg%, rising from 16 mg%, and 1.5 mg% one week earlier.  Control patient H had a BUN 18 mg%, and creatinine 1.1 mg%.  Clotted blood was centrifuged and the serum was separated and stored at $-20^{\circ}C$ until studied.

METHODS

Ultrafiltration.  Ten ml of serum were filtered through an Amicon YCO5 ultrafilter using an Amicon stirred cell (Model 8010) under 550 Kpa of nitrogen.  The ultrafiltrate was then lyophilized.

High Pressure Liquid Chromatography.  The lyophilized ultrafiltrates were reconstituted in 300ul of degassed, deionized, and distilled water.  This solution was centrifuged in a Beckman Airfuge Centrifuge at 110,000 x g and 20ul of the clear supernatant was injected into the Ternary Gradient Liquid Chromatograph (IBM, Model LC 533) which was fitted with a C18 reverse phase column (IBM, 5 um, 4.5x250mm).  The gradient elution was obtained with a mobile phase of 0.15% trifluoroacetic acid (TFA) in water and 0.15% TFA in acetonitrile at a flow rate of 0.8 ml/min.  The column effluent was monitored at 220nm.  Fractions were collected manually, lyophilized and kept frozen at $-4^{\circ}C$ until amino acid analysis was done.

Amino Acid Analysis.  The lyophilized residues of HPLC fractions were hydrolyzed in 6N hydrochloric acid at $120^{\circ}C$ for 22 hours in a sealed evacuated ampule.  The hydrolysate was evaporated and dissolved in 300ul of lithium citrate dilution buffer (Beckman) containing thioldiglycol, pH 2.2, 0.15N, and amino acids were separated with a Beckman-Spinco amino acid analyzer model 119 CL.  The final concentration of amino acids were calculated with a Beckman Integrator Model #126.

RESULTS

HPLC patterns of patients with CRF were similar but not identical and often contained more than 20 peaks.  The CRF chromatograms were clearly different from those of normals.  In general the pattern of ARF resembled that of the normal subjects and contained only one or two prominent peaks.  Following acid hydrolysis, each peak yielded three or more amino acids.  Amino acid content varied from fraction to fraction.

The normal HPLC pattern (Figure 1, patient H) had only one prominent peak with the same retention time as found in other normals.  The earlier fractions (H5, H6, H8) contained leucine, isoleucine, tyrosine, glycine, and serine.  (H5) had 45% leucine.  Of the later fraction (H10) had 73% phenylalanine; (H12) and (H14) were composed of more amino acids including glycine, isoleucine and serine.  Glycine accounted for no more than 25% of any fraction.

The HPLC pattern in ARF (Figure 1, patient JJ) was similar to the normal.  The chromatogram had one prominent peak with the same retention time.  Fraction (JJ3) was composed principally of glycine, phosphoserine and serine.  (JJ5) was 54% leucine, 21% isoleucine, and 21% tyrosine.  (JJ6) had 61% glycine, and (JJ8) had 30% glycine and 70% serine.  Among the later fractions (JJ10) had 84% phenylalanine, (JJ12) and (JJ14) had 30% glycine, and about 20% phosphoserine, and (JJ18) had 44% phosphoserine and 37% glutamic acid.

198

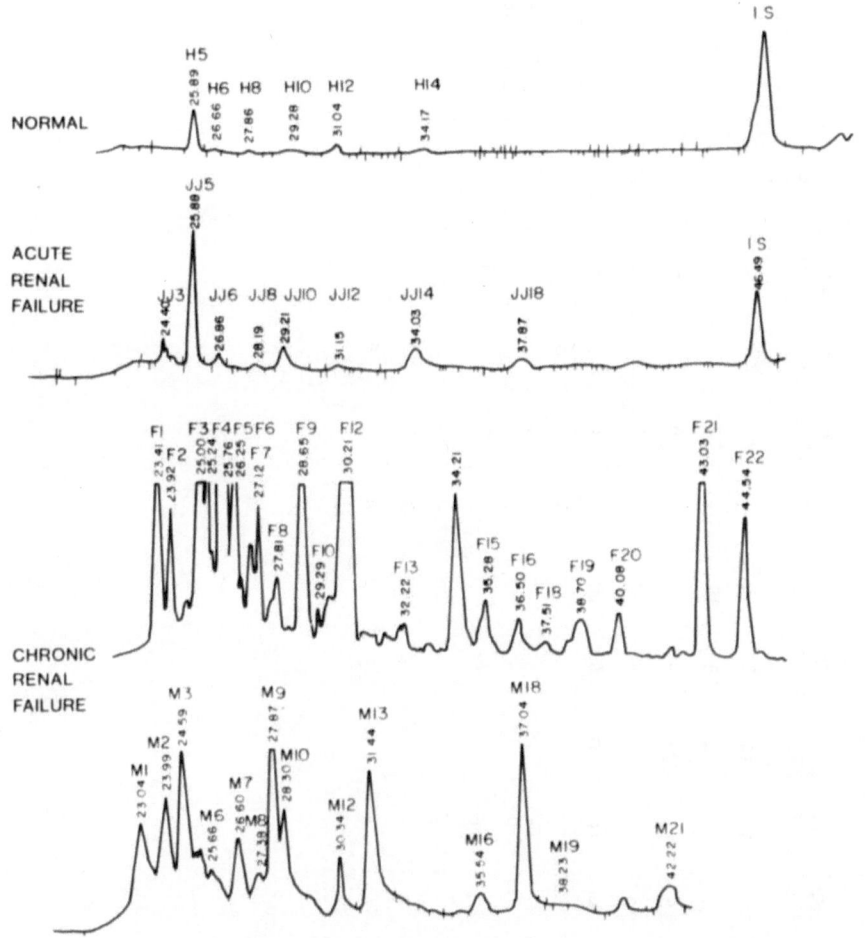

Figure 1.  HPLC patterns observed in N control H, ARF patient JJ, and CRF patients F and M.

TABLE 1. Amino Acid Composition after Acid Hydrolyis of Corresponding HPLC Peaks of CRF Patient F[b].

| CRF PATIENT F HPLC PEAK RETENTION TIME | F1 23.4 | F2 23.9 | F3 25.0 | F4 25.2 | F5 25.8 | F6 26.3 | F7 27.1 | F8 27.8 | F9 28.7 | F10 29.3 | F11 [a] |
|---|---|---|---|---|---|---|---|---|---|---|---|
| GLYCINE | 36 | 39 | 57 | 34 | 28 | 5 | 27 | 45 | 22 | 18 | |
| LEUCINE | 4 | 4 | 6 | 4 | 38 | 38 | 36 | 21 | 7 | 5 | |
| SERINE | 18 | 15 | 13 | 22 | 9 | | 10 | 6 | | 9 | |
| PHOSPHOSERINE | | 20 | | | 6 | | 3 | 4 | | 55 | |
| GLUTAMIC ACID | 8 | 9 | 10 | 10 | 4 | 10 | | 8 | | 4 | |
| PHENYLALANINE | | | | | | | 3 | | 58 | | |
| ASPARTIC ACID | 7 | 7 | 8 | | 3 | | | 4 | | | |
| ORNITHINE | 9 | | 6 | 7 | 4 | | | 3 | 4 | 6 | |
| ISOLEUCINE | | | | | | 16 | 5 | 5 | | | |
| TYROSINE | 4 | 7 | | | 4 | 31 | 3 | | | | |
| ALANINE | 8 | | | 7 | 3 | | 8 | 3 | 6 | | |
| THREONINE | 4 | | | 14 | | | | | | | |
| PHOSPHOETHANOLAMINE | | | | | | | | | | | |
| LYSINE | | | | 3 | | | 5 | | 3 | | |
| GLUTAMINE | | | | | | | | | | | |
| HALF CYSTINE | | | | | | | | | | | |
| B-ALANINE | | | | | | | | | | | |
| HISTIDINE | | | | | | | | | | | |
| HYDROXYPROLINE | | | | | | | | | | | |
| PROLINE | | | | | | | | | | | |
| VALINE | | | | | | | | | | | |

a = peak absent; b = results are shown as per cent of total amino acid content of each fraction.

200

| CRF PATIENT F HPLC PEAK RETENTION TIME | F12 30.2 | F13 32.2 | F14 34.2 | F15 35.3 | F16 36.5 | F17 a | F18 37.5 | F19 38.7 | F20 40.1 | F21 43.0 | F22 44.5 |
|---|---|---|---|---|---|---|---|---|---|---|---|
| GLYCINE | 61 | 33 | 32 | 30 | 30 | | 30 | 29 | 36 | 33 | 31 |
| LEUCINE | | 6 | 6 | 6 | 7 | | | 6 | 8 | 21 | 7 |
| SERINE | 21 | 21 | 11 | 14 | 16 | | 16 | 13 | 20 | 6 | 17 |
| PHOSPHOSERINE | | | 8 | | | | 29 | 37 | | 11 | 8 |
| GLUTAMIC ACID | 23 | 9 | 9 | 11 | 13 | | 8 | 7 | 13 | 10 | 14 |
| PHENYLALANINE | 16 | 3 | 8 | 7 | 5 | | 9 | | | | |
| ASPARTIC ACID | | 7 | 6 | 7 | 9 | | | 8 | 9 | 8 | 8 |
| ORNITHINE | | 10 | 8 | 6 | 6 | | 7 | | 9 | 10 | |
| ISOLEUCINE | | 3 | | 3 | | | | | | | |
| TYROSINE | | | 6 | | | | | | | | 3 |
| ALANINE | | | 3 | 8 | 4 | | | | | | 4 |
| THREONINE | | 4 | | 3 | 4 | | | | 5 | | 3 |
| PHOSPHOETHANOLAMINE | | | | | | | | | | | |
| LYSINE | | | 4 | 4 | 5 | | | | | | |
| GLUTAMINE | | | | | | | | | | | |
| HALF CYSTINE | | | | | | | | | | | |
| B-ALANINE | | | | | | | | | | | |
| HISTIDINE | | | | | | | | | | | |
| HYDROXYPROLINE | | 3 | | | | | | | | | 3 |
| PROLINE | | | | | | | | | | | |
| VALINE | | | | | | | | | | | |

201

TABLE 2

SUMMARY OF AMINO ACIDS FOUND IN HPLC FRACTIONS OF CRF PATIENTS F, M, B
AND J FOLLOWING ACID HYDROLYSIS[a]

|  | CRF PATIENT F | CRF PATIENT M | CRF PATIENT B | CRF PATIENT J |
|---|---|---|---|---|
| GLYCINE | 20/20 | 13/13 | 9/9 | 8/8 |
| LEUCINE | 18/20 | 10/13 | 5/9 | 6/8 |
| SERINE | 17/20 | 10/13 | 8/9 | 5/8 |
| PHOSPHOSERINE | 10/20 | 8/13 | 8/9 | 1/8 |
| GLUTAMIC ACID | 18/20 | 12/13 | 9/9 | 4/8 |
| PHENYLALANINE | 7/20 | 4/13 | 3/9 | 2/8 |
| ASPARTIC ACID | 15/20 | 10/13 | 6/9 | 4/8 |
| ORNITHINE | 14/20 | 6/13 | 2/9 | 3/8 |
| ISOLEUCINE | 5/20 | 4/13 | 4/9 | 4/8 |
| TYROSINE | 7/20 | 3/13 | 3/9 | 2/8 |
| ALANINE | 10/20 | 6/13 | 4/9 | 3/8 |

[a]Results are shown as number of fractions in which amino acid was present/
number of fractions analyzed.

many more prominent peaks than either ARF or normal chromatograms. The amino acid composition corresponding to the HPLC fractions of CRF patient F is shown in Table 1. Although there were some similarities, no two corresponding fractions had the same proportion and amino acid content in the four CRF analyzed. Glycine was found in all of the fractions in each of the patients. Leucine, serine, phosphoserine, glutamic acid, phenylalanine, and aspartic acid were most frequently present (Table 2).

DISCUSSION

High pressure liquid chromatography, only recently used to analyze uremic retention products, is rapid, sensitive, and yields reproducible chromatographic patterns of serum ultrafiltrates of patients with CRF and ARF. As others we have observed some variability in patterns, which may perhaps reflect differences in residual renal clearance or variable metabolic states (3). We observed fewer peaks in our ARF chromatograms. It is likely that most CRF peaks represent substances which accumulate slowly during the progression of renal failure, and that their accumulation has little correlation with the abrupt decrements in renal clearance that are reflected by rising serum urea nitrogen or creatinine levels. We postulate that these substances arise from catabolism of higher molecular weight proteins. Our finding of amino acids in the HPLC fractions would suggest that many of the peaks found on the HPLC chromatogram are peptides, though this would by no means rule out the additional presence of conjugates and low molecular weight non-protein substances.

Our results may be compared to previous reports. Using different separation methods, several other investigators have also isolated fractions that on subsequent analysis yielded high levels of glycine, glutamic and aspartic acid after hydrolysis (4-6). We also found such high levels of the same amino acids in previous studies using thin layer chromatography (1). Recently, Gallice et al., used sephadex gel chromatography followed by HPLC, to separate what appears to be a small peptide which contained aspartic acid, serine, glutamic acid, and alanine (7). This same combination of amino acids was also present in various proportions in twelve of our CRF fraction. At present, we do not know if we are looking at the same peptide which they isolated, or perhaps at remnants of a high molecular weight protein which has been broken at different sites. Abiko et al., have isolated a peptide with histidine, glycine, and lysine, and also a proline containing peptide (8-9). We found histidine only in our ARF pattern in fraction JJ8, and were not able to identify a proline containing peptide with our separation. We found glycine, leucine, serine, phosphoserine, glutamic acid, phenylalanine, and aspartic acid most frequently present in the hydrolysates of HPLC fractions of CRF. Tryptophan, which is destroyed by acid hydrolysis, may have been underestimated. We are presently using amino acid sequencing to fully characterize these peptides.

REFERENCES

1. J. Bergstrom and P. Furst, Uremic Toxins, pp 368-390, in: "Replacement of Renal Function by Dialysis," W. Drukker, F. M. Parsons, and J. M. Maher, Eds., Martinus Nujhoff, Boston (1983).
2. M. Shaykh, A. Dubin, G. Dunea, B. Mamdani, and S. Ahmed, Separation, isolation and amino acid composition of uremic peptides, Clin. Physiol. Biochem. 2:1-13 (1984).
3. A. Schoots, H. R. Homan, M. M. Gladdines, C. A. M. G. Cramers, R. De Smet, S. M. G. Ringoir, Screening of uv-absorbing solutes in uremic serum by reversed phase HPLC-change in blood levels in different therapies, Clin. Chim. Acta. 146:37-51 (1985).

4.  E. Burzynski, Bound amino acids in serum of patients with chronic renal insufficiency, Clin. Chim. Acta. 25:231-237 (1969).
5.  Z. Czerniak, N-Substituted amino acids in serum of patients with chronic renal insufficiency, Clin. Chim. Acta. 28:493-508 (1976).
6.  J. Menyhart and J. Grof, Many hitherto unknown peptides are principal constituents of uremic 'middle molecules', Clin. Chem. 27:1712-1716 (1981).
7.  P. Gallice, N. Fournier, A. Crevat, M. Briot, R. Frayssinet, and A. Murisasco, Separation of one uremic middle molecule fraction by high performance liquid chromatography, Kidney Int. 23:764-766 (1983).
8.  T. Abiko, M. Kumikawa, M. Ishizaki, H. Takahashi, and H. Sekino, Identification and synthesis of a tripeptide in ecum fluid of an uremic patient, Biochem. Biophys. Res. Commun. 83:357-364 (1978).
9.  T. Abiko, M. Kunikawa, H. Higuchi, and H. Sekino, Identification and synthesis of a heptapeptide in uremic fluid, Biochem. Biophys. Res. Commun. 84:184-192 (1978).

BIOCHEMICAL ELUCIDATION AND HPLC FRACTIONATION OF FLUORESCENT PEPTIDES

IN PATIENTS WITH CHRONIC RENAL FAILURE

Robert H. Williams[1], Alvin Dubin[1], Mashouf Shaykh[2],
Theodore Musiala[1], Sarosh Ahmed[2], Nadine Bazilinski[2,3],
and George Dunea[2,3]

[1]Department of Biochemistry, Rush University and Hektoen
Institute for Medical Research, Chicago, IL 60612 (USA)

[2]Division of Nephrology, Cook County Hospital and Hektoen
Institute for Medical Research, Chicago, IL 60612 (USA)

[3]University of Illinois at Chicago, College of Medicine
Chicago, IL 60612 (USA)

## ABSTRACT

We evaluated the biochemical characteristics of endogenous fluorescent substances, Ex 380 nm/Em 440 nm and Ex 400 nm/Em 460 nm, present in
sera of patients with chronic renal failure (Clin. Chem. 31:1988, 1985).
Sera from 23 patients with chronic renal failure (CRF) and from 10 normal
subjects were filtered through ultrafiltration membranes (cutoff limit of
500 Da).  Fluorescence intensity of the aforementioned substances was
significantly elevated as compared to normals ($p < 0.001$).  Fluorescence
characteristics of these substances remained unaltered after ultrafiltration and treatment with beta-glucuronidase.  Extraction of these fluorescent compounds with organic solvents (dichloromethane, ethyl acetate,
chloroform:methanol) could not be achieved after ultrafiltrates were subjected to 6N hydrochloric acid (HCl) hydrolysis.  In addition, treatment
with 6N HCl enhanced fluorescence intensity without altering fluorescence
excitation/emission maxima.  Removal of fluorescence could be accomplished
in toto by adsorption onto activated charcoal with subsequent recovery
from charcoal by treatment with sodium hydroxide, pH 12 (Ex 380 nm:
51.1%, Ex 400 nm: 91.8%).  Analysis of alkali-treated specimens by high
performance liquid chromatography demonstrated that peptides associated
with these fluorescent substances were denatured, although fluorescence
at these previously described excitation/emission maxima persisted.  Our
studies indicate that the unique fluorescence observed in the sera of
patients with CRF is not an intrinsic characteristic of a specific peptide
or its amino acids, but rather an inherent property of fluorescent molecules which may bind to these peptides.

## INTRODUCTION

Numerous fluorescent substances are present in the serum of patients
with chronic renal failure[1]; many of these compounds are in the middle
molecular weight range[2].  Various chromatographic techniques have been

used in an attempt to elucidate the biochemical nature of these substances; however, only a few compounds have been isolated[3]. Recently, we reported the presence of endogenous fluorescence in the sera of patients with chronic renal failure with characteristic excitation and emission maxima not previously described[4]. This fluorescence eluted in Sephadex G15 fractions between relative molecular mass ($M_r$) 800 to 1400 Da. Fluorescence with excitation/emission maxima of 380 nm/440 nm eluted in the vitamin $B_{12}$ marker region ($M_r \simeq 1356$ Da), and in smaller quantities, in the region of the NADPH marker ($M_r \simeq 833$ Da). Fluorescence with excitation/emission maxima of 400 nm/460 nm eluted in the vasopressin marker region ($M_r \simeq$ 1056 Da), and also in the region of NADPH, but to a lesser degree. We also demonstrated that the fluorescence remained unaltered after serum was passed through ultrafiltration membranes with molecular weight cutoffs from 10000 to 500 Da. Here, we describe some of the biochemical properties of these fluorescent substances.

MATERIALS AND METHODS

Patients. Blood samples were collected by venipuncture from 23 CRF patients just prior to hemodialysis. Blood samples were also obtained from 10 healthy volunteers. Sera were separated from clotted blood and subsequently stored at $-20^oC$ until ultrafiltration was performed.

Ultrafiltration. An aliquot of serum was filtered through an Amicon YCO5 diaflo ultrafiltration membrane (cutoff limit 500 Da), using an Amicon stirred cell (Model 8010) under 380 kPa (55 psi) of nitrogen. The ultrafiltrate was then lyophilized to dryness.

Fluorescence Emission Spectroscopy. Lyophilized ultrafiltrates were reconstituted 1:10 with distilled-deionized water. One hundred microliters of each sample was added to 3.5 ml of 0.1M potassium phosphate buffer, pH 7.7. Samples were analyzed at room temperature. Fluorescence measurements were performed as previously described[4].

High Performance Liquid Chromatography (HPLC). Analysis was performed as reported elsewhere in these proceedings[5].

Fluorescence Removal With Charcoal. One hundred microliters of ultrafiltrate was added to 3.5 ml of 0.1M potassium phosphate buffer, pH 7.7, along with 20 mg of Norit A charcoal. Each sample was mixed on a rotator for 15 minutes and subsequently centrifuged for 10 minutes at 2500 rpm. The supernatant was removed and then analyzed for fluorescence and by HPLC as previously described[4,5].

Fluorescence Recovery With Alkali. The charcoal pellet (above) was treated with 3.5 ml of 0.01M sodium hydroxide, pH 12. The suspension was then centrifuged for 10 minutes at 2500 rpm. The pH of the supernatant was adjusted to 7.0 with glacial acetic acid and subsequently lyophilized to dryness. After lyophilization, the sample was reconstituted with 3.5 ml of 0.1M potassium phosphate buffer, pH 7.7. Each sample was then analyzed for fluorescence and by HPLC as previously described[4,5].

Enzyme Hydrolysis Using Beta-Glucuronidase. Two hundred microliters of the reconstituted ultrafiltrate were hydrolyzed with 400 Units/mg of β-glucuronidase (Sigma Chem. Co.) for 24 hours at $37^oC$. Fluorescence measurements were performed as previously described[4].

Hydrochloric Acid (HCl) Hydrolysis. One hundred microliters of the lyophilized ultrafiltrate were hydrolyzed with 6N constant boiling HCl (Pierce Chem. Co.) for 24 hours at $110^oC$. Fluorescence measurements and HPLC analysis were performed as previously described[4,5]. Acid hydro-

206

lysates were also extracted with organic solvents, dichloromethane, ethyl acetate, and chloroform:methanol (2:1) at pH 7.7. The aqueous phase was subsequently analyzed for fluorescence as aforementioned.

RESULTS

Results of the fluorescence emission spectroscopy indicated a clear difference ($p < 0.001$) between fluorescence intensities in the sera ultrafiltrates (500 Da cutoff filter, YC05) obtained from 23 CRF patients on hemodialysis and from 10 normal subjects (Table 1). The fluoresence intensity measured in the sera ultrafiltrates of CRF patients was increased about 13-fold at both excitation wavelengths; however, among individual samples, the quantity of fluorescence at Ex 400 nm was always greater than at Ex 380 nm. Since glucuronide conjugates are increased in plasma ultrafiltrates from CRF patients[6], we treated YC05 ultrafiltrates with β-glucuronidase and noted the effects of enzyme hydrolysis on fluorescence intensity and emission maxima at excitation 380 nm and 400 nm. No consistent change in intensity or shift in peak emission was observed at pH 7.7 (Table 2). We also analyzed seraultrafiltrates by HPLC. A characteristic HPLC chromatogram of sera ultrafiltrates from a normal subject (A) and a patient with CRF (B) is depicted in Figure 1. Only one prominent peak was found in the normal chromatogram, however, the CRF chromatogram showed several peaks. These peaks (Figure 2A) and the characteristic fluorescence at Ex 380 nm and Ex 400 nm (Table 2) can be removed completely by prior treatment of the ultrafiltrate with activated charcoal. Even though recovery of fluorescence from activated charcoal can be achieved by alkalinization with sodium hydroxide to pH 12 (Table 3), there is virtually no change in the HPLC chromatogram (Figure 2B). Furthermore, as shown in Table 2, the maximum recovery of fluorescence at each excitation wavelength was considerably different (Ex 380 nm: 51.1%, Ex 400 nm: 91.8%). The acid hydrolysates of YC05 ultrafiltrates from sera of CRF showed no shift in peak emission at either excitation wavelength (Table 4), but the fluorescence intensity was considerably enhanced. Extraction of the acid hydrolysates by dichloromethane, ethyl acetate, and chloroform:methanol did not significantly reduce the fluorescence at pH 7.7.

DISCUSSION

Several investigators have reported an increase in endogenous fluorescence in patients with chronic renal failure. Recently, Mabuchi et al.[1] using HPLC demonstrated numerous endogenous fluorescence substances at Ex 322 nm/Em 415 nm in the sera and urine of CRF patients. Several endogenous compounds are known to have fluorescent excitation maxima in the far ultraviolet region[7]. However, the endogenous fluorescence we have hitherto demonstrated is quite unique, since the excitation/emission maxima does not occur in the highly fluorescent ultraviolet region, but rather in the near ultraviolet and visible part of the spectrum[4]. Mabuchi et al.[1] also concluded from their HPLC studies that some of the fluorescent peaks were probably peptidic substances. However, the compounds of interest do not appear to be peptides, since acid hydrolysis does not reduce fluorescence intensity or change the fluorescent properties. Furthermore, we have shown that the fluorescence intensity at Ex 380 and Ex 400 is significantly enhanced, which suggests that peptides or other substances destroyed by hydrolysis are quenching the fluorescence. Schwertner et al.[8], reported a fluorescent species with an emission maximum at 415 nm, which he contended was due to fluorescent ligands bound tightly to albumin. The fluorescent substance was highly water soluble and could be removed from serum by charcoal. These chemical properties are in accordance with our recent findings. However, the fluorescence reported here has a much higher emission maxima. Our

TABLE 1.  COMPARISON OF FLUORESCENCE IN YCO5 ULTRAFILTRATES FROM
NORMAL SUBJECTS AND PATIENTS WITH CHRONIC RENAL FAILURE (CRF)

| GROUP (n) | [a]RELATIVE FLUORESCENCE ($\mu$mol/L) | | | |
| | Ex 380 nm/Em 440 nm | | Ex 400 nm/Em 460 nm | |
| | Mean | SD | Mean | SD |
| --- | --- | --- | --- | --- |
| NORMAL (10) | 0.49 | 0.36 | 2.14 | 0.76 |
| CRF (23) | 6.15 | 1.96 | 29.42 | 11.54 |
| p-value | < 0.001 | | < 0.001 | |

[a]determined by comparison with fluorescence of 13.3 $\mu$mol/L
quinine standard.

TABLE 2.  EFFECT OF BETA-GLUCURONIDASE TREATMENT ON YCO5
ULTRAFILTRATES FROM CRF PATIENTS (n=6)

| EXCITATION WAVELENGTH 380 nm | | EXCITATION WAVELENGTH 400 nm | |
|---|---|---|---|
| [a,b] RELATIVE FLUORESCENCE μmol/L | | [a,b] RELATIVE FLUORESCENCE μmol/L | |
| YCO5 | β–GLUC | YCO5 | β–GLUC |
| 6.22 | 8.72 | 35.32 | 40.49 |
| 7.43 | 7.59 | 43.29 | 35.08 |
| 7.26 | 8.10 | 42.79 | 39.01 |
| 7.99 | 6.89 | 43.62 | 33.12 |
| 8.09 | 17.72 | 45.59 | 78.76 |
| 8.14 | 14.90 | 40.36 | 62.46 |

[a]determined by comparison with fluorescence of 13.3
μmol/L quinine standard.

[b]no change in peak emission at either excitation
wavelength was observed.

TABLE 3. ACTIVATED CHARCOAL EXTRACTION OF FLUORESCENCE IN YCO5 ULTRAFILTRATES FROM CRF PATIENTS WITH SUBSEQUENT REMOVAL BY 0.01 M SODIUM HYDROXIDE, pH 12 (n=4)

| SAMPLE NO. | MODE OF TREATMENT | EXCITATION/EMISSION WAVELENGTH | | | |
|---|---|---|---|---|---|
| | | 380/440 nm | | 400/460 nm | |
| | | [a]REL. FLUOR. $\mu$mol/L | RECOVERY (%) | [a]REL. FLUOR. $\mu$mol/L | RECOVERY (%) |
| 1 | YCO5 | 6.98 | – | 24.77 | – |
| | CHARCOAL | – | – | – | – |
| | NaOH, pH 12 | 3.84 | 55.1 | 23.93 | 96.6 |
| 2 | YCO5 | 5.93 | – | 30.23 | – |
| | CHARCOAL | – | – | – | – |
| | NaOH, pH 12 | 2.78 | 46.1 | 26.04 | 86.1 |
| 3 | YCO5 | 8.72 | – | 65.52 | – |
| | CHARCOAL | – | – | – | – |
| | NaOH, pH 12 | 4.10 | 47.0 | 58.19 | 88.8 |
| 4 | YCO5 | 8.90 | – | 69.46 | – |
| | CHARCOAL | – | – | – | – |
| | NaOH, pH 12 | 4.94 | 55.5 | 66.34 | 95.5 |
| MEAN RECOVERY (%) | | (51.1) | | (91.8) | |

[a]determined by comparison with fluorescence of 13.3 $\mu$mol/L quinine standard.

TABLE 4.   EFFECT OF ACID HYDROLYSIS AND SUBSEQUENT ORGANIC SOLVENT
EXTRACTION ON FLUORESCENCE IN YCO5 ULTRAFILTRATES FROM CRF PATIENTS

| SAMPLE NO. | MODE OF TREATMENT | [a,b]RELATIVE FLUORESCENCE μmol/L | |
|---|---|---|---|
| | | EXCITATION WAVELENGTH | |
| | | 380 nm | 400 nm |
| 1 | YCO5 | 4.35 | 28.03 |
| | 6N CH1 | 38.82 | 244.74 |
| | DICHLOROMETHANE | 38.63 | 237.60 |
| | ETHYL ACETATE | 35.37 | 216.94 |
| | CHLOROFORM:METHANOL | 36.08 | 226.33 |
| 2 | YCO5 | 4.26 | 26.35 |
| | 6N HCl | 75.40 | 523.10 |
| | DICHLOROMETHANE | 73.85 | 484.15 |
| | ETHYL ACETATE | 67.83 | 455.61 |
| | CHLOROFORM:METHANOL | 70.04 | 471.26 |
| 3 | YCO5 | 4.87 | 23.82 |
| | 6N CH1 | 118.35 | 1968.30 |
| | DICHLOROMETHANE | 112.33 | 1690.62 |
| | ETHYL ACETATE | 116.61 | 1962.78 |
| | CHLOROFORM:METHANOL | 109.96 | 1931.09 |

[a]determined by comparison with fluorescence of 13.3 μmol/L
quinine.

[b]no change in peak emission at either excitation wavelength
was observed.

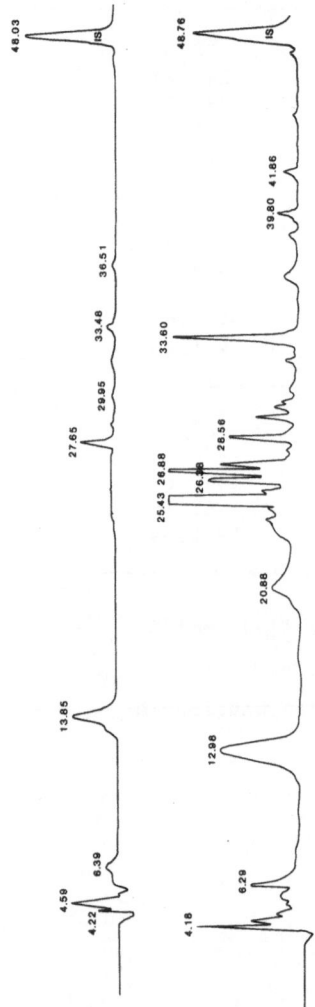

Figure 1. HPLC chromatograms of YCO5 ultra-filtrates from representing: A) normal sub-ject and B) patient with chronic renal fail-ure. Arabic numbers indicate the retention times, and (IS), the internal standard.

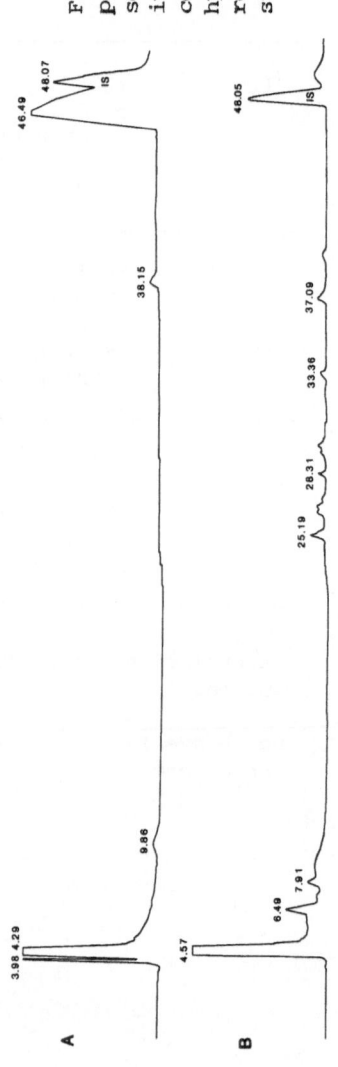

Figure 2. HPLC chromatograms of: A) su-pernatant after YCO5 ultrafiltrate from serum of CRF patient was treated with act-ivated charcoal, B) supernatant after char-coal pellet was treated with 0.01M sodium hydroxide, pH 12. Arabic numbers represent retention times, and (IS), the internal standard.

recovery studies indicate that the fluorescence we have observed in the sera of CRF patients is probably due to at least two distinct compounds. Furthermore, based on our preliminary studies using ultrafiltration membranes[4], these compounds do not appear to be bound solely to high molecular weight proteins such as albumin. Glucuronic acid conjugates have been reported to be increased in the fraction of uremic toxins often referred to as the uremic middle molecules[6]. However, our βglucuronidase studies appear inconslusive at the present time. Although no significant decrease in fluorescence intensity was observed with enzyme hydrolysis, an enhancement was noted in some cases. Therefore, these findings do not preclude the possibility that these fluorescent substances are bound to glucuronic acid. Nevertheless, these compounds are markedly elevated in patients with chronic renal failure. Consequently, we are pursuing further purification and structural identification of these endogenous fluorescent substances and the mechanism of their retention in chronic renal failure.

REFERENCES

1. H. Mabuchi, and H. Nakahashi, Liquid-chromatographic profiling of endogenous fluorescent substances in sera and urine of uremic and normal subjects, Clin. Chem. 29:675-677 (1983).
2. H. A. Schwertner, Isolation and chromatographic analysis of unidentified fluorescence in biological fluids of patients with chronic renal disease, Nephron 31:209-211 (1982).
3. J. S. Swan, E. Y. Kragten, and H. Veening, Liquid-chromatographic study of fluorescent materials in uremic fluids, Clin. Chem. 29:1082-1084 (1983).
4. M. Shaykh, N. Bazilinski, D. S. McCaul, S. Ahmed, A. Dubin, T. Musiala, and G. Dunea, Fluorescent substances in uremic and normal serum, Clin. Chem. 31:1988-1992 (1985).
5. N. Bazilinski, M. Shaykh, S. Ahmed, T. Musiala, R. H. Williams, A. Poulos, A. Dubin, and G. Dunea, Amino acid composition of uremic middle and low molecular weight retention products, Proc. Symposium Uremic Toxins (1986).
6. J. P. Monti, P. Gallice, A. Crevat, and A. Murisasco, Identification by nuclear magnetic resonance and mass spectrometry of a glucuronic acid conjugate of o-hydroxybenzoic acid in normal urine and uremic plasma, Clin. Chem. 31:1640-1643 (1985).
7. O. S. Wolfbeis and M. Leiner, Mapping of the total fluorescence of human blood serum as a new method for its characterization, Anal. Chim. Acta 167:203-215 (1985).
8. H. A. Schwertner and S. B. Hawthorne, Albumin-bound fluorescence in the serum of patients with chronic renal failure, Clin. Chem. 26:649-652 (1980).

SEPARATION FROM UREMIC BODY FLUIDS AND

NORMAL URINE OF $Na^+$, $K^+$-ATPase INHIBITOR

Philippe Gallice[+], Jean-Pierre Monti[+], Diane Braguer[+],
Claude Durand[++], Hervé Manchon[+], Antoine Murisasco[++] and
Aimé Crevat[+]


[+]Laboratoire de Biophysique, Faculté de Pharmacie,
27 Boulevard Jean Moulin, 13385 Marseille Cedex 5, France
[++]Service de Néphrologie, Hôpital Sainte-Marguerite
Marseille, France

INTRODUCTION

Previous studies [1,3] have reported that in some uremic patients
there in an abnormally high intra-erythrocytic Na concentration. Thus
sodium transport abnormalities in uremia may be demonstrated in human
erythrocytes.

This finding was classically related to a defect in $Na^+$, $K^+$-ATPase
activity. These observations indirectely led to the suggestion that an
endogenous inhibitor of the $Na^+$, $K^+$-ATPase may accumulate in the plasma
of uremic patients with chronic renal failure treated by hemodialysis [4,5].

More recently by means of [23]Na Nuclear Magnetic Resonance, with
use of aqueous shift reagent to resolve distinct signal for intra and
extracellular sodium in intact erythrocytes, we studied the intra-
erythrocytic Na in uremic patients using living red blood cells of
these patients [6]. Our results for uremic patients agreed well with
other reports, in which flame photometry was used on lysed cells, and
we showed that some uremic patients exhibits high intra-erythrocytic
Na concentrations. This is thought to be the result of an unpaired
Sodium potassium pump associated with a defect in membrane
$Na^+$, $K^+$-ATPase which may serve as a link between the pump and its energy
supply.

In the present study we try to isolate, by means of liquid chromatography techniques, the compound present in uremic body fluids and
in normal urine in the fraction or uremic toxins of so-called "uremic
middle molecules" that could be responsible for the decrease in
$Na^+$, $K^+$-ATPase activity and that accumulates in the plasma of uremic
patients with chronic renal failure treated by hemodialysis.

MATERIALS AND METHODS

## Sample preparation

Plasmatic ultrafiltrates from three chronically uremic patients (with high intra-erythrocytic Na concentration) and from urines obtained from two healthy volunteers were examined for the presence of a $Na^+$, $K^+$-ATPase inhibitor. The subjects received no medication at least 2 weeks prior to the study.

Plasmatic ultrafiltrates were obtained at the beginning of dialysis by applying negative pressure in the dialyser compartment of a capillary dialyser (Cordis Dow Corp. Miami, Florida) fitted with a cellulose acetate membrane before the dialysis fluid ran through the dialyser.

Urines were collected from healthy volunteers and filtered before use.

## Chromatographic isolation

Biological fluids were separated using a chromatographic technique previously described [7]. The main steps of this separation are : gel filtration of uremic plasma ultrafiltrates or normal urines on Sephadex G 15 yields 8 peaks numbered in order of elution. Next, peak 2 (corresponding to compounds MW 200-2000) was separated by anionic exchange into 6 fractions on DEAE A 25 and numbered 2-1 to 2-6 in order of elution. Finally each fraction was twice dessalted by gel filtration on Sephadex G 15. The fractions thus isolated were freeze-dried and stored at -20°C.

## Test of $Na^+$, $K^+$-ATPase

Dog kidney $Na^+$, $K^+$-ATPase (EC 3.6.1.3, Sigma) was incubated at 37°C for five minutes in the following medium : 100 mM Tris-Hcl, 1 mM EGTA, 3 mM ATP (vanadate free, Sigma), 20 mM $MgCl_2$, 100 mM NaCl, 20 mM KCl, pH 7.4. Various doses (0 to 3 mg/ml) of chromatographic fractions (2-1 to 2-6) from normal urines and from uremic ultrafiltrates were added in an incubating medium volume of 1 ml. Blanks were realized by parallel incubation performed in the presence of 1 mM ouabain and without chromatographic fractions. The reaction was stopped by adding cold perchloric acid (0.75 M final concentration). The resulting solutions were analysed for their inorganic phosphorus content according to Hurst's method [8]. The difference between results obtained after incubation in the presence of ouabain was considered as the $Na^+$, $K^+$-ATPase activity. The enzymic activities obtained with chromatograhic fractions were compared with those obtained with ouabain and expressed as a percentage of its inhibitory effect. Results are given as mean value of results of three experiments ± SD. Due to the accuracy of the method, an inhibitory effect higher than 10 % is considered as significant.

RESULTS

Figure 1A shows that with a final concentration of 0.6 mg/ml, fraction 2-3 and 2-6 exert an inhibitory effect on $Na^+$, $K^+$-ATPase (27 ± 5 and 23 ± 5 % respectively, mean value results of three experiments ± SD) whatever their origin (urines from healthy subjects or uremic plasma). The other fractions at the same concentration (0.6 mg/ml) showed no significant effect. Futhermore the inhibitory effect of 2-3 fraction was dose dependent (Figúre 2).

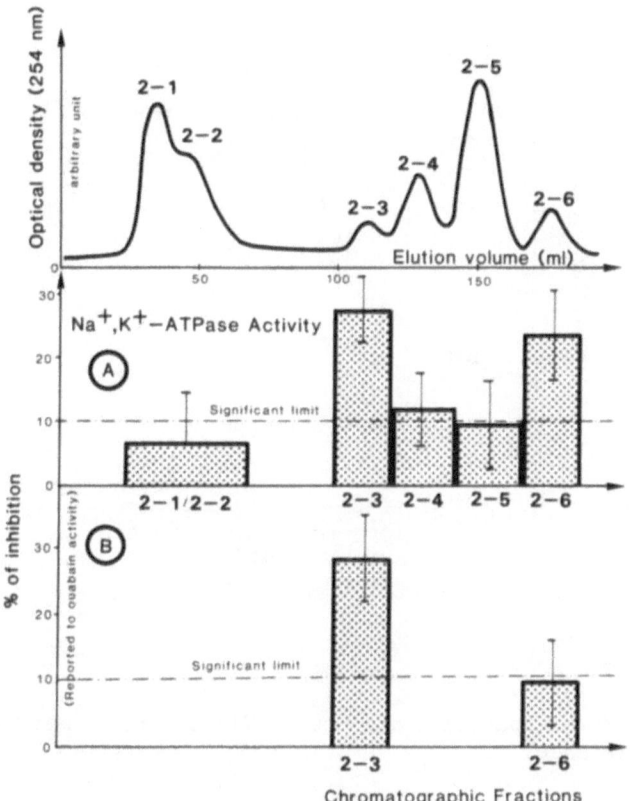

Figure 1. Upper : typical chromatogram on
Sephadex DEAE 25 column.
Lower : A) Inhibitory effect of chromatographic fractions at same concentration (0.6 mg/ml).  B) Inhibitory
effect of 2-3 and 2-6 fractions
taking into account their respective
concentration in body fluids.
Results are given as mean values of
3 experiments ± SD.

DISCUSSION

Results indicate that two chromatographic fractions inhibit
Na$^+$, K$^+$-ATPase activity. But this finding must be reported to the respective concentrations of these fractions in body fluids. Indeed, weighting
of the dried fractions after their separation demonstrates that the concentration of 2-3 fraction was about 3 fold higher than that of 2-6. So when
this difference is taken into account, only the 2-3 fraction exhibits a
significant inhibitory effect (Figure 1B).

This effect is dose dependent as illustrated in Figure 2. Dose response
curves show that the specific activity of 2-3 fraction was not the same from
one  patient to another, suggesting that the 2-3 fraction is a mixture in
which the concentration of Na$^+$, K$^+$-ATPase inhibitor varies according to
its origin. moreover in urine obtained from healthy subjects the same
 fraction exerts a similar inhibitory effect. So a possible physiologic
role for this substance may be postulated.

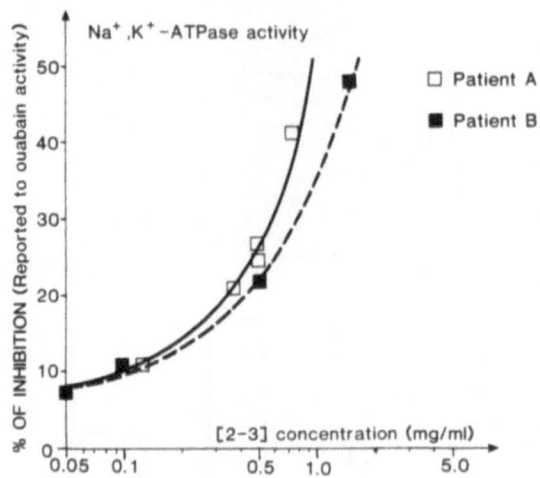

Figure 2. Dose response curves of 2-3
fractions from patient A ( □ )
and patient B ( ■ ) on
Na⁺, K⁺-ATPase activity.

ACKNOWLEDGEMENT

We thank Mr H. Bouteille and Mrs M. Vidalin for their skilfull technical assistance.

REFERENCES

1. C. H. Cole,  Decreased ouabain-sensitive adenosine triphosphate activity in the erythrocyte membrane of patients with chronic renal
   desease,  Clin. Sci. Mol. Med.  45 : 775 (1973).
2. E. Kinsey, M. Smith and L.G. Welt,  The red blood cell as a model for
   the study of uremic toxins,  Arch. Intern. Med. 126 : 827 (1970).
3. L.G. Welt, J.R. Sachs and T.J. Mc Manus, An ion transport defect in
   erythrocytes from uremic patients,  Trans. Assoc. Am. Physic.,
   77 : 169 (1964).
4. H.J. Kramer, J. Pennig, D. Klingmüller, J. Kipnowski, K. Glänzer and
   R. Düsing,  Digoxin-Like Immunoreacting Substances(s) in the serum
   of patients with chronic Uremia , Nephron. 40 : 297 (1985).
5. R. Dzurik, B. Lichardus, V. Spustova, J. Ponec, A. Gerykova, and
   P. Bakos,  A procedure for the isolation of an inhibitor of sodium
   transport from the urine of ureamic patients . Physiol. Bohemoslov.
   31 : 573 (1982).
6. J.P. Monti, P. Gallice, A. Crevat, M. El Mehdi, C. Durand and
   A. Murisasco, Intra-erythrocytic Sodium in Uremic Patients, as determined by "High resolution" ²³Na Nuclear Magnetic Resonance ,
   Clin. Chem. 32 : 104 (1986).
7. P. Gallice, N. Fournier, A. Crevat, M. Briot, R. Frayssinet and
   A. Murisasco,  Separation of one uremic middle molecules fraction by
   High Performance Liquid Chromatography,  Kidney Int. 23 : 764 (1983)
8. R.O. Hurst,  The determination of nucleotide phosphorus with a Stannous
   chloride-Hydrazine sulfate reagent, Canad. J. Biochem. 42 : 287 (1964).

# ISOLATION OF SIALYLCOMPOUNDS FROM HEMOFILTRATE OF CHRONIC UREMIC PATIENTS AND IDENTIFICATION BY NUCLEAR MAGNETIC RESONANCE

Gerhard Weißhaar[1], Helmut Brunner[2], Horst Friebolin[1],
Wolfgang Baumann[1], Helmut Mann[2], Hans-Jesef Opferkuch[3],
and Heinz-Günther Sieberth[2]

[1] Organisch-Chem. Institut der Unversität Heidelberg
Im Neuenheimer Feld 270, D-6900 Heidelberg, F.R.G.
[2] Abteilung Innere Medizin II der R.W.T.H. Aachen
Pauwelsstr., D-5100 Aachen, F.R.G.
[3] Zentrale Arbeitsgruppe Spektroskopie, Deutsches
Krebsforschungszentrum, Im Neuenheimer Feld 280
D-6900 Heidelberg, F.R.G.

## INTRODUCTION

Sialylcompounds in the molecular weight range between 500 and 2000 D
(sialyloligosaccharides, -glycopeptides) are structurally related to the
carbohydrate components of serum- and tissue glycoconjugates, which are
involved in numerous biological processes[1]. Many of them occur in cow
colostrum, human milk and urine; structures and amount of excreted compounds
vary under different physiological and pathological conditions. In serum
they are usually not detectable. Since they are proposed to originate from
biosynthesis and degradation of sialoglycoconjugates, their identification
in uremic body fluids may contribute to a better knowledge of their
metabolism.

## MATERIAL AND METHODS

### Preparative Fractionation of Hemofiltrate

Hemofiltrates of patients undergoing regular dialysis treatment were
desalted and concentrated by modified reverse osmosis (membrane DDS-930,
nominal cut-off 500 D)[2]. For a representative preparative isolation eight
hemofiltrates (156 l) of the same patient were pooled, desalted, lyophilized
and fractionated by gel chromatography (Sephadex G-15 and Biogel P-6;
0.05 mol/l $NH_4HCO_3$), ion-exchange chromatography (SP-Sephadex C-25;
0.01 mol/l $CH_3COONa$; DEAE-Sephacel, 0.01-1 mol/l $NH_4HCO_3$). High-performance
liquid chromatography was achieved with Nucleosil 5 $NH_2$ (65% $CH_3CN$/0.005
mol/l $NH_4HCO_3$).

### Analytical Methods

Sialic acid was quantitatively determined according to Warren[3] after
mild acid hydrolysis (0.05 mol/l $H_2SO_4$, 1 h, 80 °C). 250, 300 and 500 MHz
one- and twodimensional NMR spectra were recorded on Bruker spectrometers
WM 250, WH 300 and AM 500. All measurements were run at 22 °C in $D_2O$ with
trimethylsilyl-Na-propionate-$d_4$ as internal standard ($\delta = 0$).

Proton decoupled [13]C-NMR spectra were recorded in 10 mm tubes at 30 °C on a Bruker WM 250 spectrometer operating at 62.89 MHz in $D_2O$ with methanol as internal standard ($\delta$ = 49.7 relative to TMS).

## RESULTS AND DISCUSSION

The crude freeze-dried material, obtained after reverse osmosis of the hemofiltrate, was separated on Sephadex G-15 into 18-20 fractions with sialylcompounds detectable in the first six fractions. Fraction 6 ($V_e/V_o$ between 1.4 and 1.5) was subjected to a further gel filtration on Biogel P-6 and three sialic acid containing fractions were obtained (see Fig.1). Fractions 6.4, 6.5 and 6.6, indicated in Fig.1, were submitted to cation--exchange chromatography on SP-Sephadex C-25 and, in each case, the first eluted fractions were collected (6.4.1, 6.5.1 and 6.6.1). After desalting on Sephadex G-15 (water as eluent), anion-exchange chromatography using DEAE-Sephacel was performed:

Fraction 6.6.1 yielded one major fraction (6.6.1.2), eluted under starting conditions, which was identified as a mixture of the two isomers of sialyl-lactose, NeuAcα2-3Galβ1-4Glc and NeuAcα2-6Galβ1-4Glc (compounds 1 and 2) on the basis of their known [1]H-NMR spectra[4].

Fraction 6.5.1 yielded two sialic acid containing fractions (6.5.1.2 and 6.5.1.5). Fraction 6.5.1.2 was subfractionated into two fractions using Biogel P-6. While the minor fraction consisted of compounds 1 and 2, as in 6.6.1.2, [1]H- and [13]C-NMR spectra of the major fraction showed an equimolar mixture of two monosialyloligosaccharides. These were separated from each other by high-performance liquid chromatography on Nucleosil 5 $NH_2$, analysed by [1]H- and [13]C-NMR spectroscopy and identified as NeuAcα2-3Galβ1-4GlcNAc (compound 3) and NeuAcα2-6Galβ1-4GlcNAc (compound 4). Characteristic [1]H--resonances of NeuAc (H-3e, H-3a, NAc), Gal (H-1, H-3), GlcNAc (H-1α, H-1ß, NAc) and almost all [13]C-resonances of both compounds were assigned on the basis of model compounds.

Fraction 6.5.1.5 was subfractionated on Biogel P-6; the [1]H-NMR spectrum of the major fraction was identical with an enzymatic hydrolysation product of polysialic acid ɟNeuAcα2-8ɟn, which has been analysed by twodimensional relayed H,H-COSY-NMR spectroscopy[5] and identified as NeuAcα2-8NeuAc (compound 5).

By rechromatography of fraction 6.4.1 using anion-exchange chromatography followed by Biogel P-6, the pure compound 6 was isolated; its 500 MHz [1]H-NMR spectrum is shown in Fig.2. Characteristic proton resonances of NeuAc (H-3e, H-3a, NAc) indicate two different bound sialic acids. The anomeric region ($\delta$ = 4.5-5.0) shows two other sugar components; one of them

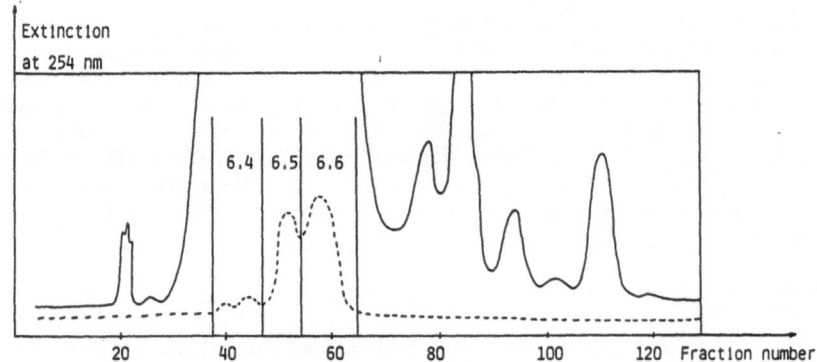

Fig.1 Gel filtration of fraction 6 (from Sephadex G-15)
using Biogel P-6/200-400 in 0.05 mol/l $NH_4HCO_3$,
column:16x111 mm, flow rate:12 ml/h, 15 min/fract.
----- sialic acid (according to Warren-assay)

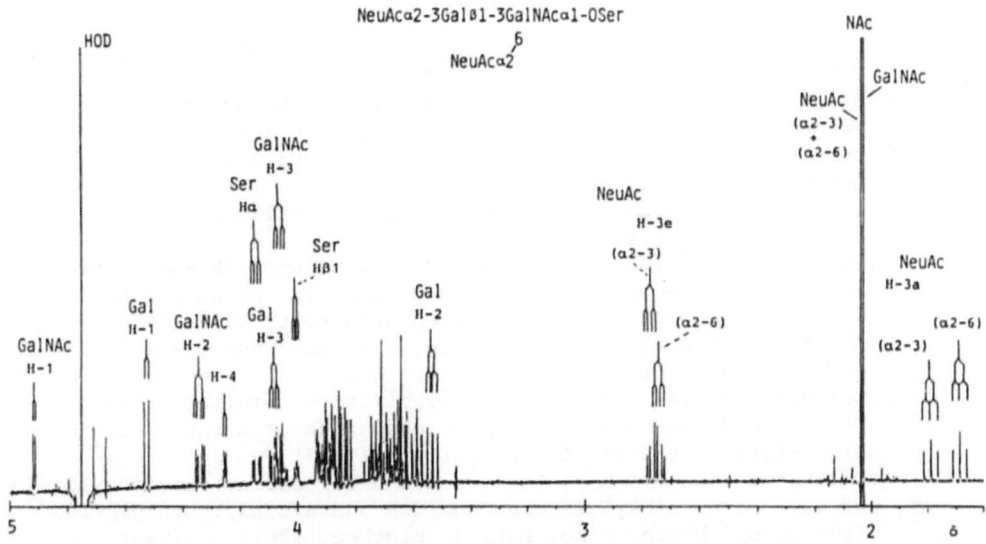

Fig.2 Resolution enhanced 500 MHz $^1$H-NMR spectrum of compound 6, dissolved in $D_2O$, recorded at 22 °C

is N-acetylated (see NAc-region δ = 2.0-2.1). The identification of Gal, GalNAc and Ser is based on the twodimensional relayed H,H-COSY-NMR spectrum. The assignments are given in Fig.2.

High resolution $^1$H-NMR spectroscopy has proved to be a powerful tool for structure elucidation of pure compounds as well as to check the purity of fractions obtained during different separation steps.

Structures, molecular weights and approximate yields of the isolated compounds are listed in Tab.1. In the mean time we found the same compounds in the hemofiltrates of different other patients[6].

While compounds 1,2,3,4 and 6 were also found in the urine of healthy subjects[7], and compounds 1 and 3 in peritoneal dialysis fluids[8], the occurence of Di-N-acetyl-D-neuraminic acid in biological material has not yet been described. Whether this compound is specific for the uremic state has to be proved in further investigations.

The reason for the accumulation of these sialylcompounds in uremic body fluids and their possible clinical relevance still have to be elucidated.

Tab.1 Structures of isolated sialylcompounds

| Compound | Structure | MW [D] | Approximate yield[a] [mg] |
|---|---|---|---|
| 1 | NeuAcα2-3Galß1-4Glc | 633 | 25.0 |
| 2 | NeuAcα2-6Galß1-4Glc | 633 | 2.5 |
| 3 | NeuAcα2-3Galß1-4GlcNAc | 674 | 10.0 |
| 4 | NeuAcα2-6Galß1-4GlcNAc | 674 | 10.0 |
| 5 | NeuAcα2-8NeuAc | 600 | 2.0 |
| 6 | NeuAcα2-3Galß1-3GalNAcα1-OSer<br>6<br>/<br>NeuAcα2 | 1052 | 2.5 |

[a] from 156 l hemofiltrate

221

# REFERENCES

1. R.Schauer (Ed.), Sialic acids, chemistry, metabolism and function; Springer Verlag, Heidelberg (1982).
2. H.Brunner, H.Mann, U.Essers, R.Schultheis, T.Byrne and R.Heintz, Preparative isolation of middle molecular weight fractions from the hemofiltrate of patients with chronic uremia, Artif. Org. 2:375 (1978).
3. L.Warren, The thiobarbituric acid assay of sialic acids, J. Biol. Chem. 234:1971 (1959).
4. U.Dabrowski, H.Friebolin, R.Brossmer and M.Supp, [1]H-NMR studies at N-Acetyl-D-neuraminic acid ketosides for the determination of the anomeric configuration, Tetrahedron Lett. 4637 (1979).
5. B.Kwiatkowski, R.Stirm, H.Friebolin, W.Baumann and H.-J. Opferkuch, publication in preparation.
6. G.Weißhaar, Isolierung und Strukturaufklärung sialinsäurehaltiger Verbindungen aus dem Hämofiltrat chronisch urämischer Patienten Dissertation, Universität Heidelberg (1987).
7. J.Parkkinnen and J.Finne, Isolation and structural characterization of five major sialyloligosaccharides and a sialylglycopeptide from normal human urine, Eur. J. Biochem. 136:355 (1983).
8. G.Le Moel, G.Strecker, S.Troupel, M.Dolegeal, C.Jacobs, A.Galli and J.Agneray, Carbohydrate content of middle molecular weight substances (MMWS) in uremic patients: preliminary results, Artif. Org. 4 (Suppl.):37 (1980).

# IDENTIFICATION OF TWO UREMIC TOXINS BY NUCLEAR

# MAGNETIC RESONANCE AND MASS SPECTROMETRY

Jean-Pierre Monti[+], Philippe Gallice[+], Diane Braguer[+],
Claude Durand [++], Antoine Murisasco[++] and Aimé Crevat[+]

[+]Laboratoire de Biophysique, Faculté de Pharmacie
27 Boulevard Jean Moulin, 13385 Marseille Cedex 5, France
[++]Service de Néphrologie, Hôpital de Sainte-Marguerite
Marseille, France

## INTRODUCTION

A few uremic toxins (UT) previously called uremic middle molecules were isolated in pure form and identified [1-5]. In a previous study, by using a three step chromatographic separation we had isolated ten uremic toxins and had identified two of them [6,7]. The first is a double conjugate of glucuronidate-o-hydroxyhippuric acid identical to Fürst's compound [1]. The second is a glucuronic acid conjugate of o-hydroxybenzoic acid. In the present paper, we report the identification by $^1$H nuclear magnetic resonance and mass spectrometry of two new uremic toxins represented by peaks 2-5-2 and 2-5-5.

## MATERIALS AND METHODS

### Isolation

UT were obtained from urine of healthy subjects, free of drugs, and from plasmatic ultrafiltrate of uremic patients. The UT subfractions were separated as follows [8,9]. The first chomatographic step is a gel filtration of plasmatic ultrafiltrate or urine on Sephadex G15 with Tris HCl 0.01 M buffer (pH 8.6) as an eluant. So is obtained crude middle molecules (fraction 2). The second step is an anion exchange chromatography of the fraction 2 on DEAE Sephadex A 25. The sample is eluted with a solution of Tris HCl 0.01 M buffer (pH 8.6) containing a sodium chloride concentration increasing from 0 to 0.15 M. The fraction 2 is so separated into seven components : 2-1 through 2-7. 2-5 fraction is then desalted by gel permeation chromatography using bidistilled water as an eluant. Finally this fraction is divided into 10 molecules (2-5-1 through 2-5-10) by means of "high performance" liquid chromatography. This last step is performed by using a reversed phase ion pairing procedure involving tetrabutylammonium phosphate (TBAP) as a counterion. Before use the counterion is removed from the molecule so separated by passing the solution through DEAE Sephadex A 25.

## Identification

<u>¹H NMR spectroscopy</u>. ¹H NMR spectra were recorded at 200.13 MHz in the pulse Fourier transform mode with a Bruker AM 200 spectrometer (Service Interuniversitaire de RMN, Faculté de Pharmacie, Marseille, France). Experimental conditions were the same as those specified previously [6], however the number of scans was 500-1000 because of the small quantity of product. Proton resonance assignments were performed by spectrum integration, homonuclear double irradiation and comparison with free compounds.

<u>Mass spectrometry</u>. The spectra were recorded with a NERMAG-RIO-10C gas chromatograph-mass spectrometer (Centre de Recherche Clin-Midy, Montpellier, France). The experimental conditions were : direct insertion probe ; electron energy, 70 eV ; ionization current, 100 µA ; ion-source temperature, 150°C ; reagent gas, ammonia ($10^{-1}$ Torr) for positive chemical ionization and methane ($5.10^{-2}$ Torr) for negative chemical ionization by electron capture.

## RESULTS AND DISCUSSION

Figure 1 shows the ¹H NMR spectrum of the 2-5-5 compound. For the aliphatic region we note the chemical shifts and the coupling features characteristic of a glucuronic acid [6,7] and an ethyl group. For the aromatic region we note a NMR spectrum of a dihydrobenzoic acid with three possibilities for the hydroxy substitution : 2-4, 2-5 or 3-4, corresponding to the coupling features observed.

The presence of a glucuronic acid, which is a bulky substituant, modifies the chemical shift of aromatic protons. A β glucuronidase enzymic hydrolysis releases the dihydrobenzoic acid and we obtain the characteristic spectrum of a 3-4 dihydrobenzoic acid. The $CH_2$ chemical shift of ethyl group is characteristic of an O-ester. Consequently, the

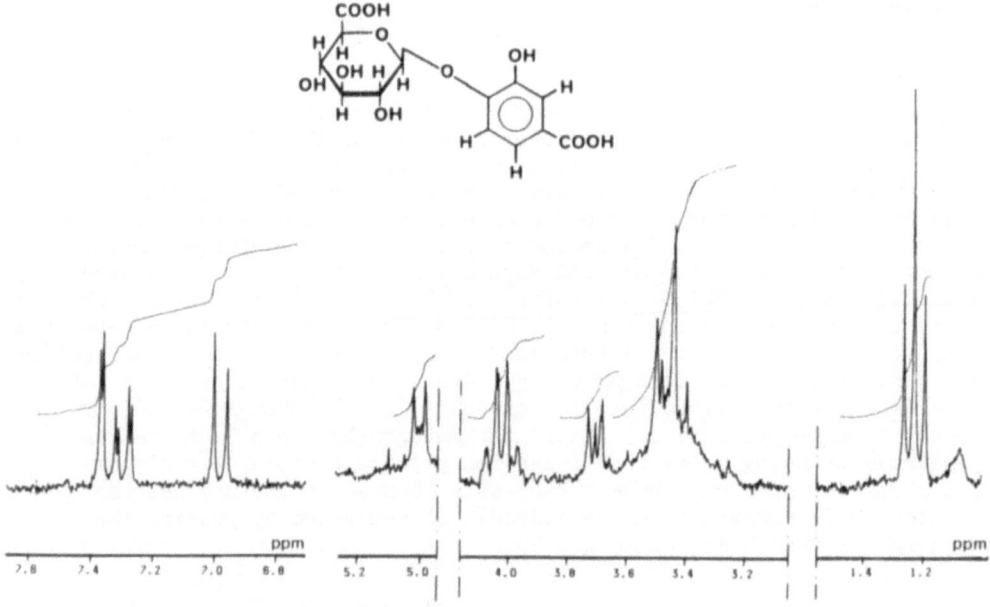

Figure 1. ¹H NMR spectrum of the 2-5-5 compound.

glucuronic substituant is linked via one of two hydroxyl groups.

By using the substituant effect of glucuronic acid on aromatic protons in the 2-5-10 molecule [6], it is possible to calculate the theoretical chemical shifts expected for the aromatic region in the 2-5-5 compound ; $H_1$ = 7.54 and 7.25 ppm, $H_2$ = 6.75 and 7.04 ppm, $H_3$ = 7.35 and 7.18 ppm for substitution on hydroxyl 3 and on hydroxyl 4 respectively. Comparison with the experimental chemical shifts : $H_1$ = 7.37 ppm, $H_2$ = 6.98 ppm and $H_3$ = 7.30 ppm indicates that the glucuronic acid is linked via the hydroxyl 4. Thus the 2-5-5 compound is a glucuronic acid conjugate of 3-4 dihydroxyethylbenzoate. This structure is confirmed by the mass spectrometry study. By using positive chemical ionization, we obtain the ion $|M + H|^+$ (m/z 359), $M^+$ (m/z 358), the glucuronide ions (m/z 194 and 177). and the aglycone ion (m/z 164). Negative chemical ionization by electron capture reveals molecular ions $M^-$ (m/z 358) and $|M-H|^-$ (m/z 357).

But this compound is not derived from a classical metabolic pathway[10]. So doubts remain about this ethyl substituant. This esterification may be due to the separation procedure ; indeed the second chromatography step is

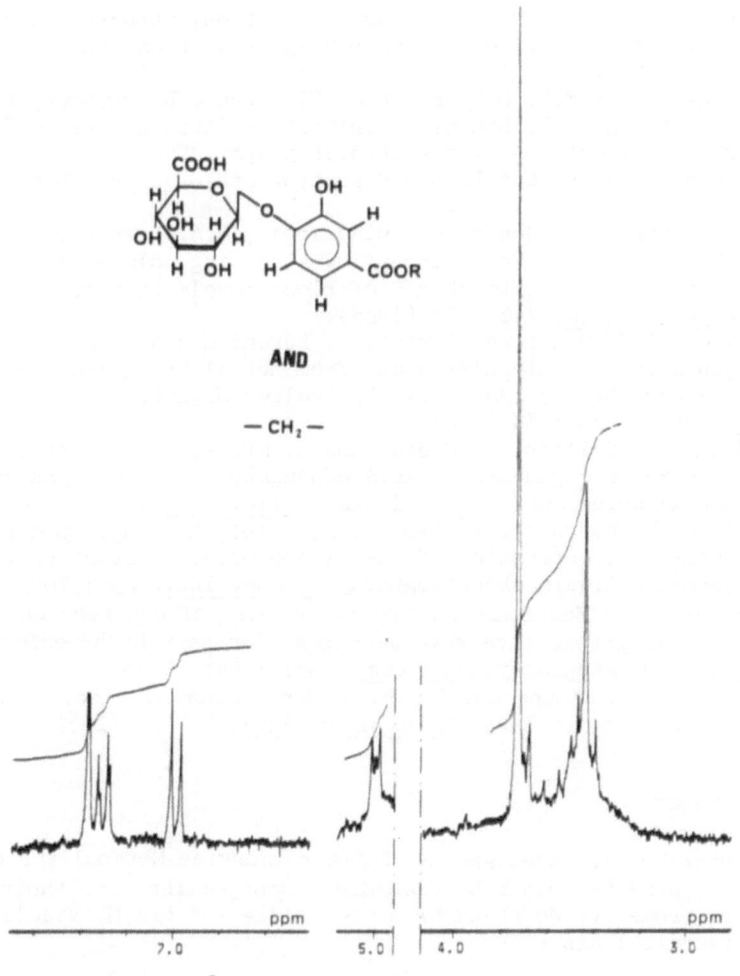

Figure 2. $^1$H NMR spectrum for 2-5-2 compound.

carried out with Sephadex DEAE. So the structure of 2-5-5 molecule is probably a glucuronic acid conjugate in position 4 of a 3-4 dihydroxybenzoic acid (Figure 1).

Figure 2 shows the [1]H NMR spectrum for 2-5-2 compound. In the aromatic region are found the same coupling features as in the 2-5-5 compound. In the aliphatic region we note the presence of a glucuronic acid and that of a singlet, corresponding to a $CH_2$ group. The chemical shift of this methylene group is identical to that of glycine as in the 2-5-10 compound [6]. However, amino-acid analysis does not show the presence of glycine. The results obtained by mass spectrometry do not allow a complete identification of the 2-5-2 structure. We propose a partial structure which is shown in Figure 2. Further experiments are being carried out to identify this structure.

REFERENCES

1. L. Zimmerman, P. Fürst, J. Bergström and H. Jörnval, A new Glycine-containing compound with a blocked amino group from uremic body fluids, Clin. Nephrol., 14 : 107 (1980).
2. L. Zimmerman, H. Jörnvall, J. Bergström, P. Fürst and J. Sjöval, Characterization of a double conjugate in uremic body fluids, FEBS Lett., 129 : 237 (1981).
3. T. Abiko, I. Onodera and H. Sekino, Isolation, structure and biological activity of the Trp containing pentapeptide from uremic fluid, BBRC., 89 : 813 (1979).
4. Ad. C. Schoots, F.E.P. Mikkers, H.A. Claessens, R. De Smet, N. Van Landschoot and M.E. Ringoir, Characterization of uremic "Middle Molecular" fractions by gaz Chromatography, Mass Spectrometry, Isotachophoresis, and Liquid Chromatography, Clin. Chem., 28 :45 (1982).
5 Ad. C. Schoots, H.R. Homan, M.M. Gladdines , C.A. Cramers, R. De Smet, and M.E. Ringoir, Screening of UV-Absorbing solutes in uremic serum by reversed phase HPLC-change of blood levels in different therapies, Clin. Chim. Acta, 146 : 37 (1985).
6. P. Gallice, J.P. Monti, A. Crevat, C. Durand and A. Murisasco, A compound from uremic plasma and from normal urine isolated by liquid chromatography and identified by Nuclear Magnetic Resonance, Clin. Chem., 31 : 30 (1985).
7. J.P. Monti, P. Gallice, A. Crevat and A. Murisasco, Identification by NMR and MS of a glucuronic acid conjugate of o-hydroxybenzoic acid in normal urine and uremic plasma, Clin. Chem., 31 : 1640 (1985).
8. P. Gallice, N. Fournier, A. Crevat, M. Briot, R. Frayssinet and A. Murisasco, Separation of one middle molecule fraction by high performance liquid chromatography. Kidney Int., 23 : 764 (1983).
9. P. Gallice, J.P. Monti and A. Crevat, Removal of counter-ion reagent after semi-preparative reversed-phase ion-pair high-performance liquid chromatography, J. Chrom., 331 : 445 (1985).
10. J.M. Hicks, D.S. Young and I.D.P. Wooton, Abnormal blood constituents in acute renal failure, Clin. Chim. Acta, 7: 623 (1962).

ACKNOWLEDGEMENT

We thank Drs Y. Sales and M. Berther (Service Metabolisme et Pharmaco-cinétique -Centre de Recherche Clin-Midy, Montpellier) for their assistance in Mass spectrometry. We thank Mr H. Bouteille and Mrs M. Vidalin for their skillful technical assistance.

# KINETICS OF SOLUTE REMOVAL IN HEMODIALYSIS

Frank A. Gotch

Franklin Hospital
Hemodialysis Treatment and Research Center
San Francisco, California, USA

The basic premise of dialysis therapy for uremia is that uremic patho-physiology is dependent on abnormal solute levels in body water which can be normalized by dialysis. Consequently, the clinician is concerned with predicting changes in concentration and body content of solutes as a function of the amount of dialysis prescribed. This can best be achieved through kinetic analysis.

The first kinetic description of the patient-dialyzer system was provided by Wolf's classic paper in 1951.[1] His modeling assumptions were: body solute distribution in a single compartment equal to total body water (V,L); first order diffusive transport across the dialyzer at rate defined by dialysance, D, L/min, into a dialysate bath volume (Vd,L); and conservation of mass or constant total solute content of V + Vd. He considered V and Vd to be constant, thus he neglected fluid transfer from V to Vd in the system and considered but did not include a metabolic source of solute generation such as urea generation during dialysis. Two simultaneous rate equations describing mass balance in the patient-dialyzer system result from these formulations:

$$(1) \qquad V \frac{dC}{dt} = -D(C-Cd)$$

$$(2) \qquad Vd \frac{dCd}{dt} = D(C-Cd)$$

where C is blood concentration and t is time.

Solution of Eqs (1) and (2) for the boundary conditions t=0, C=C(o), resulted in:

$$(3) \qquad C(t) = \left[ \frac{V}{V+VD} \right] \left[ 1 + \frac{Vd}{V} e^{-\frac{D(V+Vd)t}{V \cdot Vd}} \right] C(o)$$

Equation (3) provided the first quantitative description of the change in blood concentration and body solute content over time as a function of V, Vd and D, which can be considered to define the dose of dialysis prescribed with a batch dialysate system. It can also be noted that Eq (3) could be written as

$$(4) \qquad V \frac{dC}{dt} = - KC$$

where K is the actual dialyzer clearance. Solution of this expression results in

$$(5) \qquad C(t) = Coe^{-\frac{Kt}{V}}$$

Wolf pointed out that K is not constant but is constantly decreasing over time in a batch system due to increasing CD. However, Eq (5) can be rearranged to find the effective or average clearance ($\overline{Ke}$) with a batch system if C(o), C(t), t and V are known in accordance with

$$(6) \qquad \overline{Ke} = \frac{1}{t} \ln\left(\frac{Co}{Ct}\right) V$$

Kinetic modeling and principles of mass balance for analysis of hemodialysis data were clearly described thirty-six years ago by Wolf. Although kinetic considerations and mass balance analysis can be used to rigorously define the dose of dialysis it has not been so easy to demonstrate the utility of kinetic modeling to guide dialysis therapy of uremia. In large part this is because the relative importance of the various retained metabolites in the pathogenesis of the uremic syndrome continues to be poorly defined. Several models of therapy have been devised in attempts to address this perplexing problem.

## The Square-Meter-Hour and Dialysis Index Models

In the early years of chronic dialysis, Scribner observed that prevention of neuropathy appeared to be more dependent on treatment time than BUN levels and that peritoneal dialysis seemed to provide better control of neuropathy but lower levels of small solute clearance than hemodialysis. These observations led Scribner to hypothesize that larger molecular weight toxins poorly cleared by conventional cellulosic membranes were responsible for uremic neuropathy.[2] These theoretical uremic neurotoxins were considered to be middle molecules (MM) and removal rates were permeability limited as compared to small molecules which were significantly blood and/or dialysate flow limited over the clinical blood flow operational range in dialyzers.

The square-meter-hour kinetic model[3] was developed as a result of these clinical observations and was based on the following assumptions: (1) dialyzer MM clearance (KdMM) was nearly equal to the dialyzer permeability-area product, KoA, and permeability (Ko) was constant for the membranes in clinical use and, therefore, dialyzer MM clearance (KdMM) was directly proportional to A; (2) KdMM was quite low so the blood level decreased slowly and the blood to dialysate diffusion gradient remained high throughout dialysis; (3) endogenous MM toxins could be realistically modeled from the transport characteristics of a marker solute, Vitamin B12.

It followed from the above assumptions that total MM removal (JMM) with two therapies employing different membrane areas (A1, A2) and treatment times (t1, t2) could be described in accordance with

$$(7) \qquad JMM1 = \overline{C1} \cdot Ko \cdot A1 \cdot t1$$

$$(8) \qquad JMM1 = \overline{C2} \cdot Ko \cdot A2 \cdot t2$$

Combining Equations (7) and (8) and solution for t2 when JMM1 = JMM2 and $\overline{C1} = \overline{C2}$ results in

$$(9) \qquad t2 = \frac{A1}{A2}\, t1$$

Clinical implementation of this model was based on the further assumption that 30 hours of dialysis weekly with the 0.7 m$^2$ Kiil dialyzer had historically provided fully adequate JMM. Substitution of these values in Eq (9) for A1 and t1 results in

$$(10) \qquad t2 = \frac{21}{A2}$$

Equation (10) indicates that 21 square-meter-hours of treatment were required for adequate MM concentration control and that the major parameter determining dialysis time was membrane area. Recognition that Ko could vary significantly in commercially available membrane-support systems led to a further refinement of this model based on in vitro dialyzer Vitamin B12 clearance (KdB12) values[4] and the contribution of residual renal function[5]. The fully developed MM model was called the dialysis index[6] and again the target for optimal treatment was based on historical experience with the Kiil dialyzer. The dialysis index (DI) was initially defined only for MM[6] but subsequently generalized to include DI values for theoretical solutes of 1000 daltons and creatinine.[7]

## The Kjellstrand Number (Nkj)

Kjellstrand and colleagues proposed a different approach to quantification of dialysis[8] based on Eq (5) above. Equation (5) when rearranged gives

$$(11) \qquad \frac{C(t)}{Co} = e^{-\frac{Kt}{V}}$$

Kjellstrand approached the relationships in Eq (11) empirically and showed that C(t)/Co for BUN, creatinine and uric acid was exponentially related to a parameter which he termed the Kjellstrand number and was defined as the product of dialyzer clearance, ml/min, and treatment time, hr, divided by body weight, Kg. Thus, the parameter Nkj had units of ml · hr/min/Kg but was an analogue of the dimensionless exponent, Kt/V, since it consisted of the product of clearance and treatment time divided by body weight and V was expected to be proportional to body weight.

The Kjellstrand number was proposed as a universal parameter to quantitatively express the amount of dialysis therapy prescribed to patients varying greatly in weight and treated with dialyzers with widely varying clearances. Kjellstrand was particularly impressed with the clinical need for such a parameter because he treated very small pediatric patients as well as adult patients. Although the Nkj parameter could not be used mathematically to directly model and prescribe dialysis therapy because of its units, it was a rational representation of the amount of dialysis prescribed. The analogue of Nkj, Kt/V, has subsequently been shown to be a good predictor of adequate dialysis as will be discussed below.

## The Urea Model

As is the case for the MM models described above, the urea model also was conceived as a result of clinical observations. We had the opportunity to compare the clinical outcome with conventional therapy to that with a five fold increase in KdB12 and were dismayed to find no significant difference in measured middle molecule concentration and no change in the clinical manifestations of the uremic syndrome.[9] Further, with the same dialyzer urea clearance and treatment time, and similar patient body weight and dietary protein prescription, we observed a two fold difference in predialysis BUN values. These observations led us to conclude that KdB12 was not controlling on uremic toxicity and that there appeared to be a two fold variation in net urea nitrogen generation and protein intake relative to body size in patients with similar diet prescriptions. In view of the dependence of clinical uremia in part on the rate of protein catabolism, it appeared reasonable to postulate that the basic dialysis requirement is proportional to the protein catabolic rate (pcr) relative to the lean body mass. Since our MM studies over a five fold range of clearance were negative, we concluded that the amount of dialysis required for adequate therapy was a function of low molecular weight solute clearances. Thus, the conceptual basis of the urea model was that low molecular weight uremic toxin generation was a function of pcr and urea could be used to quantify pcr and serve as a generic marker solute to quantify low molecular weight toxic solute removal by dialysis.

The urea kinetic model was formulated so that urea generation rate and urea distribution volume (equivalent to total body water) could be measured from readily obtained clinical data and thus the net protein catabolic rate determined for individual patients.[10] The original model assumed a constant volume[11] but was subsequently refined to the more rigorous variable volume model.[12,13]

Similarly to the MM models, the original implementation of the urea model was based on the historical Kiil data which was interpreted to indicate that adequate thrice weekly dialysis resulted in a midweek predialysis BUN value of 80 mg/dL. Therefore, a midweek BUN of 80 mg/dL was targeted with measured pcr = 1.1 gm/Kg using the urea model. In clinical use of this model, the dietary protein intake (dpi, gm/Kg/day), considered equal to pcr in steady state dialysis therapy, was prescribed to be 1.1 ± 0.3 gm/Kg/day and midweek BUN was targeted at 70 mg/dL for pcr = 0.8 and at 90 mg/dL at pcr = 1.4 gm/Kg/day. A sizeable body of clinical data was accumulated using this model and showed good overall clinical outcome[14,15] with a reduction in mean treatment time from 6.0 to 3.8 hours thrice weekly.

These clinical results indicated that the urea model could be used to reduce treatment time but they did not shed light on the minimum adequate level of treatment since the total urea clearance provided was not reduced substantially below safe levels demonstrated historically. The treatment was individualized to body size (V) and pcr and made more efficient. The U.S. National Cooperative Dialysis Study (NCDS) was designed to compare clinical outcome with conventional levels of therapy to that with markedly reduced therapy.[16] This study was guided by the variable volume urea model and clinical outcome analyzed both statistically[17] and mechanistically[16] with somewhat different conclusions. The mechanistic analysis will be primarily considered here and a detailed comparison to the statistical analysis can be found elsewhere.[16]

Analysis of the urea model shows that with thrice weekly dialysis therapy the midweek predialysis BUN will increase linearly with pcr if the level of treatment expressed as urea Kt/V is held constant.[16] These relationships are depicted in Fig 1 where the linear dependence of BUN on pcr is shown for Kt/V values ranging from 0.3 to 1.5 which encompass the range of therapy included in the NCDS. Analysis of clinical outcome in the NCDS as a dependent variable of Kt/V showed high incidence of morbidity for all Kt/V values less than 0.8. There was a sharp, step function decrease in morbidity in the range of Kt/V = 0.8 to 0.9 and morbidity remained low and constant as Kt/V increased further to 1.5, the upper bound on therapy in the study. Thus, it was concluded that Kt/V < 0.8 was clearly inadequate therapy, 0.8 < Kt/V < 0.9 was a region of transition from high to low morbidity and Kt/V ≥ 1.0 was required to assure adequate therapy for all patients with pcr ≤ 1.1 gm/Kg/day. The dialysis requirement for pcr > 1.1 was not well defined due to the design of the NCDS and the adequate dialysis line in Fig 1 for this region simply represents the upper bound on the study since morbidity was low and constant in this region. Recent reports suggest the adequate domain may extend into the triangle with a question mark.[18,19]

The NCDS results show that the dialysis requirement is not proportional to pcr, as originally postulated, at low levels of intake. The minimum level of adequate dialysis requires a Kt/V = 1.0 even at very low dpi and suggests there is generation of low molecular weight toxins independent of pcr which must be removed by dialysis. To the extent that the modeling line for adequate dialysis is correct with pcr > 1.1, it would indicate that Kt/V must increase in proportion with pcr and would suggest that uremic toxicity is pcr dependent in this region.

The NCDS data also demonstrated that the BUN alone is useless as a criterion to define an adequate dialysis prescription. An adequate level of dialysis is predicted with BUN ranging from 30 to 95 mg/dL depending on the pcr. Clinical assessment of adequacy of the combined nutritional (protein intake) and dialytic components of therapy requires knowledge of pcr, Kt/V urea and BUN.

We have prescribed therapy with the urea model for over 10 years now and have used the NCDS results to target therapy the past 4 years. All patients are modeled and undergo kinetic evaluation at monthly intervals. The means of the last four monthly measurements of BUN, pcr and Kt/V for 66 patients on thrice weekly dialysis are depicted in Fig 2. The therapy coordinates of these patients closely follow the NCDS adequate dialysis modeling line and none are distributed in the domain of inadequate dialysis defined by the study. The dashed lines represent ± 10% confidence limits on prescribed Kt/V and it is apparent that over 95% of the patients fall within ± 10% of the targeted therapy.

## Kinetic Modeling in Clinical Practice

Kinetic modeling has not been widely applied in clinical practice up to the present time. However, a case can be made that modeling has had a substantial impact on clinical practice. The middle molecule and urea models are both based on the fundamental law of conservation of mass and the concept that uremic toxicity is solute concentration dependent. The models require that the total amount of solute generated between dialyses (GT) must equal the total flux or removal of solute (JT) during dialysis, and, for comparable treatment, mean solute concentration ($\overline{C}$) must remain constant. These fundamental relationships can be simplistically expressed by the equation

$$(12) \qquad GT = JT = \overline{C}(K)(t)$$

where K is dialyzer clearance and t is treatment time. When GT and $\overline{C}$ remain constant it follows that for two different treatments K1 · t1 and K2 · t2, the relationship of t2 to t1 is

$$(13) \qquad t2 = t1\left(\frac{K1}{K2}\right)$$

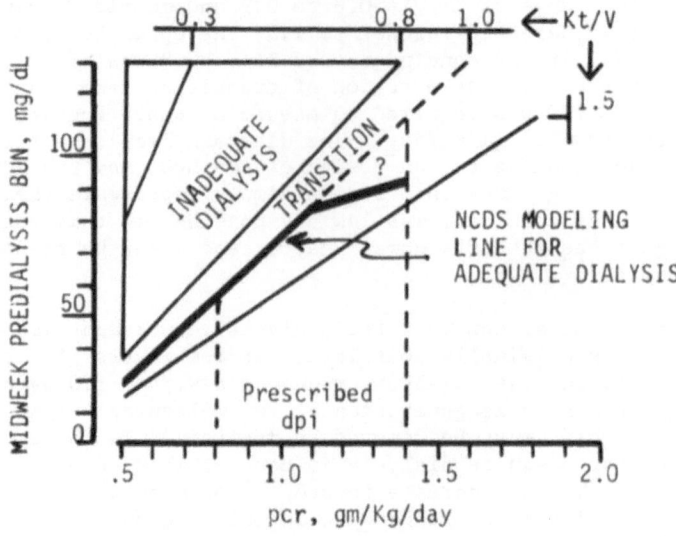

Figure 1.   Clinical Outcome in the NCDS.

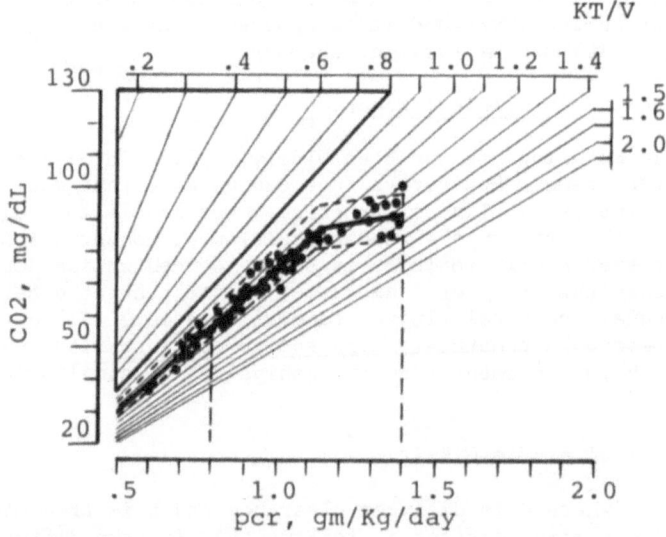

MEAN VALUES OVER 4 MONTHS OF KINETICALLY
PRESCRIBED THERAPY FOR 67 PATIENTS

Figure 2.   Mean Values Over 4 Months of Kinetically
Prescribed Therapy for 67 Patients.

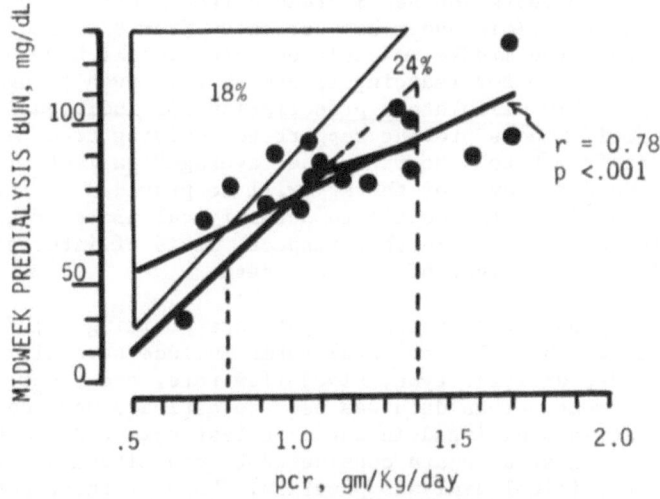

Figure 3.  A Small Random Sample of Current
Clinical Dialysis Prescription.

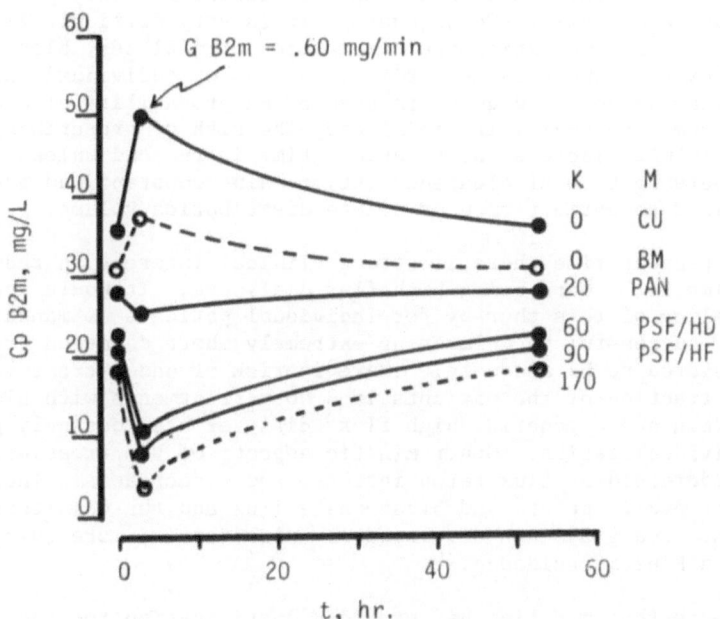

PLASMA B2m PROFILES CALCULATED FROM
TWO POOL VARIABLE VOLUME MODEL

Figure 4.  Calculated B-2 Microglobulin
Profiles.

Over the past 20 years KdU has increased from 80 to 160 ml/min and KdB12 from 20 to 40 ml/min and t has decreased from 8 to 3 or 4 hours. As shown by Eq (13), the middle molecule and urea models have both provided reassuring justification for reducing treatment time even though they are rarely used to actually calculate a prescription for individual patients. Although the kinetic models provide support for halving treatment time over the past 20 years from 8 to 4 hours in the "average" patient, it does not follow that an adequate level of therapy will be provided to individual patients without applying the models to actually calculate prescriptions with reduced treatment time. In this respect, it is of interest to examine clinical practice in the light of the NCDS data.

We dialyze a substantial number of patients visiting San Francisco during the summer months. The referral forms include several monthly predialysis BUN values, dialyzer type, blood flow rate, treatment time and frequency of treatment. This data was used to estimate pcr and Kt/V urea on 17 visiting patients dialyzed in our unit last summer with results plotted in Fig 3. These data are considered to constitute a small random sample of current clinical dialysis practice. None of these patients had model guided prescription calculations and the treatment time ranged from 3 to 5 hours with a mean of 3.7 hours. Inspection of the data points in Fig 3 shows 18% of patients with $Kt/V < 0.8$ and 24% with $0.8 < Kt/V < 1.0$ so that 42% had Kt/V less than the minimum concluded to be an adequate level in the NCDS. These data illustrate the difficulties with empiric prescription of the dialysis dose. The 7 patients with $Kt/V < 1.0$ all have clinically acceptable BUN levels ranging from 70 to 93 and 6 of the 7 have acceptable pcr levels of 0.8 to 1.1 gm/Kg/day. The model generated NCDS data, however, indicate that these patients are at significantly increased risk for uremic morbidity.

The uremic syndrome is complex, multifactorial and only partly responsive to adequate dialysis. Consequently, it is very difficult to determine from clinical observation the combination of dialyzer, blood flow rate, dialysate flow rate and treatment time required for individual patients to provide therapy which is adequate to assure the probability of dialysis responsive uremic morbidity is minimized. The risk of prescribing inadequate dialysis will increase as treatment time is reduced unless care is taken to assure that total clearance (Kt) remains constant and adequate when normalized to patient size or solute distribution volume.

At the present time there is strong clinical interest in reducing treatment time to 2 hours using high flux dialyzers. It would appear that kinetic modeling of this therapy for individual patients is mandatory if we are to be successful in delivering extremely short duration treatments which are assured to be adequate. There is risk of under treatment in a substantial fraction of the patients if 2 hour treatments with blood flow rate 400 ml/min and a generic "high flux" dialyzer are routinely prescribed without individualization. Other kinetic aspects of the treatment may also have to be addressed as flux rates increase and t decreases. The rates of Na and H20 removal, acetate and bicarbonate flux and the magnitude of transcellular urea gradients developed may also require more quantitative attention and kinetic guidance.[20]

It appears that modeling has provided justification for and resulted in substantial reduction in treatment time over the past 20 years but is rarely used to assure a defined, adequate amount of treatment for individual patients. Consequently, it is probable that for some fraction of patients the dialysis prescription may be inadequate to assure the probability of dialysis responsive uremic morbidity is minimized. The probability of prescribing an inadequate dose of dialysis is likely to increase if treatment time is shortened to 2 hours without quantification and kinetic model guidance of therapy.

## Beta-2 Microglobulin (B2M) Kinetics

The above discussion of the dialysis prescription has been concerned with the kinetics of removal of low and middle molecular weight toxins. Recently evidence has been obtained indicating that B2M (molecular weight 12,400 daltons) may be an important toxin in dialysis patients. It has been shown that amyloid accumulating in synovial and osseous tissue is commonly comprised of B2M[21,22] and has been suggested that this substance may be responsible for the high incidence of carpal tunnel syndrome and cystic bone disease in long term dialysis patients. Consequently there is a strong need to define the removal kinetics of this material in order to determine whether it may be possible with dialysis therapy to ameliorate toxic effects resulting from its accumulation.

Although numerous measurements of blood levels in dialysis have been reported, to date there is only one kinetic study of this material reported.[23] This single kinetic study indicates that B2M appears to be distributed in two compartments with volumes approximating plasma and interstitial fluid. The generation rate was calculated from this study and found to average 0.12 mg/min and appears to range from 0.05 to 0.20 mg/min. When renal function goes to zero, it appears that B2M can be removed by extra renal mechanisms but at a very slow rate with whole body clearance $\sim$ 3.5 ml/min.

These kinetic parameters were used in an attempt to tentatively analyze reported B2M blood levels in dialysis patients and to tentatively predict what changes might occur with increased dialyzer clearances of B2M. The following B2M kinetic model was formulated: two pool distribution in plasma ($V_p$) and interstitial fluid ($V_i$) with initial volumes of 3.0 and 11.5 L respectively. Total fluid removal was 2.5 L, distributed over $V_i$ and $V_p$ and treatment time was 180 min. The capillary mass transfer coefficient was estimated to be 0.2 L/min.[24] B2M generation rate (G) was taken to be 0.12 mg/min[23] and a thrice weekly treatment schedule assumed. Renal clearance was assumed to be zero and extra renal clearance 3.5 ml/min.

The results of this analysis are shown in Fig 4 where several B2M plasma concentration profiles are shown. The top profile has been reported for cuprophane membranes[25] which have zero clearance of B2M. A striking rise during dialysis is typically seen which would require a large increase in generation rate to 0.6 mg/min based on the kinetic model formulated above.

The second curve would be predicted for a dialyzer with zero clearance but no increase in G during dialysis.

The third curve was calculated for a dialyzer with K = 20 ml/min and no increased G during dialysis. This curve approximates data reported for polyacrylonitrile (PAN) dialyzers quite well.[26]

The lower three curves were calculated for dialyzers with B2M clearances ranging from 60 to 170 ml/min and no increase in G during dialysis. These curves reasonably well approximate some of predialysis concentration reported with polysulfone membrane (PSF) dialyzers.[25]

The time averaged concentration (TAC) of B2M in the top curve in Fig 4 would be 44 mg/L while in the bottom curves it would be $\sim$ 15 mg/L. Thus the kinetic model would suggest that it might be possible to reduce TAC B2M by 66% with a biocompatible open membrane dialyzer.

The curves in Fig 4 must be considered highly theoretical at this time. They do, however, point up the need for kinetic studies of this material in order to gain understanding of the removal kinetics of B2M and develop a data base which could be used to quantify the dose of dialysis with respect to control of B2M accumulation in dialysis patients. Long term clinical correlations would also be required to assess the clinical value of enhanced removal of B2M.

## Conclusion

In conclusion it must be noted that the kinetic models reviewed here must be viewed as very primitive kinetic formulations with respect to the molecular etiology of uremia. There is still very limited understanding of the relationships between uremic abnormalities and specific solute concentrations. The urea model does provide for quantification of protein intake and hence serves as a guide to the rate of generation of the protein catabolites hydrogen ion, inorganic phosphorous and urea. However, its validity for guidance of therapy rests on empiric correlations[16] between the probability of uremic morbidity and the level of therapy prescribed. It does not shed light on the basic molecular abnormalities resulting in the uremic syndrome. Our understanding of the patient-dialyzer system kinetics is still far from complete.

## References

1.  A. V. Wolf, D. G. Remp, J. E. Kiley, and G. D. Currie, Artificial kidney function: Kinetics of hemodialysis, J Clin Invest 30:1062-1070 (1951).
2.  B. H. Scribner, Discussion, Trans Amer Soc Artif Intern Organs 11:29 (1965).
3.  A. L. Babb, R. P. Popovitch, T. G. Christopher, and B. H. Scribner, The genesis of the square-meter hour hypotehsis, Trans Amer Soc Artif Intern Organs 17:81-91 (1971).
4.  A. L. Babb, P. C. Farrell, D. A. Uvelli, and B. H. Scribner, Hemodialyzer evaluation by examination of solute molecular spectra, Trans Amer Soc Artif Intern Organs 18:98-106 (1972).
5.  A. L. Babb, P. C. Farrell, M. J. Strand, D. A. Uvelli, J. Multinovic, and B. H. Scribner, Residual renal function and chronic hemodialysis therapy, Proc Clin Dialysis & Transplant Forum 2:142-149 (1972).
6.  A. L. Babb, M. J. Strand, J. Multinovic, and B. H. Scribner, Quantitative description of dialysis treatment: A dialysis index, Kid Int 7:S23-S30 (1975).
7.  A. L. Babb, M. J. Strand, D. A. Uvelli, and B. H. Scribner, The dialysis index: A practical guide to dialysis treatment, Dialysis & Transplantation 6:9-13 (1977).
8.  C. M. Kjellstrand, R. L. Evans, R. J. Petersen, L. W. Rust, J. Shideman, T. J. Buselmeier, and L. T. Rozelle, Consideration of the middle molecule hypothesis, Proc Clin Dialysis & Transplant Forum 2:127-142 (1972).
9.  F. A. Gotch, J. A. Sargent, and J. H. Peters, Studies on the molecular etiology of uremia, Kid Int 7:S276-280 (1975).
10. M. F. Borah, P. Y. Schoenfeld, F. A. Gotch, J. A. Sargent, M. Wolfson, and M. H. Humphreys, Nitrogen balance during intermittent dialysis therapy of uremia, Kidney Int 14:491-500 (1978).

11. F. A. Gotch, J. A. Sargent, M. L. Keen, M. Seid, and R. Foster, Comparative treatment time with Kiil, Gambro and Cordis-Dow kidneys, Proc Clin Dialysis & Transplant Forum 3:217-229 (1973).

12. J. A. Sargent, and F. A. Gotch, Principles and biophysics of dialysis, in: "Replacement of Renal Function by Dialysis", W. Drukker, F. Parsons, J. Maher, eds., The Hague: Martinus Nijhoff, 2nd edition, 53-96 (1983).

13. P. C. Farrell, and F. A. Gotch, Dialysis therapy guided by kinetic modeling: Application of a variable-volume single-pool model of urea kinetics, Trans Second Australian Conference on heat and mass transfer, 29-37 (1977).

14. F. A. Gotch, J. A. Sargent, and M. Keen, Clinical results of intermittent dialysis therapy guided by ongoing kinetic analysis of urea metabolism, Trans Amer Soc Artif Intern Organs 22:175-189 (1976).

15. F. A. Gotch, A quantitative evaluation of small and middle molecule toxicity in therapy of uremia, Dialysis & Transplantation 9:183-192 (1980).

16. F. A. Gotch, and J. A. Sargent, A mechanistic analysis of the National Cooperative Dialysis Study (NCDS), Kidney Int 28:526-534 (1985).

17. N. M. Laird, C. S. Berkey, and E. G. Lowrie, Modeling success or failure of dialysis therapy: The National Cooperative Dialysis Study, Kidney Int 23:S101-S107 (1983).

18. J. I. Shapiro, W. P. Argy, T. A. Rakowski, A. Chester, A. S. Siemsen, and G. E. Schreiner, The unsuitability of BUN as a criterion for prescription dialysis, Trans Amer Soc Artif Intern Organs 29:129-135 (1983).

19. F. A. Gotch, Discussion, Trans Amer Soc Artif Intern Organs 29:133-134 (1983).

20. F. A. Gotch, Dialysis of the future, Kidney Int (in press).

21. F. Gejyo, S. Odani, T. Yamada, N. Honma, H. Saito, Y. Suzuki, Y Nakagawa, H. Kobayashi, Y. Maruyama, Y. Hirasawa, M. Suzuki, and M. Arakawa, Beta-2 microglobulin: A new form of amyloid protein associated with chronic hemodialysis, Kidney Int 30:385-390 (1986).

22. A. Z. Genves, M. Emmett, M. G. White, G.Greenway, and D. B. Michaels, Carpal tunnel syndrome with cystic bone lesions secondary to amyloidosis in chronic hemodialysis patients, Amer Jour Kidney Dis 7:130-134 (1986).

23. C. Vincent, N. Pozet, and J. P. Revillard, Plasma Beta-2 microglobulin turnover in renal insufficiency, Acta Clinica Belgica 35:S10 2-13 (1980).

24. E. M. Landis, and J. R. Pappenheimer, Section II, Circulation, in: "Handbook of Physiology", American Physiology Society, 2:1008

25. J. Bommer, H. P. Seelig, R. Seelig, and E. Ritz, Beta-2 microglobulin levels in hemodialyzed patients, Abstract, EDTA (1986).

26. J. M. Vandenbroucke, C. van Ypersele de Strihou, Relationship between membrane characteristics and dialysis induced changes in Beta-2 microglobulin levels, Abstract, EDTA (1986).

AN ADVANCED, USER-FRIENDLY MICROCOMPUTER

PROGRAM FOR HEMODIALYSIS KINETICS

Tom Buur

Department of Nephrology
University Hospital of Linkoping
Linkoping, Sweden

## INTRODUCTION

This project has been undertaken as a study in medical informatics. The aim is to develop a multipurpose program for hemodialysis kinetics with the following characteristics: 1. Can be used on widespread computer systems. 2. Is easy to use 3. Can treat the general, unsimplified case, if desired 4. Can be customized for individual requirements.

At some stage of program development potential users should be consulted in order to optimize user-friendliness and investigate their actual requirements.[1] To initiate such a dialogue preliminary results are presented here.

## METHODS

To achieve a more general solution most equations were developed de novo by solving mass balance differential equations. Where necessary various iterative procedures were constructed to solve the most general case possible. Simplified equations/algorithms were then treated as special cases of their general counterparts. PASCAL was used as the programming language and the programming method was that of modular structuring and so-called "top-down design".[2] Guidelines for user-friendliness were followed.[1]

## RESULTS

### Program description

A 4500-line program has been produced. The compiled version will run on an IBM PC or highly compatible computer with a minimum of 256 K RAM.

In the present version, which is limited to single pool kinetics, a total of 71 genuine equations/algorithms have been implemented for the calculation of 21 different variables of interest to the nephrologist. Well-known equations from major texts on hemodialysis kinetics are incorporated,[3,4] but several algorithms represent new developments.

The computations performable cover a range from the simplest calculation of dialyzer clearance, to a complex algorithm rendering unequally

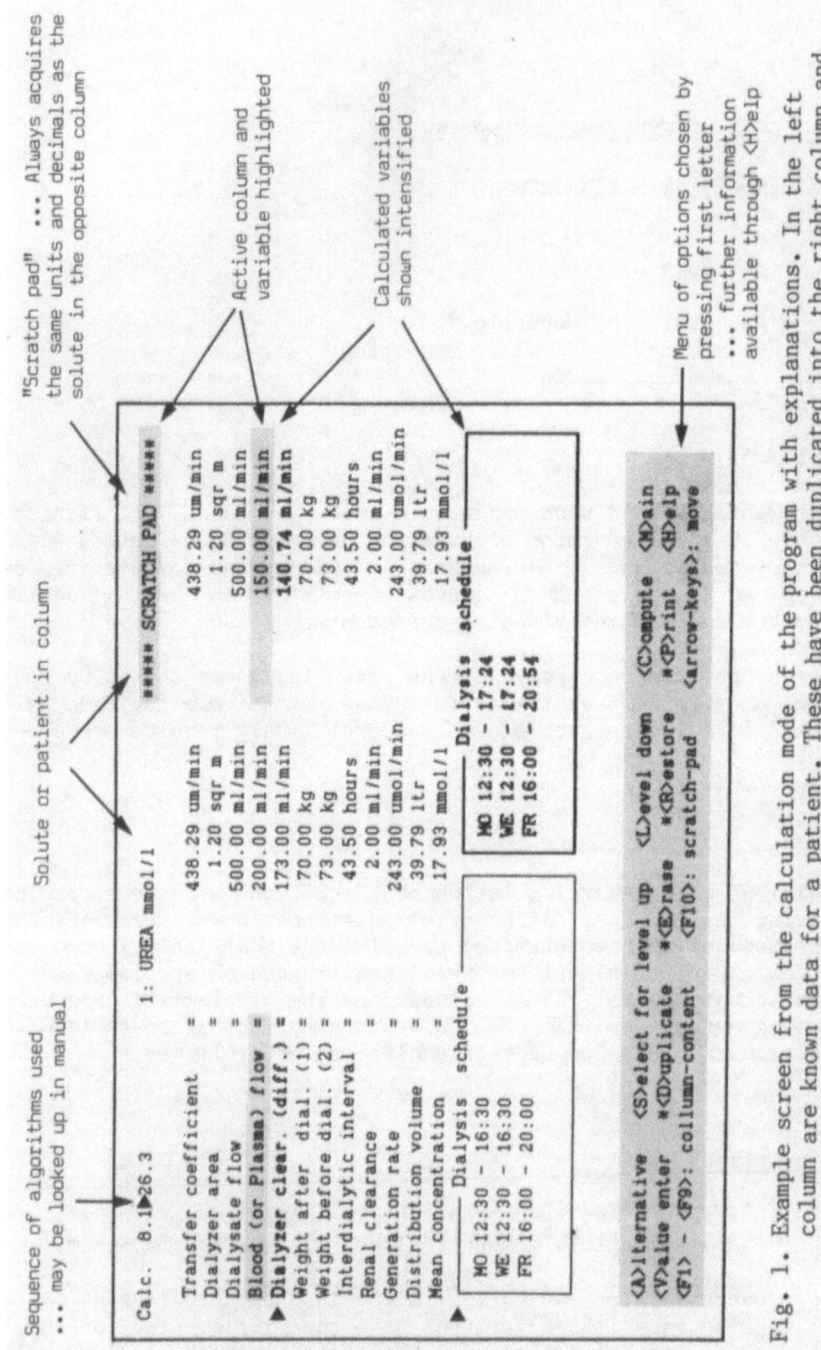

Sequence of algorithms used ... may be looked up in manual

Solute or patient in column

"Scratch pad" ... Always acquires the same units and decimals as the solute in the opposite column

Active column and variable highlighted

Calculated variables shown intensified

Menu of options chosen by pressing first letter ... further information available through <H>elp

```
Calc.  8.1▶26.3          1: UREA mmol/l          ***** SCRATCH PAD *****

 Transfer coefficient       =      438.29 um/min         438.29 um/min
 Dialyzer area              =        1.20 sqr m            1.20 sqr m
 Dialysate flow             =      500.00 ml/min         500.00 ml/min
 Blood (or Plasma) flow     =      200.00 ml/min         150.00 ml/min
▶Dialyzer clear. (diff.)    =      173.00 ml/min         140.74 ml/min
 Weight after dial. (1)     =       70.00 kg              70.00 kg
 Weight before dial. (2)    =       73.00 kg              73.00 kg
 Interdialytic interval     =       43.50 hours           43.50 hours
 Renal clearance            =        2.00 ml/min           2.00 ml/min
 Generation rate            =      243.00 umol/min       243.00 umol/min
 Distribution volume        =       39.79 ltr             39.79 ltr
 Mean concentration         =       17.93 mmol/l          17.93 mmol/l
     ─── Dialysis  schedule            Dialysis  schedule
 MO 12:30 - 16:30                 MO 12:30 - 17:24
 WE 12:30 - 16:30                 WE 12:30 - 17:24
 FR 16:00 - 20:00                 FR 16:00 - 20:54


 <A>lternative    <S>elect for level up   <L>evel down   <C>ompute   <M>ain
 <V>alue enter    *<D>uplicate  *<E>rase  *<R>estore     *<P>rint    <H>elp
 <F1> - <F9>: collumn-content    <F10>: scratch-pad      <arrow-keys>: move
```

Fig. 1. Example screen from the calculation mode of the program with explanations. In the left column are known data for a patient. These have been duplicated into the right column and the consequences of a fall in blood flow are investigated. Algorithm 8.1 calculates the resulting fall in dialyzer clearance, which in turn is used by algorithm 26.3 to render the dialysis schedule, necessary to maintain mean concentration constant.

spaced dialysis schedules from time averaged solute concentrations, considering weight-gain, charged solute and the presence of solute in dialysate.

Up to 4 algorithms may be linked, so that  the results from one are used for calculation in the next. At all times the user is clearly shown what is calculated, from which parameters, and how. Editing functions make it simple to compare the results from changing one or more variables (Fig. 1.). During the linking process the program will ascertain that none of the parameters required for the algorithm being added is also calculated by a succeeding algorithm.

Some of the more commonly used combinations of algorithms, such as those used for urea kinetics[5,6], may be accessed as a single package calculating all relevant variables.

To further add to the flexibility of the program the value for a variable may be entered, not only as the mere value, but also as a mathematical expression containing transcendental functions such as logarithms and exponentials.

The program can be run with either 9 different solutes or 9 different patients simultaneously in memory for comparison purposes. But further data may be saved to, or loaded from diskette. Also, the user has the option of making a version of the program with solute names, units and decimals according to his own choice.

User-friendliness has been achieved by extensive error-checking and the use of menus and help-screens in a manner similar to that used in commercial programs. A manual has been written with an appendix describing the algorithms, and a tutorial teaching the use of the program in a few hours.

## Speed of performance

The complex algorithms for calculating a dialysis schedule are executed in 15 - 20 seconds on the slowest IBM compatible computers. All other computations, including  fairly complex algorithms for time averaged concentration at steady state, are executed in less than 2 seconds.


## DISCUSSION

### Technical considerations

From a programming point of view modular structuring has proved very useful. When adding to this the principle of generalizing the solutions as far as possible, the actual implementation of equations and algorithms has become considerably facilitated.

Extensions are easy to create, since the modules work independently of each other (Fig. 2). The generalizing approach has the advantage of reducing program length, but the drawback is, that some algorithms, assuming simplifications, are executed at a slower speed, since they are treated as a special case of the general solution.

### Areas of application

The program can be used for a variety of applications, but is especially suitable for answering "what-if" questions (Fig. 1). The present version of the program assumes the user has a basic understanding of hemodialysis kinetics, to correctly exploit the ability to link algorithms. However, it demands little knowledge about computers, explaining even simple matters

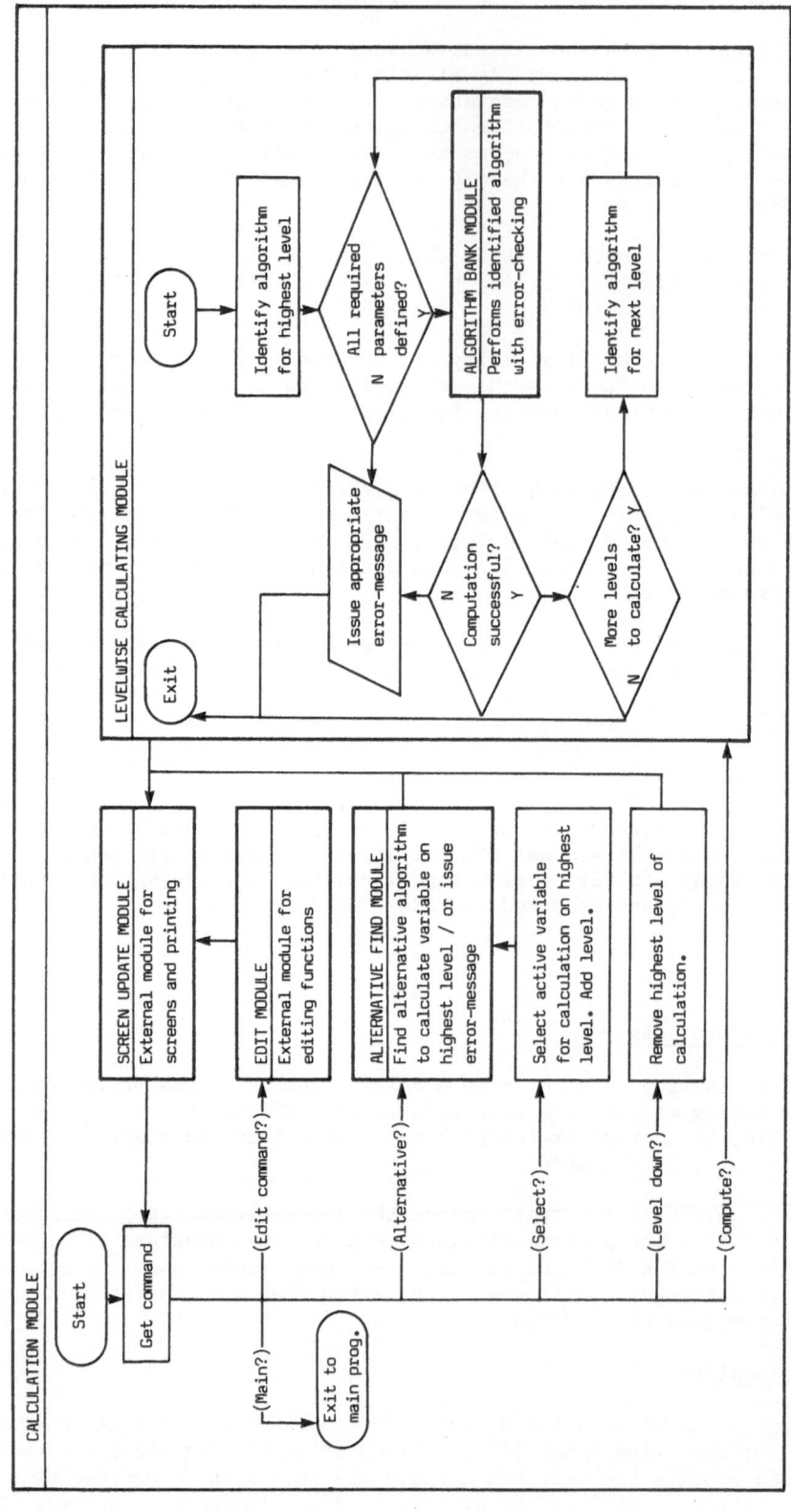

Fig. 2. Illustration of the modular structure of the program. The calculation module, shown here, is itself nested on a third level in other modules. The figure summarizes the actions taken by the program, when the user chooses an option from the menu seen in Fig. 1. Notice the algorithm bank module. New algorithms can be added here without any changes to the rest of the program, except for the addition of an identification code.

242

with the tutorial and help screens. Lesser skill in the kinetics field does not preclude the use of the packaged algorithms for urea kinetics. Also, the program may be used for theoretical experiments, while studying the literature on hemodialysis kinetics.

For large scale studies, where only a few well-defined problems have to be solved and results are merged with other data for statistical analysis etc.,[6,7] it may be more practical to have a computer program written for that specific purpose. A program of the present type could, however, be used for pilot-studies and prototyping of models.

## Future developments

Future expansions that would be possible are the incorporation of modules for two-pool kinetics, sodium balance simulation, heparin kinetics etc. The program could be interfaced with an expert system to help the user find ways to solve a particular problem, thus minimizing the requirements on his skills in the kinetics field.

Already the program is quite powerful, and before making further changes, potential users should be able to try it first and express their opinion. This would optimize user-friendliness and facilitate the construction of a true multipurpose tool for hemodialysis kinetics, much in analogy with the program packages available for statistical analysis.

## AVAILABILITY

From the author's viewpoint the project is undertaken as a study in medical informatics. The participants are provided with the program and updates free of charge, except for the materials used, but in return are expected to answer eventual questionnaires concerning their use of the program and kinetic models in general.

Send three diskettes and self-adhesive label with name and address to Dr. Tom Buur, Dept. of Nephrology, University Hospital of Linkoping, S-581 85 Linkoping, Sweden.

## ACKNOWLEDGEMENT

The author expresses his appreciation to Flemming Hermansen, M.D., systems programmer, for much advice during the development of the program.

## REFERENCES

1.  A. T. Johnson, User friendliness in microcomputer programs, Comp. Prog. Biomed. 19:127 (1985).
2.  G. M. Schneider, S. C. Bruell, "Advanced programming and problem solving with PASCAL," John Wiley & Sons, New York (1981).
3.  C. K. Colton, E. G. Lowrie, Hemodialysis: Physical principles and technical considerations, in: "The Kidney (2nd ed.)," B. M. Brenner, F. C. Rector, ed., Saunders, Philadelphia (1980).
4.  J. A. Sargent, F. A. Gotch, Principles and biophysics of dialysis, in: "Replacement of renal function by dialysis (2nd ed.)," W. Drukker, F. M. Parsons, J. F. Maher, ed., Martinus Nijhoff Publishers, Dordrecht (1983).
5.  E. G. Lowrie, B. P. Teehan, Principles of prescribing dialysis therapy: Implementing recommendations from the National Cooperative Dialysis Study, Kidney Int. 23:S113 (1983).

6.  J. A. Sargent, Control of dialysis by a single-pool urea model: The National Cooperative Dialysis Study, <u>Kidney</u> <u>Int.</u> 23:S19 (1983).
7.  F. A. Gotch, J. A. Sargent, A mechanistic analysis of the National Cooperative Dialysis Study (NCDS), <u>Kidney</u> <u>Int.</u> 28:526 (1985).

IS SERUM UREA A GOOD INDEX FOR PREDICTING MORBIDITY IN HEMODIALYSIS (HD)?

R. Marcén, S. Lamas, C. Quereda, L. Orofino, J.L. Teruel,
R. Matesanz, and J. Ortuño

Department of Nephrology. Hospital Ramón y Cajal
Madrid, Spain

INTRODUCTION

Blood urea nitrogen (BUN) has been considered a useful guide in the prescription and assessment of diet and dialysis therapy [1] since a direct relationship has been shown between the former and hospitalization rate [2,3]. On the other hand, a decline in morbidity and mortality has been observed with the increase in BUN and protein intake [4]. The aim of the present work was to study the possible link between the clinical situation and the predialysis serum urea.

PATIENTS AND METHODS

A total of 51 patients on HD, 33 male and 18 female, was studied over a year period. Age varied between 17 and 70 years old, time on HD between 6 and 94 months. All patients underwent dialysis 3 times a week for 3-4 hours using a Cuprophan hollow fiber dialyser (1-1.2 sqm). The lenght of dialysis and dialyser features were prescribed empirically in accordance with the clinical conditions of each patient. In every case, instructions were given concerning diet (35 Kcal/Kg/day and 1.2 g of protein/Kg/day) but strict control was not carried out. The interdialysis symptoms, incidence of symptomatic dialysis, excessive weight gain, infections, other intercurrent disorders and hospitalization were collected prospectively. Blood was drawn at monthly intervals to determine hemoglobin, hematocrit, leucocyte and platelet counts (Hemalog-6000), urea, creatinine and electrolytes (Astra 8 autoanalyzer) prior to one first weekly dialysis. On 3 occasions the protein catabolic rate (PCR) was calculated as previously described [1]. At the same time the triceps skinfold thickness (TSF) and midarm muscle circumference (AMC) were measured. The patients were classified arbitrarily into two groups depending on the mean concentration of serum urea; Group I - serum urea above 260 mg/dl (226.0 $\pm$ 19.5 mg/dl), 26 patients, and Group II below that level (179.5 $\pm$ 17 mg/dl), 25 patients. Chi squared and student's t-test for unpaired data were used for the comparison between the groups.

RESULTS

The age and percentage of males were higher in Group II (41.1 + 16.5 vs 53.8 + 11.6 years; 50% vs 80%) although there was no statistical significance. The PCR was above 0.8 g/Kg/day, in 49 out of 51 patients but greater in Group I (1.3 + 0.3 vs 1.1 + 0.2 g/Kg/day, $p < 0.01$). Considering anthropometric measures, the relative body weight was also better preserved in Group I (96.7 + 14.2% vs 88.7 + 11.4%, $p<0.01$). In 12 patients from Group I (46.2%) and in 10 from Group II (40%) the TSF was below the normal range as was the AMC in 12 (46.2%) and 18 (72%) patients respectively. The latter was more commonly decreased in males (72.7% vs 38.3%, $p<0.05$). There were no differences in the incidence of accompanying processes: hypertension, heart disease, degree of osteodistrophy and chronic liver disease. The interdialysis and HD symptoms, fluid overload, infection, and hospitalization rate were similar. The number and causes of hospitalization are expressed in Table 1. The number of days in the hospital was higher in the Group II (29 vs 215 days) and one patient in this group accounted for the 60% of hospitalizations. All deaths took place in Group II and the cause was not related with uremia: liver disease, cancer, sudden death. With the exception of urea and creatinine the remainder of the analytical figures were comparable in both Groups.

Table 1.- Number and Causes of Hospitalization

|  | Group I | Group II |
|---|---|---|
| Cardiovascular | 3 | 5 |
| Infections | 1 | 4 |
| Malnutrition | - | 5 |
| Gastrointestinal | - | 2 |
| Confusion | - | 1 |
| Cancer | 1 | - |
| Other | - | 2 |
| Total | 5* | 19** |

* 5 patients;        ** 7 patients.

DISCUSSION

The highest levels of urea were not consistent with an increase in the morbidity of the patient on HD, in contrast to earlier published results [2,3]. On the contrary, the association of a higher PCR and a better maintenance of some anthropometric parameters suggest a better nutritional status. The greatest reductions in weight, body fat and muscle mass were observed in males [5] who dominated the group with lower urea. Our results are similar to those of Avram et al. [6] who found few differences in the degree of rehabilitation of the patients in relation to the pre-dialysis biochemical values. In our experience, variations in the concentration of urea ranking between 150-260 mg/dl were not accompanied by significant differences in the patient clinical situation. These findings would undermine the usefulness of analytical determinations in the prescription of dialysis therapy.

REFERENCES

1.   E. G. Lowrie. B. P. Teehan, Principles of prescribing dialysis
     therapy: Implementing recommendations from the National Cooperative
     Dialysis Study. Kidney Int. 23 (Suppl. 13): 113 (1983).

2.   E. G. Lowrie, N. M. Laird, T. F. Parker, J. A. Sargent, Effect of the
     hemodialysis prescription on patient morbidity, N. Engl. J. Med.
     305: 1176 (1981).

3.   T. F. Parker, N. M. Laird, E. G. Lowrie, Comparison of the study
     groups in the National Cooperative Dialysis Study and a
     description of morbidity, mortality and patient withdrawal,
     Kidney Int 23 (Suppl. 13): 42 (1983).

4.   S. R. Acchiardo. L. W. Moore. P. A. Latour, Malnutrition as the main
     factor in morbidity and mortality of hemodialysis patients, Kidney
     Int. 24 (Suppl. 16): 199 (1983).

5.   P. Y. Schoenfeld, R. R. Henry, N. M. Laird, D. M. Roxe, Assessment of
     nutritional status of the National Cooperative Dialysis Study
     population, Kidney Int 23 (Suppl. 13): 80 (1983).

6.   M. M. Avram, P. A. Slater, A. Gan, M. Iancu, A. N. Pahilan, D. Okanya,
     K. Rajpal, S. K. Paik, M. Zouabi, P. A. Fein, Predialysis BUND and
     creatinine do not predict adequate dialysis, clinical rehabilitation
     or longevity, Kidney Int. 28 (Suppl. 17): 100 (1985).

PREDICTED TIME OF DIALYSIS IN PATIENTS WITH DIFFERENT

VOLUMES OF UREA DISTRIBUTION - THE USE OF MATHEMATICAL MODEL

Ninoslav Ivanovski and Georgi Masin

Department of Nephrology,Medical Faculty
University "Ciril and Methodius"
Skopje, Yougoslavia

## INTRODUCTION

On the basis of a confirmed evidence that urea,as a major product of protein catabolism, is a good indicator for the "uremic toxycity",there were worked out mathematical models which would try to define the parametres of the individual approach of the dialysis therapy.[2,3]

The kinetic urea model based on the principle of the conservation of mass,well known as a Fick's principle[4] deals with individual patient parametres such a residual urea clearence $(K_r)$,the volume of urea distribution (V) and the urea generation rate (G).

The individual volume of urea distribution,numerically close to the whole body water,[5] and its relation with the time of dialysis and the time of interdialysis has been analysed from other authors.[6,7] The aim of our paper is a presentation of the influence of the volume of urea distribution, as a individual patient parameter,on the predicted time of dialysis in order to hold the predialysis urea concentration $(C_0=C_2)$ to values on 23.3 mmol/1 and postdialysis $(C_1)$ on 11.6 mmol/1 for all patients,or TACurea (time avereged concentration)= 17.45 mmol/1.

## METHODS

There were analysed 30 well controled patients on the chronic dialysis program with twice and trice a week regular dialysis,all on capilary dialysator "Hemomed 1.3" and Gambro AK-10 monitors.The renal diseases were:Chronic glomerulonephritis (6),Chronic pyelonephritis (4),Nephroangiosclerosis (4),Balkan endemic nephropathy (5),Diabetic nephropathy (5),Interstitiopathy (2) and Lupus nephropathy (1).

### Single pool kinetics

There was used a pharmacokinetic principle of folowing kinetics of urea in one well mixed single pool which mathematic form is described earlier:[2,3,6]

$$C = C_0 e^{-KT/V} + G/K \ (1 - e^{-KT/V}) \qquad /1/$$

Table 1. Calculated values for G,V,TD and TID

| PATIENTS | G-mgr/min | V-ml | TD-min | TID-min |
|---|---|---|---|---|
| 1. | 7.87 | 57.962 | 254.4 | 2414 |
| 2. | 7.31 | 50.562 | 165 | 2304 |
| 3. | 6.67 | 54.288 | 206.6 | 2184 |
| 4. | 8.69 | 64.347 | 331.3 | 3161.8 |
| 5. | 7.67 | 60.770 | 146 | 1646 |
| 6. | 8.11 | 45.842 | 186.7 | 2179 |
| 7. | 8.53 | 37.759 | 137.1 | 2952 |
| 8. | 6.18 | 40.487 | 148.1 | 2132 |
| 9. | 4.20 | 46.641 | 235 | 3942 |
| 10. | 6.64 | 40.860 | 193 | 2029 |
| 11. | 8.96 | 53.462 | 267.13 | 1952 |
| 12. | 5.49 | 45.856 | 183.9 | 2950 |
| 13. | 5.05 | 55.654 | 208 | 4149 |
| 14. | 5.98 | 56.191 | 371 | 3339 |
| 15. | 10.1 | 57.835 | 251.7 | 2537 |
| 16. | 4.65 | 40.745 | 221 | 3559 |
| 17. | 6.81 | 42.544 | 189 | 2263 |
| 18. | 5.67 | 55.916 | 205.4 | 5271 |
| 19. | 6.19 | 48.856 | 208.6 | 2540 |
| 20. | 5.43 | 41.962 | 132 | 2952 |
| 21. | 6.40 | 36.453 | 181 | 1832 |
| 22. | 4.90 | 27.310 | 143 | 1817 |
| 23. | 5.48 | 32.260 | 136 | 1906 |
| 24. | 5.76 | 41.270 | 130 | 2310 |
| 25. | 6.43 | 44.846 | 145.7 | 2248 |
| 26. | 5.68 | 34.590 | 135.2 | 1990 |
| 27. | 8.93 | 75.891 | 305 | 2879 |
| 28. | 9.38 | 41.020 | 155.7 | 1520 |
| 29. | 5.03 | 32.038 | 143 | 2052 |
| 30. | 7.04 | 58.251 | 241 | 3672 |

G= Generation rate for urea     TD = Time of dialysis
V= Volume of urea distribution    TID= Interdialysis interval

## Time of dialysis and interdialysis interval

The predicted time of dialysis (TD) and interdialysis interval (TID) are mathematic form expressed through a rearrangement of the equation /1/:

$$TD = - \frac{V}{(Kr+Kd)} \ln \frac{C_1 (Kr+Kd)-G}{C_0 (Kr+Kd)-G} \qquad /2/$$

$$TID = - \frac{V}{Kr} \ln \frac{C_2 Kr-G}{C_1 Kr-G} \qquad /3/$$

The dialisator urea clearence (Kd) were from 170 to 191 ml/min, while residual renal clearence (Kr) from 0 to 5.1 ml/min. We used, conditionally, the "non-toxic" urea profil according to our comprehension, TACurea=17.45 mmol/1 ($C_0$=$C_2$=23.3 and $C_1$= 11.6 mmol/1).

<u>Urea generation rate (G) and volume of distribution (V)</u>

The values of G and $V_0$ were realised according to iterative algorithm[9] using computer technique. G was expressed as a mgr/min, while V in ml. Statistical analysis was performed with Student's t test. Coeficient of correlation and linear regression analysis by the standard technique.

RESULTS

The values for G and V,TD and TID for a values of "non-toxic" urea concentration (TACurea=17.45 mmol/l) are presented on Table 1.
According to the values for V,patients were devided into two groups: I group with $V_1$= 39.670$\overline{+}$1.41 ml (n=17) and predicted $TD_1$=154.41$\overline{+}$5.93 min, and II group with $V_2$= 53.9$\overline{+}$2.73 (n=13) and predicted $TD_2$=262.1$\overline{+}$18.3 min.

DISCUSSION

From the analysis of the mutual influence between the volume of urea distribution and predicted (calculated) time of dialysis,we could decide obout the evident and statistically confirmed influence expressed with coeficient of correlation r=0.84 or TD= 5.316 V- 47.423 (Fig. 1).
In other words,an increased volume of distribution means a prolonged period of dialysis,under the conditions used in our analysis.Patients with lower volume of urea distribution need a shorter time of dialysis under the same conditions.The similar results were confirmed from other authors.[6] The increased body mass,perheaps,is the explenation for this fenomenon.

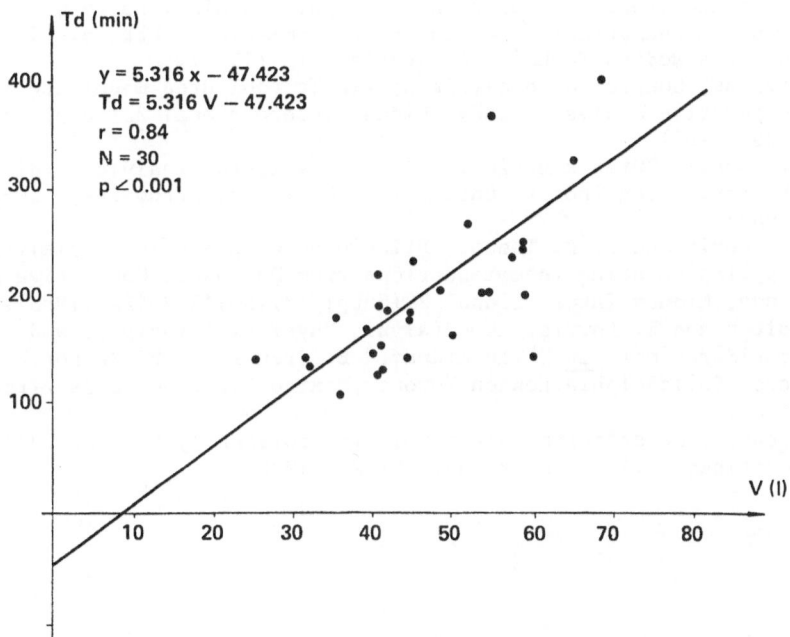

Fig. 1.  Correlation between time of dialysis (Td) and
         volume of urea distribution (V)

We should note that the value for "urea non-toxic" profil was used conditionally and it is more an expression of an experiens than a relevant scientific evidence.

Our results were confirmed in the every day practice,but thay were used in hospital conditions for a determination of the individual inter-dialysis interval.We consider the relevant value as identical with calcu-lated one,if the differance between tham was no bigger than 10%.


CONCLUSION

Trying to calculate the time of dialysis on the basis of urea kinetics separately for each patient,reaching in that way conditionally "non-toxic" urea profil (TACurea=17.45 mmol/l),we consider it as an attempt for deter-mination of the individual dialysis dose.The procedure is applicable at hos-pital dialysis and also at home dialysis when patient could modulate the therapy by himself,according to this individual parametres.


REFERENCES

1.  J. Bergström and P. Fürst, Uraemic toxins, in: "Replacement of Renal Function by Dialysis," W. Drukker, W. Parsons, and J. Maher ed., Martinus Nijhoff Medical Division, The Hague, Boston, and London (1983).
2.  J. Sargent and F. Gotch, Mathematic modeling of dialysis therapy, Kidney International 18(Suppl):S-2-S-10 (1980).
3.  F. Gotch, J. Sargent, M. Keen, M. Lam, M. Prowitt, and M. Grady, Clinical results of intermittent dialysis therapy (IDT) guided by ongoing kinetic analysis of urea metabolism, Trans Amer·Soc Artif Int Organs XXII:175 (1975).
4.  D. Riggs, "The mathematical approach to physiological problems," The M.I.T. Press Cambridge, Massachusetts, London, (1976).
5.  R. Swenson, D. Sanfelipo, D. Hall, W. Walker, Analysis of changing plas-ma concentration of urea and creatinin in dialysis by means of a semple mathematical model and a programmable pocket calculator, Opuscula Medico-Technica Lundensia XVII (1977).
6.  J. Sargent, Control of dialysis by single pool urea model:The Nationale Cooperative Dialysis Study, Kidney International 23(Suppl 13):S-19-S-25 (1983).
7.  N. Ivanovski, "Urea kinetic model as a basis for individual dialysis therapy, M.Sc Thesis, University "Ciril and Methodius," Skopje (1985)
8.  E. G. Lowrie and P. B. Teehan, Principles of prescribing dialysis the-rapy:Implementing recommendations from Nationale Cooêrative Dialysis Study, Kidney International 23(Suppl 13):S-113-S-122 (1983).
9.  C. Colton and E. Lowrie, Hemodialysis:Physical principles and technical considerations, in: "The Kidney," B. Brener and F. Rector, ed.,Saun-ders, Philadelphia,London,Toronto,Mexico City,Rio de Janeiro,Tokyo (1981).
10. R. Brent, Some efficient algorithms for solving systema of nonlinear equations, SIAM J Numer Audl 10:327 (1973).

# MEMBRANES AND THEIR REMOVAL OF UREMIC TOXINS

Nicholas A. Hoenich

Department of Medicine
Medical School
University of Newcastle upon Tyne,  UK.

## INTRODUCTION

The human kidney performs several functions continuously:-  the excretion of metabolic waste products or uremic toxins (in the form of urine), the maintenance of body fluid, electrolyte and acid base homeostasis, the selective reabsorption and secretion of a variety of plasma constituents, the production or activation of hormones, the biodegradation of circulating substances and the metabolic production and consumption of amino acids.

Uremia, or urine in the blood, is a clinical syndrome of multiple systemic dysfunction and is a consequence of renal failure or insufficiency. However, due to the number of functions that the kidney has, some of the symptoms and signs of uremia may be the result of endocrine disturbances rather than inadequate uremic toxin removal.

Although a large number of metabolites are elevated as a consequence of renal insufficiency (Table 1), their role in the aeteology of uremia and the breakthrough in our understanding of the nature of uremic toxicity resulting from their elevation remains elusive.  Bergstrom and Furst (1) have made an extensive study of the subject and suggested that a solute could be considered a uremic toxin if it fulfils the undermentioned criteria:-

(a)  the compound should be positively identified; specific and accurate analytical methods in biological fluids should be available.

(b)  Plasma level and/or tissue concentrations of the compound should be elevated in uremic subjects.

(c)  high concentrations should be related to specific uremic symptoms which disappear when the concentration is reduced.

(d)  toxic effects should be obtained in human subjects, and experimental animals at concentrations comparable to those found in body fluids of uremic patients.

It has been established for some time that uremia, acute or chronic, is reversible by appropriate treatment.  In the early stages of chronic renal failure, treatment may be conservative, protein restriction which reduces azotemia and is associated with the diminution of acidosis as well

Table 1. Compounds known to be retained in uremia.

Organic Substances

Urea
Creatinine
Methylglutanidine
Guaninosuccinic acid
Other guanidines

Products of Nucleic Acid Metabolism

Uric acid
Cyclic Amp
Pyridine

Amino Acids and Dipeptides

Amines

Indoles

Phenols

Carbohydrate Derivatives

Other Metabolites

Inorganic Substances

Water
Sodium
Potassium
Hydrogen Ions
Magnesium
Phosphate
Sulphate
Trace elements

as partial relief of some uremic symptoms such as nausea and vomiting. When
conservative treatment is no longer able to control the symptoms of renal
insufficiency (generally when endogenous creatinine clearance is below
10 ml/min) and the patient is unable to lead a "normal" life, alternative
therapies are instigated. These include haemodialysis, haemofiltration,
haemodiafiltration and intermittent or continuous ambulatory peritoneal
dialysis. All these modes of treatment involve a combination of diffusive
and convective transport of uremic toxins across an artificial semi-
permeable membrane or the peritoneal membrane.

The purpose of this paper is to discuss the uremic toxin transport
properties of such membranes.

MEMBRANES AVAILABLE FOR THE REMOVAL OF UREMIC TOXINS

Human Membranes

The peritoneum in an adult has a surface area of approximately $2m^2$,
and covers the visceral organs. It is continuous and forms a closed sack.
The space within it contains small amounts of fluid. In adults, it may be

comfortably enlarged by the instillation of two or more litres of fluid. The knowledge that the peritoneal membrane was semi-permeable has been known to physiologists before the end of the 19th century (2). It received further study in the first quarter of the 20th century and by 1923 intermittent peritoneal dialysis had been used for the treatment of uremia in humans. By the late 1950's and early 1960's intermittent peritoneal dialysis became a safe and standardised procedure used both for the treatment of acute and chronic renal failure, although it was not until 1965 that the permeability of peritoneum for 'middle' molecules was rediscovered following the observations of Scribner (3); since which time its use in the intermittent (IPD) and continuous (CAPD) peritoneal dialysis mode of treatment of renal failure has grown.

## Artificial Membranes

### Membranes Based on Cellulose

Cellulose based membranes form the basis of dialysis treatment. The most widely used membrane is Cuprophan (Enka AG - Wuppertal, Barmen, FRG) manufactured by the solubilization of cellulose in an ammonia solution of cupric oxide which is extruded into an acid bath to yield cellulose in either flat sheet, tubular or hollow fibre format.

A number of variants of this regeneration technique exist and are used by other manufacturers leading to the availability of a range of cellulosic membranes (Table 2).

Table 2. Production of Cellulose Based Membranes

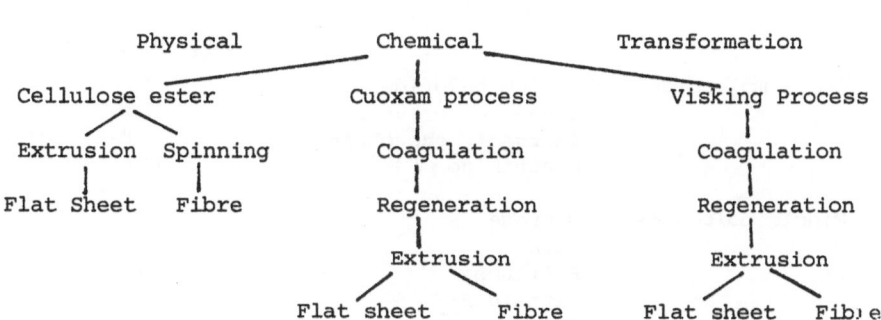

### Synthetic Membranes

The advent of the middle molecular hypothesis (4) led investigators to search for new membranes with increased ability to remove middle weight uremic toxins since it was considered that removal of uremic toxins in the molecular weight range 300-2000 daltons by cellulose based membranes was inadequate.

Although a number of new membranes were developed, of which only a few have evolved to clinical use. Among these are:- polyacrylonitrile (PAN) originally produced by Rhone Poulenc (France) and now also produced by Asahi Medical (Japan); polymethylmethacrylate (PMMA) (Toray Industries, Japan); polysulfone (Amicon Division WR Grace and Co., MA, USA) and Fresenius AG, Bad Homburg, FRG); polycarbonate (Gambrane), (Ab Gambro - Sweden); and ethylene-vinylalcohol (EVAL) (Kuraray Co. Japan). Many of these membranes, in addition to having increased ability to remove middle weight uremic toxins, have a high hydraulic permeability and cannot be used as dialysers with conventional equipment, and in consequence, their clinical use has been confined to haemofiltration and associated techniques. The availability of a new generation of haemodialysis equipment offering ultrafiltration control is likely to lead to an increased utilization of such membranes for use in conventional haemodialysis, not only due to their enhanced uremic toxin removal, but also because of such membranes enhanced biocompatibility compared with the widely used Cuprophan (5-6).

## MASS TRANSFER PROPERTIES OF MEMBRANES

### The choice of solutes for characterization

The primary function of treatment is to remove various solutes from the patient. The solutes whose removal rates are most commonly studied are urea (MW 60), a mild uremic toxin which in high concentrations is associated with nausea, vomiting, headache and fatigue; creatinine (MW 113) a compound routinely measured clinically and used as an indicator of renal function; inorganic phosphate (MW 98) which causes increased parathyroid activity and carries the risk of deposits in organs and tissues at high concentrations.

These solutes are all of low molecular weight and as discussed above, during treatment a wide range of solutes need to be removed. Unfortunately, there are very few large molecules that can be used conveniently to characterise middle weight uremic toxin removal. Vitamin $B_{12}$ (MW 1355) and Inulin (MW 5175) are frequently used as such indicators but there is no evidence that either of these molecules is toxic. Other metabolites used for the characterization of membranes are shown in Table 3.

Table 3. Compounds suitable for solute removal characterization

|  |  | Molecular Mass |
|---|---|---|
| Small molecules | Sodium chloride | 58 |
|  | Urea | 60 |
|  | Inorganic phosphate | 98 |
|  | Creatinine | 113 |
| Middle molecules | Sucrose | 342 |
|  | EDTA | 380 |
|  | Raffinose | 504 |
|  | Vitamin $B_{12}$ | 1355 |
|  | Inulin | 5200 |
| Large molecules | Cytochrome C | 13400 |
|  | Haemoglobin | 68000 |
|  | Albumin | 69000 |

Although such characterization is often undertaken in vitro, it may not be a guide to in vivo performance due to the presence of proteins (and thus protein binding) in the latter, while the removal of large molecules even in vitro, may be subject to factors other than molecular size such as electrical charge.

As renal function deteriorates there is a deterioration in not only the excretory function of the kidney, but also in its ability to maintain homeostatic control, water overload results and this is controlled to some extent in chronic renal failure by the ability of the membrane to remove water and, coupled with the fact that the removal of large molecular weight solutes is governed by convective flow or water removal, thus water removal rates must also be included in solutes used for the characterization of membranes.

THEORETICAL ASPECTS

Solute Removal

Solute removal across a membrane is by two mechanisms:- by passive diffusion from the blood to the dialysate across a concentration gradient and by convection or ultrafiltration which augments diffusive solute transport.

The relative importance of these two components varies depending upon the modality of treatment. In conventional haemodialysis and peritoneal dialysis, diffusive processes predominate, while in haemo-filtration, the solute removal is principally by convective means.

Clearance is the most important and familiar clinical parameter of membrane performance. It is a special case of the more generalised dialysance expression. It may be defined as the solute removed from the blood per unit time divided by the incoming concentration and provides a measure of the volume of blood completely cleared of the solute per unit time. It is analogous to renal clearance and for a dialyser in which there is measurable ultrafiltration, it may be expressed as :-

$$K_D = Q_B \frac{(A - V)}{A} + Q_F \frac{V}{A}$$

where  $Q_B$  =  blood flow
       $Q_F$  =  ultrafiltration
       $A$    =  metabolite concentration at inlet to the dialyser
       $V$    =  metabolite concentration at outlet to the dialyser

Peritoneal clearance ($K_p$) as measured clinically is not truly analogous to clearance defined above. It is commonly defined as :-

$$K_p = \frac{C.V_D}{p.t}$$

where  $C$    =  concentration in peritoneal fluid
       $V_D$  =  volume of the exchange
       $t$    =  exchange time
       $p$    =  plasma concentration

As the metabolite concentration in the dialysis fluid rises, thus the clearance deteriorates and the above in effect is a measure of the average rate of removal.

More complicated relationships have been developed to measure "instantaneous" peritoneal clearance (called peritoneal dialysance), one such relationship being :-

$$\text{Plasma peritoneal dialysance} = \frac{-V_B \, V_D}{V_B + V_D} \quad \text{In} \left[ \frac{1 - C_D \frac{(V_D + V_B)}{C_B \, V_B}}{t} \right]$$

where  $V_D$  =  dialysate volume at end of exchange
       $V_B$  =  distribution volume in body
       $C_B$  =  solute concentration in plasma measured at mid-point of exchange
       $C_D$  =  solute concentration in dialysate at conclusion of exchange
       $t$    =  time for exchange

Peritoneal dialysance, therefore, is truly analogous to haemo-dialysis clearance and haemofiltration clearance. However, in the latter, modifications to the widely used formulas defining clearance are necessary since using this technique solute is cleared from the blood by the physical removal of plasma water. Plasma clearance in haemofiltration ($K_{HF}$) may be expressed as :-

$$K_{HF} = \frac{Q_F . \, C_F}{C_{pi}}$$

where   $Q_F$ = ultrafiltrate flow
$C_F$ = solute concentration in ultrafiltrate
$C_{pi}$ = solute concentration in plasma entering device

If the effects of the exclusion volume of the hydrated proteins and the exclusion of larger molecular weight solutes from the electrolytes are neglected, the relationship simplifies to :-

$$QF. \quad S$$

where  S = sieving coefficient.

Figure 1 compares the solute removal characteristics of peritoneal dialysis, haemodialysis and haemofiltration.

For diffusion based processes, small molecular weight solutes are removed more rapidly than large molecular weight solutes. In contract, in convectively based processes there is no size discrimination until the size of the solute is so large that it is physically held back by the membrane pore.

Fluid Removal

Fluid removal across membranes used in haemodialysis and haemo-filtration has two components. One due to the hydrostatic pressure gradient, the other due to the osmotic pressure gradient. The rate of hydrostatic fluid removal being a function of membrane permeability, membrane area and applied pressure.

This relationship in vitro for aqueous solutions will be linear with increasing pressure, but for protein containing solutions a deviation from linearity will be observed due to protein polarization at the surface of the membrane at high rates of fluid removal.

In peritoneal dialysis, fluid removal is accomplished osmotically - glucose being most commonly used to create the osmotic gradient.

Quantitation of Uremic Toxin Removal

Quantitation of uremic toxin removal may be accomplished by a number of different ways, clearance or dialysance offered over time being central to any such comparison.

Clearance of small and middle molecular weight metabolites measured in vitro are summarised in Table 4 for both cellulose based and synthetic membranes while the ultrafiltration capacity of the membranes is also shown. Clearances in peritoneal dialysis, notably for small molecules, are appreciably lower than those shown (7). Large molecules on the other hand, are cleared in a comparable manner to that observed for haemodialysers. Table 5 compares haemodialysis and haemofiltration when using the same membrane. Haemofiltration offers a lower clearance of small molecules but for higher molecules its removal efficiency is significantly higher than for haemodialysis.

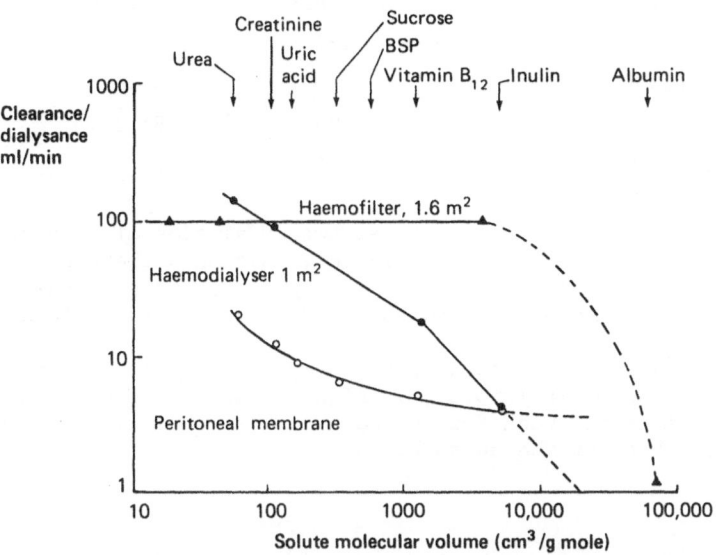

Figure 1.  Clearance or dialysance curves are plotted against solute molecular volume.  Dotted lines represent likely extrapolations of the curves.

Table 4.  In vitro removal of small and middle molecular solutes by haemodialysers utilising cellulose based and synthetic membranes

Blood flow  200 ml/min
Dialysate flow  500-530 ml/min
Temperature  38°C
Ultrafiltration 0 ml/min

| Dialyser | Membrane | Surface Area $(m^2)$ | Clearance (ml/min) | | | | |
|---|---|---|---|---|---|---|---|
| | | | Urea (60) | Creatinine (113) | $PO_4$ (136) | Vit.$B_{12}$ (1355) | Ultra-filtration ml/hr/mmHg |
| Sorin 11.08 | Cuprophan [1] 8 micron | 1.06 | 164 | 144 | 123 | 35.8 | 5.0 |
| Cordis 90 SCE | SCE 30 micron | 1.1 | 147 | 118 | - | 26.0 | 2.9 |
| Cordis CDAK 3500 | Cellulate 40 micron | 0.9 | 115 | 91 | - | 30.6 | 5.9 |
| Hospal Filtral | AN 69 HF 45 micron | 1.15 | 164 | 151 | 131 | 78.2 | 40.2 |
| Toray B1-100 | PMMA 25 micron | 1.2 | 149 | 129 | 110 | 49.5 | 11.8 |
| Bellco PS1 | Polysulfone 40 micron | 0.7 | 145 | 128 | 121 | 68.1 | 90.3 |

[1]  membrane thickness

Table 5.  In vitro comparative clearance in haemodialysis and haemofiltration using polyacrylonitrile membrane at a blood
flow of 200 ml/min.

|  | Haemo-<br>dialysis | Haemo-<br>filtration | Human<br>Kidney |
|---|---|---|---|
| Urea (60) | 135 | 70 | 136 |
| Vitamin $B_{12}$ (1355) | 60 | 65.8 | - |
| Inulin (5200) | 23 | 54.4 | 218 |
| Ultrafiltration rate (ml/min) | 0 | 70 | |

Haemodiafiltration combines these two modalities of treatment and
was first introduced by Leber in 1978 (8).  Although the uremic toxin
removal capacity of such a technique is superior, its clinical application
has been limited by sophisticated equipment required to perform it.

In quantification of therapy, not only the removal rate but also the
duration of treatment is important.  Figure 2 shows the clearances
computed in litres per week for the more common modalities of treatment
compared with that for the human kidney.

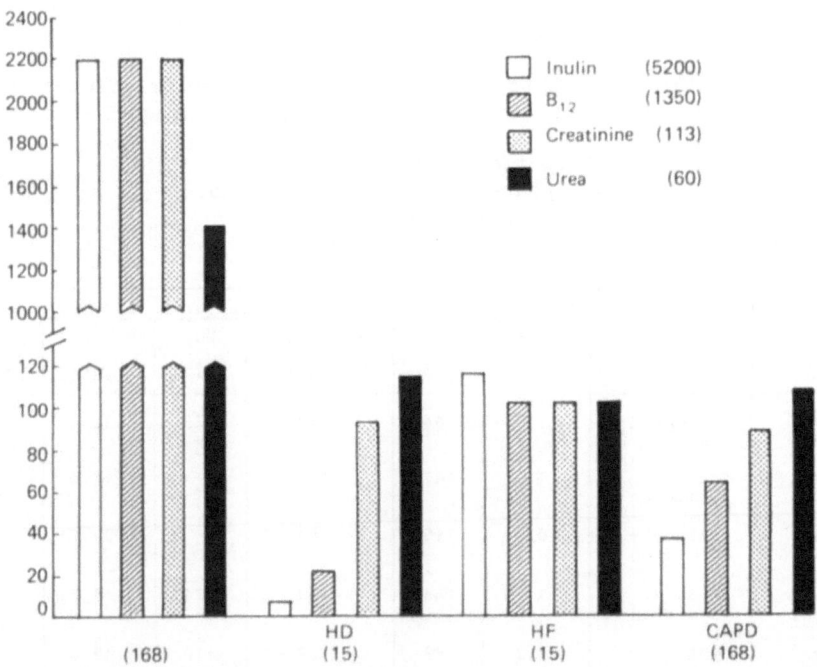

Figure 2.  Clearances computed in litres per week for solutes of different
size showing comparative clearance profiles for haemodialysis,
haemofiltration (HF), CAPD and a pair of normal human kidneys.
The figure in parentheses is the usually employed operating
time in hours per week.

Clearly the aim of any therapy is to offer a level of uremic toxin removal achieved by the human kidney. However, such comparisons are unrealistic in the sense that whereas the human kidney performs its function continuously, other treatment modalities (with the exception of continuous ambulatory peritoneal dialysis) are intermittent and in order to achieve the same level of uremic toxin removal, either the treatment time or efficiency of removal need to be increased. Both these options represent specific problems and may in the immediate future by unattainable.

## CONCLUSIONS

It is well recognised that uremia can be effectively treated and that the uremic toxin removal capabilities of the modalities of treatment are inferior to that achieved by the human kidney. Furthermore, such treatments also remove indiscriminately vital solutes such as amino acids and vitamins (9,10).

In order to improve efficiency of solute removal, a positive identification of uremic toxins and their role in the pathogenesis of uremia is required. Once this has been achieved, a better and more selective removal becomes a possibility.

## REFERENCES

1.  Bergstrom, J., Furst, P. Uraemic toxins (Chapter 19), in:-
    "Replacement of Renal Function by Dialysis". ed. Drukker, W.,
    Parsons, F.M., Maher, J. 1983. Martinus Nijhoff, Boston.

2.  Drukker, W. Peritoneal Dialysis - A Historical Review (Chapter 21),
    in:- "Replacement of Renal Function by Dialysis". ed. Drukker, W.,
    Parsons, F.M., Maher, J. 1983. Martinus Nijhoff, Boston.

3.  Scribner, B.H. 1965, Discussion. Trans Am Soc Artif Intern Organs
    15: 87.

4.  Babb, A.L., Popovich, R.P., Christopher, T.G., Scribner, B.H. 1971.
    The genesis of the square meter hour hypothesis. Trans Am Soc
    Artif Intern Organs. 17: 81-91.

5.  Jacob, A.I., Gavellas, G., Zarco, R., Perez, G., Bourgoignie, J.J.
    1980. Leukopenia, hypoxia and complement function with different
    hemodialysis membranes. Kidney Int. 18: 505-509.

6.  Hoenich, N.A., Johnston, S.R.D., Woffindin, C., Kerr, D.N.S. 1984.
    Haemodialysis leucopenia: the role of membrane type and reuse.
    Contr Nephrol. 37: 120-128.

7.  Robson, M., Oreopoulos, D.G., Izatt, S., Ogilvie, R., Rapoport, A.
    de Veber, G.A. 1978. Influence of exchange volume and dialysate
    flow rate on solute clearance in peritoneal dialysis. Kidney Int.
    14: 486-490.

8.  Leber, H.W., Wizemann, V., Goubeaud, G., Rawer, P., Schutterle, G.
    1978. Simultaneous hemofiltration/hemodialysis: An effective
    alternative to hemofiltration and conventional hemodialysis in
    the treatment of uremic patients. Clin Nephrol. 9: 115-121.

9.   Lasker, N., Harvey, A., Baker, H.  1963.  Vitamin levels in hemo-
     dialysis and intermittent peritoneal dialysis.  <u>Trans Am Soc Artif
     Intern Organs</u>.  9:  51-55.

10.  Giordano, C., De Santo, N.G., Capoticasa, G., Di Leo V.A.  1980.
     Amino acid losses during CAPD. <u>Clin Nephrol</u>.  14:  230-237.

# REPLACEMENT OF EXCRETORY KIDNEY FUNCTION BY HIGH-EFFICIENCY HEMODIAFILTRATION (HDF) WITH A PEPTIDE-PERMEABLE MEMBRANE

V. Wizemann, F. Techert, S. Brüning, H.-W. Birk and
K. Mueller

Justus-Liebig-University, Department of Internal Medicine
Klinikstr. 36
D-6300 Giessen, F.R.G.

## INTRODUCTION

Despite considerable progress in dialysis technology in the last decade clinical progress in chronic dialysis patients was absent or neglectable. The National Cooperative Dialysis Study reveilled that morbidity in pts on MHD increased when removal of small uremic solutes was inadequate[1]. The lower limit for dialysis adequacy, as measured by urea kinetics, was a clearance times time product divided by urea distribution volume (Kt/V), which was 0.8 or lower. Thus, morbidity appears to be interrelated to a definable degree of urea removal. On the other hand, such a relationship cannot be demonstrated when solely an improved removal of "middle molecules" (index: Vit. $B_{12}$) was obtained[2].

Although dialysis therapy has been the most successful organ replacement therapy the clinical effects of this therapy have been far from being satisfactory. Analyzing the causes, the current dose of dialysis therapy (quantity) might be still too small and/or the quality of detoxification might be insufficient. As for the latter, the majority of artificial membranes used for dialysis purposes, can remove approximately one % of the molecular weight spectrum of potential uremic toxins, as compared to the glomerular filter.

To overcome, at least in part, the deficiencies of standard dialysis and hemofiltration, we developed a high-efficiency hemodiafiltration (HDF) and performed an ABA study to assess the clinical and biochemical efficiency of such a method. In principle, a patient was treated for a comparatively long time (3 x 4 hrs/week) by a high dose of diffusive transport (dialysis over a 2.5 m surface, 500 ml/min) and by a simultaneous high dose of convective transport (concomitant hemofiltration and substitution of 60 l of bicarbonate solution in a pre- and -mid dilution mode). To improve the efficiency of detoxification in the high molecular range, two polysulphone F60 in line were used as hemodiafilters, which allow permeation of substances up to a molecular weight of 150 000 daltons.

## ACUTE EFFECTS OF HIGH-EFFICIENCY HDF

At a blood flow of 300 ml/min, clearances (ml/min) were obtained in

10 pts for urea (275 $\pm$ 8.5), creatinine (240 $\pm$ 22), phosphate 225 $\pm$ 22), inulin (123 $\pm$ 15) and ß$_2$-microglobulin (100 $\pm$ 36). In 4 hrs, from dialysate and filtrate, a mean of 30 g urea, 4 g creatinine, 26 mM phosphate and 550 mg of ß$_2$-microglobulin could be removed per patient. Kt/V was 1.47.

In comparison to standard hemodialysis performed in the same 10 pts, there was decrease in pre/post HDF serum values for ß$_2$-microglobulin from 44.4 to 17.9 mg/l (p < 0.0005), for PTH from 3.25 to 0.76 ng/ml (p < 0.01), for myoglobin from 444 to 327 ng/ml (p < 0.01) and for amylase (MW 50 - 55 000 daltons) activity from 72 to 50 U/l (p < 0.001).

A (HD) - B (HDF) - A (HD) STUDY

To assess the clinical effects of high efficiency HDF an ABA study was performed in 7 pts, consisting of a 6 month treatment by standard hemodialysis (3 x 4 hrs, 1.3 m$^2$ cuprophane dialyser, blood flow 200 ml/min) followed by a 6 month HDF period and a second HD control period of further 6 months. Atpresent, parts of the data can only be presented for the AB part of the study.

Probably due to bicarbonate buffer in dialysate and substituate, there was an excellent control of renal acidosis during the HDF period: Pre-HDF bicarbonate was 26.5 + 2.7 mM/l as compared to period A$_1$ (HD, 18.7 + 1.1) and A$_2$ (HD, 19.4 + 3.1). Pre-treatment blood pressure was unchanged throughout the periods (A$_1$ = 140/79, B = 146/84, A$_2$ = 146/84 mmHg) as was post-treatment weight and weight gain between dialysis. There was no tendency in the number of hypotensive episodes in the study periods, no difference in the number of nurse interventions or in number of hypertensive episodes.

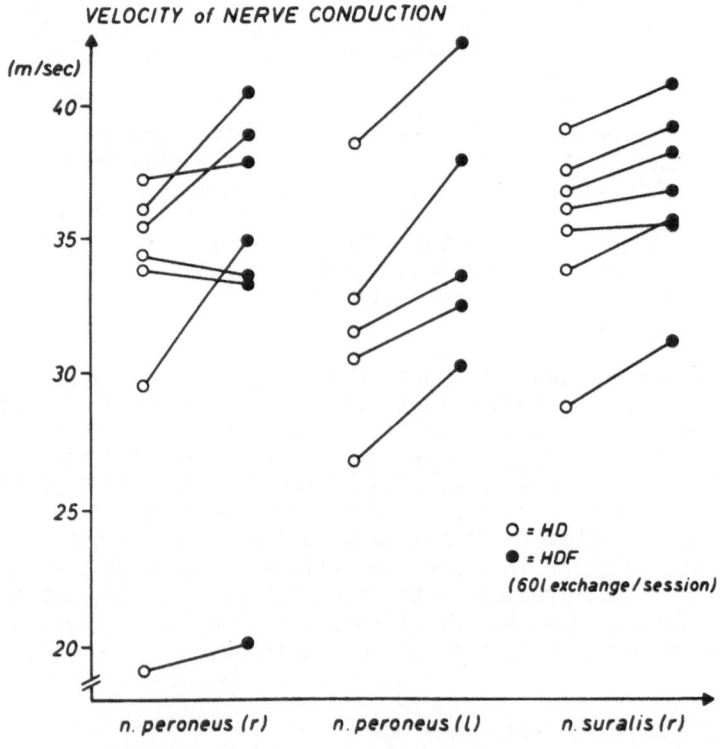

Fig. 1. Velocity of nerve conduction after 6 months of HD and HDF in 7 patients

At the end of the HDF period (B) there was a slight but significant ($p < 0.05$) improvement in velocity of nerve condiction, as compared to the preceeding HD period ($A_1$, figure 1). Renal anemia was unchanged in period B and $A_1$.

Serum biochemistry: Between period A (HD) and B (HDF) there was no difference in the pre-treatment levels of urea ($158 \pm 51$ vs. $136 \pm 29$ mg/dl), creatinine ($13.5 \pm 2.1$ vs. $13.3 \pm 1.2$ mg/dl), phosphate ($1.77 \pm 0.55$ vs. $1.71 \pm 0.41$ mM/l) and total protein ($69 \pm 11$ vs. $68 \pm 8$ g/l). However, albumin concentration was significantly increased after 6 months of HDF ($42 \pm 3$ g/l) as compared to HD ($34 \pm 4$ g/l, $p < 0.001$, figure 2). A further indication, that protein synthesis (or nutritional status) might be improved, can be indicated by an increase of serum transferrin during the HDF period (2.23 g/l) in comparison to the HD period (1.86 g/l). Adverse reactions such as fever, increase in leucocyte counts, or sings of sepsis were not observed during HDF.

DISCUSSION

To our knowledge, we applied the most efficient renal replacement therapy described so far. Despite acute improvement of pathological biochemics, long term treatment with high efficiency HDF had remarkably few beneficial effects on the clinical status of the patients. Balancing the pros and cons, one has to consider the potential dangerous effects of infusing 180 l/week of fluid into a patient. Although we did not observe adverse effects related to HDF and we were unable to demonstrate presence of endotoxins, long term adverse effects cannot be ruled out. Thus, high efficiency HDF is not suitable for replacing routine HD. In single patients, suffering from severe poly-

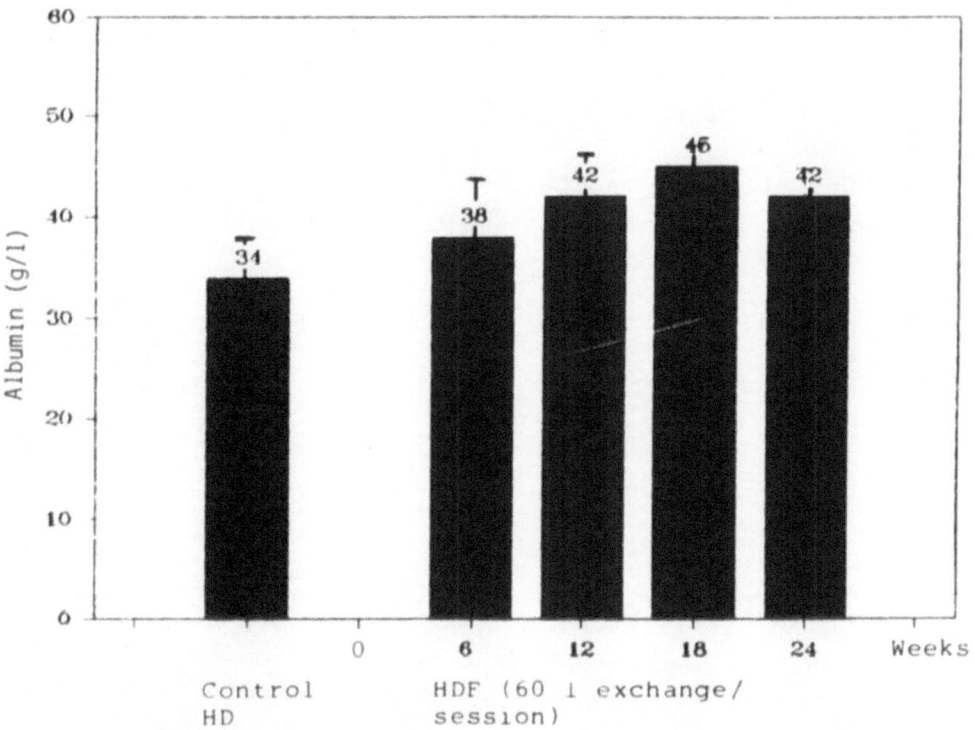

Fig. 2. Increase of pre-treatment albumin content in the serum following HDF treatment

neuropathy, HDF might be indicated. One question remaines to be answered: How can the detoxification quality of renal replacement therapy be improved and thereby the clinic of uremic intoxication? The answer might be that the current membranes - although improved in biocompatibility and sieving characteristics - are still far from the qualtity of the glomerular filter and secondly, that the duration and dose of therapy, especially in the higher molecular range, might be further augmented.

REFERENCES

1.  T. F. Parker, N. M. Laird, and E. G. Lowry, Comparison of the study groups in the National Cooperative Dialysis Study and a description of morbidity, mortality and patient withdrawl, Kidney Int. 23, Suppl. 13, 1-122 (1983).
2.  F. Gotch, Advances in nephrology and dialysis, 1st Satellite Congress of the EDTA-ERA Parma, September 26th, 1986.

# PAIRED FILTRATION-DIALYSIS (PFD) : A SEPARATE CONVECTIVE-DIFFUSIVE SYSTEM FOR EXTRACORPOREAL BLOOD PURIFICATION IN URAEMIC PATIENTS

Ghezzi P.M.*, Zucchelli P.°, Ringoir S.+, Nigrelli S.*,
Santoro A.°, Gervasio R.#, Sanz Moreno C.^, Vanholder R.+,
and Botella J.^

*Santa Croce Hospital, Cuneo (Italy) - ° M. Malpighi Hospital, Bologna (Italy) - + University Hospital, Gent (Belgium) - ^Puerta de Hierro Hospital, Madrid (Spain) - #Sorin Biomedica, Saluggia (Italy)

## INTRODUCTION

In haemodiafiltration (HDF) mass transfer of solutes through a membrane depends on both convection and diffusion. Diffusion has major importance in removing low-molecular weight solutes which would require a greater filtration flow if only convection were employed. Convection is essential for mass transfer of high-molecular weight solutes, since diffusive membrane permeability markedly decreases as molecular weight increases. The total clearance efficacy of HDF is less than the sum of the two single procedures since diffusion and convection do not take place subsequently but simultaneously and are mutually affected (1). In fact, simultaneous diffusion and convection with the same membrane determine convective clearance as low as contemporary diffusive clearance is high (2-4). The following equation, allowing calculation of convective clearance ($K_{conv}$), explains this fact.

$$K_{conv} = Q_{uf} \cdot \frac{C_o}{C_i} \cdot T$$

where:

$Q_{uf}$ = ultrafiltration rate

$C_o$ = outlet solute concentration

$C_i$ = inlet solute concentration

$T$ = transmittance

The less than one the $C_o/C_i$ ratio, the less than $Q_{uf}$ the convective clearance share. In short, the higher the diffusive efficacy of a membrane for a given solute, the less the contemporary convective removal of that solute.

## MATERIALS AND METHODS

In order to improve the HDF purifying performance, a two-chamber system was conceived, able to perform convection and diffusion simultaneously
but separately. As first practical application, a small-surface area
(0.4 m²) polysulphone hollow fibre haemofilter was connected upstream of
a hollow fibre Cuprophan haemodialyzer (effective surface area: 1.1 m²;
membrane thickness: 8 /um). In this two-chamber system, the polysulphone
haemofilter provides the convective process of blood purification (at
constant $Q_{uf}$) and the body weight control, while the Cuprophan haemodialyzer provides the diffusive process (at TMP as close as possible to
0 mmHg). Between the two components packed in the same housing, afterdilu-
tion may occur, so as to avoid excessive increase in blood viscosity
leading to haemodynamic problems in the haemodialyzer, and to obtain a
partial replacement of the ultrafiltrated plasma water in relation to
required body weight reduction. In patients with low haematocrit levels
and/or high blood flows, total or partial reinfusion downstream of the
haemodialyzer may be performed. As replacement solution, saline (0.9% NaCl)
was infused in case of bicarbonate dialysis. This two-chamber system was
named Paired Filtration Dialysis or PFD (Fig.1).
In order to determine clinically the system features, a study was performed  on 26 chronic uraemic patients undergoing maintenance haemodialysis by Cuprophan®flat plate filters (15 cases) or hollow fibre filters
(11 cases) with 8-11.5 /um thick and 1-1.3 m² effective surface area membrane. Out of 26 patients 18 were males and 8 females; the mean age was
54.3 ± 14.4 years (max 68, min 19). The diagnosis of renal failure was
chronic glomerulonephritis (9cases), chronic pyelonephritis (5 cases),
nephroangiosclerosis (3 cases), polycystic kidney disease (4 cases),
diabetes (3 cases) and analgesic nephropathy (1 case). In 3 cases the
exact diagnosis was unknown. The mean dialytic age was 61.4 ± 23.5 months
(max 101, min 7). the mean residual diuresis was 256 ± 311 ml/24 hours
(from total anuria in 11 cases to 1250 ml/24 hours). Informed consent
according to the principles of the declaration of Helsinki was obtained in
all cases prior to the study.

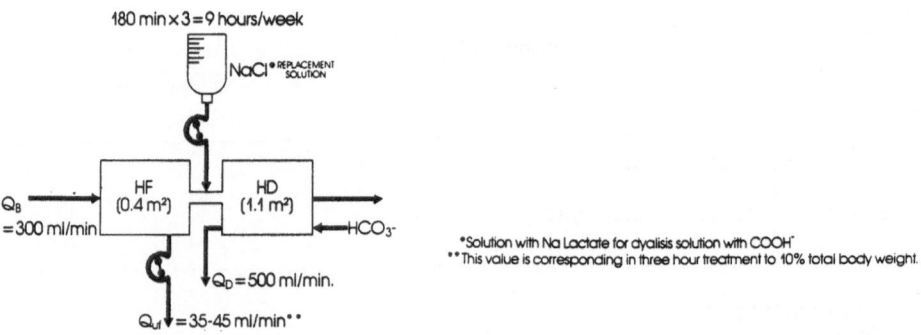

Fig.1. PAIRED FILTRATION DIALYSIS - TWO CHAMBER TECHNIQUE (PFD-TCT)

In each patient determinations were made on whole blood 60 min after start of PFD session, such as clearances of urea, creatinine, uric acid, phosphates and endogenous vitamin $B_{12}$ (RIA) at 300 ml/min $Q_b$, 450 ml/min $Q_d$, 41 ± 2 ml/min $Q_{uf}$ and 22 + 3.8 ml/min replacement solution rate. In each patient, during 180 min-long sessions, per cent extractions (start/ end ratio) were taken of urea, creatinine, uric acid and phosphates; for these solutes the convective share was also estimated, assaying their concentration in the ultrafiltrated plasma water.

By reversed-phase high-power liquid chromatography (HPLC) extraction of solutes of peaks 1, 2, 3a, 3b, 4, 5, 6, 8 was determined in 5 patients. Results were compared by the paired t - test between a group of patients undergoing 180 min-sessions of PFD and a second group of patients under-going 240 min-sessions of HD (same $Q_b$ and $Q_d$).

Serum levels of beta$_2$-microglobulin were assayed by radioimmunoassay in 11 anuric patients at the time when their treatment was switched from HD to PFD and 3 months thereafter. Results were compared by the paired t - test.

Moreover, for each patient, at the beginning and 3, 6, 12 months after regular PFD treatment performed three times a week for 180 min/ session, the following biochemical and clinical parameters were determined at start of session, after the longest interval: sodium, potassium, cal-cium and phosphates; BUN, blood creatinine, blood uric acid, total pro-teins, albumin and haemoglobin; pH, arterial $pO_2$ and $pCO_2$ and bicarbon-ates; body weight and increase in interdialytic body weight; systolic and diastolic arterial pressure.

Finally the mean interdialytic body weight increase (4415 ± 521 g) was further studied on 6 critically ill patients undergoing bicarbonate HD sessions lasting 265 ± 26 min. On these patients a comparative evalua-tion was performed, between 57 successive HD sessions (4 months) and 56 successive PFD sessions (4 months). Studied parameters were body weight at start and end of session and the corresponding weight loss; duration of each session; weight loss per minute and per hour; effective urea clearance per session and per week (2) and Kt/V according to Gotch and Sargent (5). Results were compared by the paired t - test.

RESULTS

In vivo clearances resulted to be 220 ± 18 ml/min for BUN, 191 ± 16 ml/min for creatinine, 177 ± 12 ml/min for uric acid, 188 ± 8 ml/min for phosphates and 85 + 9 ml/min for endogenous vitamin $B_{12}$ (determinations were made on whole blood 60 min after start of PFD session, at 300 ml/min $Q_b$, 450 ml/min Qd, 41 ± 2 ml/min $Q_{uf}$ and 22 ± 3.8 ml/min replacement solution rate).

Per cent extractions for the same patients (start/end session ratio) were 57 ± 9% for BUN, 55 ± 8% for creatinine, 62 ± 11% for uric acid and 53 ± 11% for phosphates. The real extractions for convection only, calcu-lated by ultrafiltrate concentrations, were 5.42 ± 0.8 g for urea, 1.08 ± 0.1 g for creatinine, 0.47 ± 0.1 g for uric acid and 0.49 ± 0.13 g for phosphates. These values were obtained with 7680 ± 789 ml mean total ultrafiltrate and 3885 ± 727 ml of replacement solution.

Tab. I

Extractions, in the same patients (5), during HD (240 min) and PFD (180 min). Study on UV-absorbing solutes in the serum by HPLC. Data in %

| Pick n | Solute | Extractions % PFD | HD | p |
|---|---|---|---|---|
| 1 | Creatinine | 60 ± 6 | 57 ± 6 | NS |
| 2 | Ps. Uridine | 49 ± 6 | 45 ± 7 | < 0.05 |
| 3a | Uric acid | 65 ± 4 | 66 ± 5 | NS |
| 3b | Xanthine | 73 ± 3 | 74 ± 3 | NS |
| 4 | Unknown | 59 ± 17 | 63 ± 9 | NS |
| 5 | Unknown | 51 ± 11 | 46 ± 8 | NS |
| 6 | Unknown | 62 ± 6 | 62 ± 6 | NS |
| 8 | Hippuric acid | 66 ± 6 | 64 ± 7 | NS |
| | Total UV-solutes | 57 ± 6 | 59 ± 7 | NS |

NS = non-significant

Extraction of solutes of peaks 1, 2, 3a, 3b, 4, 5, 6, 8 by HPLC are reported in Table I.

Serum levels of Beta$_2$-microglobulin in 11 anuric patients resulted to be 43.79 $\pm$ 22.47 mg/l before regular PFD treatment, and 32.71 $\pm$ 14.04 mg/l after start of PFD (Student's t test for paired data, t = 2.272, p 0.05).

Results of determination of biochemical and clinical parameters are reported in Table II before regular PFD treatment and 3, 6, 12 months thereafter. Results of comparative evaluation between HD and PFD on 6 critically ill patients are reported in Table III.

DISCUSSION AND CONCLUSIONS

These results would indicate that:

1. Structure and geometry of the two-chamber system are simple and the device was developed easily. Small and middle molecule clearances are satisfactory. Membrane and ultrafiltrate quantity being equal, the convection mass transfer is always better in PFD than in any other HDF techniques, as convection and diffusion are performed simultaneously but in two different chambers. Likewise, by this technique there are no convective obstacles to diffusion and no diffusive obstacles to convection, as the two mass transfers are performed by different membranes.
Consequently, in PFD either the same ultrafiltrate quantity as in other HDF techniques may be kept, achieving a better convective removal, or the ultrafiltrate quantity may be reduced for similar results. This allows either shortening of session duration or improvement of clearance performance if session duration remains unchanged.
Anyhow, by reducing $Q_{uf}$, smaller quantities of replacement solution need to be infused, with advantages of volume, qualitative/quantitative contamination (trace elements such as aluminium, pyrogens, etc.) and cost.

2. Since high $Q_{uf}$ are not necessary for proper convective purification of middle molecules (witness endogenous vitamin B$_{12}$ clearance), PFD can be performed with normal $Q_b$ (250-350 ml/min) even in patients having high haematocrit levels, with no risk of haemodynamic complications due to haemoconcentration. Moreover, low $Q_{uf}$ prevents secondary membrane formation and its consequent sieving coefficient decrease, or at any rate reduces its importance and effects (6).

3. HPLC clearly shows that PFD can extract the same amounts of solutes than conventional HD in a shorter time. Again, this allows either shortening of session duration or improvement of clearance performance if session duration remains unchanged.

4. Radioimmunoassay of beta$_2$-microglobulin showed a statistically significant decrease of serum levels in 11 anuric patients 3 months after regular PFD treatment. This result is most probably due to a better convective clearance with respect to HD, as beta$_2$-microglobulin cannot be cleared by diffusion through a Cuprophan®membrane because of its

Tab. II

Biochemical and clinical parameters during conventional HD (3x240-300 min/week) and PFD (3x180 min/week)

| Parameter | HD | PFD (3 months) | PFD (6 months) | PFD (12 months) | |
|---|---|---|---|---|---|
| Sodium (mEq/l) | 140.1 ± 2.4 | 139.4 ± 2.7 | 141.5 ± 2.1 | 139.0 ± 2.4 | NS |
| Potassium (mEq/l) | 5.7 ± 1.1 | 5.7 ± 1.2 | 5.9 ± 1.3 | 5.7 ± 1.2 | NS |
| Calcium (mg/dl) | 8.3 ± 0.5 | 8.2 ± 0.7 | 8.8 ± 0.8 | 8.9 ± 0.8 | NS |
| Phosphates (mg/dl) | 5.7 ± 1.3 | 5.5 ± 1.9 | 5.1 ± 2.0 | 5.3 ± 1.8 | NS |
| BUN (mg/dl) | 105.1 ± 12.9 | 100.4 ± 12.5 | 98.7 ± 11.7 | 97.8 ± 12.1 | NS |
| Blood creatinine (mg/dl) | 14.2 ± 3.3 | 15.4 ± 3.8 | 14.5 ± 3.1 | 13.8 ± 3.1 | NS |
| Blood uric acid (mg/dl) | 8.5 ± 1.1 | 8.7 ± 1.5 | 8.4 ± 2.2 | 8.7 ± 2.1 | NS |
| Total proteins (g/dl) | 6.6 ± 0.5 | 6.4 ± 0.6 | 6.5 ± 0.5 | 6.6 ± 0.4 | NS |
| Albumin (g/dl) | 3.3 ± 0.3 | 3.3 ± 0.3 | 3.9 ± 0.2 | 4.0 ± 0.2 | NS |
| Haemoglobin (g/dl) | 8.81 ± 1.88 | 8.21 ± 2.4 | 8.23 ± 2.1 | 9.2 ± 2.2 | NS |
| Arterial pH | 7.34 ± 0.03 | 7.33 ± 0.03 | 7.33 ± 0.05 | 7.33 ± 0.04 | NS |
| Arterial $pO_2$ (mmHg) | 96.5 ± 15.4 | 95.4 ± 14.8 | 99.4 ± 13.2 | 98.3 ± 14.1 | NS |
| Arterial $pCO_2$ (mmHg) | 30.5 ± 3.65 | 31.9 ± 3.56 | 30.5 ± 4.43 | 30.4 ± 4.22 | NS |
| Bicarbonates (mEq/l) | 16.2 ± 2.11 | 16.5 ± 2.01 | 17.8 ± 2.55 | 17.7 ± 2.4 | NS |
| Body weight (Kg) | 66.2 ± 7.10 | 66.5 ± 7.11 | 65.8 ± 6.90 | 66.6 ± 6.8 | NS |
| Increase BW interdial. (Kg) | +3.9 ± 1.90 | +3.8 ± 2.11 | +3.1 ± 1.9 | +3.4 ± 1.8 | NS |
| Systolic blood pressure (mmHg) | 144 ± 26 | 140 ± 24 | 137 ± 23 | 139 ± 21 | NS |
| Diastolic blood pressure (mmHg) | 79 ± 11 | 81 ± 11 | 78 ± 12 | 80 ± 8 | NS |

Data are expressed as mean + SD. Statistical analysis by Student's t test for paired data. Initial observations: 26 patients. Drop-out: 1 after 6 months and 4 after 12 months for renal transplantation.

Tab. III

Comparison between HD and PFD in 6 "critical" uraemic patients. Data concerning 57 sessions/patient along 4 months for HD and 56 sessions/patient for PFD. Student's t test for paired data.

| Parameter | HD | PFD | diff. | diff.% | p |
|---|---|---|---|---|---|
| BW before (g) * | 72703 ± 8912 | 73395 ± 9133 | + 692 | + 0.95 | NS |
| BW after (g) * | 68288 ± 8457 | 68712 ± 8283 | + 424 | + 0.62 | NS |
| BW loss (g) * | 4415 ± 521 | 4683 ± 439 | + 268 | + 6.07 | NS |
| Session duration (min) * | 265 ± 26 | 216 ± 20 | - 49 | - 18.5 | < 0.0005 |
| Weight loss/min (g) ** | 16.66 | 21.68 | + 5.02 | + 30.1 | < 0.0005 |
| Weight loss/hr (g) ** | 999.62 | 1300.83 | + 301 | + 30.1 | < 0.0005 |
| $C_{urea}$ (ml/min) ** | 112.02 ± 25.52 | 153.90 ± 17.52 | + 41.9 | + 37.4 | < 0.0005 |
| $C_{urea}$/week(ml/min) ** | 8.79 ± 1.98 | 9.99 ± 1.77 | + 1.20 | + 13.6 | < 0.02 |
| Kt/V ** | - 0.75 ± 0.02 | - 0.86 ± 0.03 | + 0.11 | + 14.7 | < 0.05 |

*    measured
**   calculated

NS  =  non-significant

high molecular weight (11,800 daltons). PFD, like HF, seems to achieve a better control of beta$_2$-microglobulin levels. Being this evaluation performed by a Cuprophan®membrane, reduced levels of beta$_2$-microglobulin cannot be due to better membrane biocompatibility (8).

5. Better depurative efficiency of PFD with respect to HD is also witnessed by constant biochemical and clinical parameters observed in 26 patients examined before regular PFD treatment and 3, 6, 12 months thereafter.
   Such results were obtained by shortening of each session duration from 240 to 180 min. No accumulation of low-molecular weight solutes; no variations in electrolyte content or acid-base equilibrium; no anaemia, body weight increase or blood pressure variations were ever observed. The clinical picture was unchanged.

6. The study on 6 critically ill patients with elevated body weight increase allows to conclude that PFD may achieve greater weight loss in the same time, better clinical tolerance (interdialytic hypotension frequency decreased from 18.17% to 2.1%; cramps frequency decreased from 13.06% to 3.85%) and better depurative efficiency with respect to HD, Q$_b$ and diffusive surface area being equal.

7. The continuous availability of on-line plasma water, as happens also in HF, but not in HD and HDF, allows precise biochemical and electrolytic monitoring, which is a source of information for variations, including automatic, of the dialysis solution composition and flow and other treatment parameters.

REFERENCES

1. Sprenger K.G.B., Stephan H., Kratz W., Huber K., Franz H.E., 1985, Optimizing of hemodiafiltration with modern membranes? Contr. Nephrol., 46: 43 Karger, Basel.
2. Ghezzi P.M., Frigato G., Fantini G.F., Dutto A., Meinero S., Cento G., Marazzi F., d'Andria V., Grivet V., 1983, Theoretical model and first clinical results of the paired filtration-dialysis (PFD). Life Support System, Proceedings of Xth annual Meeting ESAO, Bologna, 15-17 September, 1983, W.B. Saunders Publ., London, p. 271.
3. Ghezzi P.M., Grivet V., Nigrelli S., Meinero S., Marazzi F., Canepari G., Cento G., Gervasio R., Botella J., 1985, A new dialytic strategy (paired filtration-dialysis) using a polysulphone membrane. Malta Meeting on Clinical Aspects of Polysulphone Membranes, Malta, October 24-27 in press.
4. Gupta B.B., Jaffrin M.Y., 1984, In vitro study of combined convection-diffusion mass transfer in hemodialysers. Int.J.Art.Org., 5: 263.
5. Gotch F.A., Sargent J.A., 1985, A mechanistic analysis of the National Cooperative Dialysis Study (NCDS). Kidney Inter., 28: 526.
6. Sprenger K.G.B., Kratz W., Lewis A.E., Stadtmüller U., 1983, Kinetic modeling of hemodialysis, hemofiltration and hemodiafiltration. Kidney Inter., 24: 143.
7. Floege J., Granolleras C., Deschodt G., Branger B., Oulès R., Shaldon S., Koch K.M., 1986, Beta$_2$-microglobulin kinetics during hemodialysis and hemofiltration. XXIIIrd Congress of EDTA-European Renal Associa-

tion, June 29-July 3, Budapest, Abstracts, p.122.

8. Zingraff J., Beyne P., Bardin T., Touam M., Uzan M., Man N.K., Drüeke
   T., 1986, Dialysis amyloidosis and plasma $beta_2$-microglobulin.
   XXIIIrd Congress of EDTA-European Renal Association, June 29-July 3,
   Budapest, Abstracts, p. 162.

REMOVAL OF SMALL PROTEINS - THE STEPCHILD OF DIALYSIS THERAPY
PROTEIN PERMEABILITY OF DIFFERENT HEMOFILTRATION MEMBRANES ANALYSED BY
SDS-POLYACRYLAMIDE GEL ELECTROPHORESIS (SDS-PAGE)

H.-W. Birk and V. Wizeman

Department of Internal Medicine
Justus-Liebig-University
Giessen, F.R.G.

## INTRODUCTION

The kidney plays an essential role in the catabolism of small circulating proteins. Low molecular weight proteins (LMWP) with sizes smaller than
that of albumin are filtered by the glomerulum and hydrolyzed in the proximal tubular cells after endocytotic absorption[1,2,3]. Daily filtered LMWP
loads are in the range of 5 g. The renal catabolism is responsible for 30
- 80 % of the plasma turnover of LMWP[1]. In renal failure this catabolism is
decreased leading to increased LMWP plasma levels, which might contribute
to the development of the uremic syndrome as many of the LMWP are biologically active components like enzymes or hormones.

An ideal approach for artificial kidney replacement should therefore
allow to eliminate an equivalent LMWP fraction in addition to the small
uremic solutes. It has been claimed that hemofiltration (HF) is superior to
hemodialysis (HD) in removing larger solutes[4,5,6]. However no precise data
exist on protein permeability of the different membranes employed for HF.
In an attempt to determine which HF membrane is best suited to mimic the
natural glomerular filter we have compared different HF membranes by SDS-
PAGE.

## PATIENTS AND METHODS

Four patients (3 females, 1 male, 39 - 69 years old) were treated in
succession by post-dilution HF (Haemoprocessor, Sartorius) using polyamide
(FH88, Gambro), polysulfone (F60, Fresenius), cellulose triacetate (Haemofilter, Sartorius) and polyacrylnitrile (PAN 200, Asahi medical) membranes.
Filtrate volumes were in the range of 20 - 30 liters. Aliquots of the filtrates were concentrated 50-fold in Amicon cells using YM5 membranes with a
5 kD molecular weight cut off. Total protein concentration was measured by
the Lowry method[7]. Molecular weight analysis of the concentrated filtrate
proteins was performed by SDS-PAGE[8] on linear gradients (7.5 - 20 % T,
3 % C). Separated proteins were silver stained (modification of[9]) and gel
pattern evaluated quantitatively by densitometric scanning (LKB 2202 Ultro-
scan Laser Densitometer). Molecular weights were estimated by comparison
with marker proteins (Sigma). Human standard plasma was obtained from
Behring.

Table   Protein permeability and molecular weight selectivity of different hemofiltration membranes

| | Total protein (g/l) | Molecular Weight Distribution (D) | | | Filtrate Volume Required for Removal of 5 g Protein (l) |
|---|---|---|---|---|---|
| | | ≥50.000 | ≤50.000 – ≥25.000 | ≤25.000 – ≥ 5.000 | |
| Standard Plasma | 73,000 | 87,3 % | 2,5 % | 10,? % | / |
| Polyamide-Filtrate | 0,103±0,007 | 71,3 % | 16,1 % | 12,6 % | 48,5 |
| Polysulfone-Filtrate | 0,135±0,090 | 47,1 % | 25,6 % | 27,3 % | 37,0 |
| Cellulose-triacetate-Filtrate | 0,043±0,008 | 17,0 % | 14,6 % | 68,4 % | 116,3 |
| Polyacryl-nitrile Filtrate | 0,007±0,001 | 6,1 % | 8,1 % | 85,8 % | 714,3 |

RESULTS

Molecular weight selectivity: Densitometric scans of SDS-PAGE pattern (Figure) show that polyamide and polysulfone membranes are permeable for high molecular weight proteins up to 200 kD while cellulose triacetate and polyacrylnitrile membranes are only permeable for proteins smaller than albumin. Quantitative evaluation of the densitometric scans was performed by determination of the area under the single peaks and expression as per cent of total protein (data not shown). Calculation of the percentage of proteins in the molecular weight areas > 50 kD, 50 - 25 kD and 25 - 5 kD show (Table) that more than 87 % of the plasma proteins are of molecular weight higher than 50 kD. In the filtrates the percentage of these large proteins decreases in the sequence polyamide, polysulfone, cellulose triacetate and polynitrile, while the percentage of LMWP smaller than 25 kD increases in the same sequence. The polysulfone membrane reveals the best permeability for proteins between 50 kD and 25 kD.

Total protein permeability: The polysulfone membrane showed the highest permeability (0.135 g/l) closely followed by the polyamide membrane (0.103 g/l). The permeability of the cellulose triacetate membrane was lower (0.043 g/l) and the permeability of the polyacrylnitrile membrane very low (0.007 g/l) (Table).

DISCUSSION

30 - 80 % of the LMWP plasma turnover is mediated by the kidney[1]. Reduction of the catabolic kidney function in chronical renal failure leads to increased plasma levels of LMWP as demonstrated for ß2-microglobulin, ribonuclease, lysosome, retinol binding protein, $\alpha$1-microglobulin, ß2-gluco-proteins, $\alpha$1-antitrypsin, glucagon, ACTH, insulin PTH, growth hormone and prolactin[1,4,10]. The possible role of these elevated LMWP plasma levels in the pathogenesis of the uremic syndrome is often discussed[1,4,5,10]. An effective renal replacement therapy should therefore allow to remove LMWP from the plasma in amounts equivalent to those hydrolysed by the normal kidney.

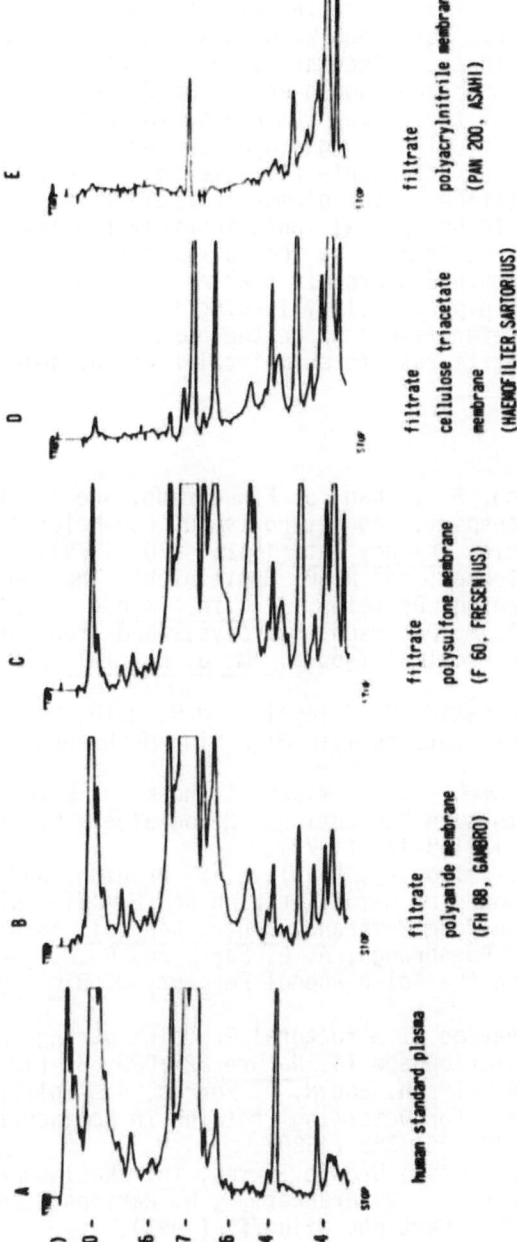

Figure 1. Determination of the molecular weight spectrum of proteins in different hemofiltrates by densitometric scanning of SDS-PAGE pattern.

From experimental data[1] one can calculate that removal of about 5 g of LMWP per day is desirable. As HD membranes are impermeable for LMWP[4,5,10], HF with highly permeable membranes seems to be the method of choice. However, no precise data exist on the protein permeability of these membranes. Characterization of protein permeabilities was yet mainly done by calculation of sieving coefficients for some selected proteins of known molecular weights on the basis of immunological methods. We have analysed the whole molecular weight spectrum of hemofiltrate proteins by SDS-PAGE, silver staining and quantitative densitometric scanning. With these methods we could demonstrate remarkable differences in the molecular weight selectivity of the tested membranes. An ideal HF membrane should on one hand mimic the protein molecular weight selectivity of the glomerulum and on the other hand provide a total protein permeability high enough to allow the removal of LMWP in the range of about 5 g with an acceptable filtrate volume. Our data show that with regard to the imitation of the glomerular selectivity the cellulose triacetate membrane would be optimal. Unfortunately the total protein permeability of this membrane is low, so that a filtrate volume of about 116 l would be required to remove 5 g protein (Table). The polysulfone membrane provides a higher protein permeability leading to an acceptable filtrate volume (37 l) and therefore seems to be the best suited membrane for the removal of LMWP despite its restricted molecular weight selectivety.

REFERENCES

1. T. Maak, V. Johnson, S. T. Kan, J. Figueiredo, and D. Sigulem, Renal Filtration, Transport, and Metabolism of Low-Molecular Weight Proteins: A Review, Kidney Int. 16:251-270 (1979).
2. T. Waldmann, W. Strober, and R. P. Mogielnicki, The Renal Handling of Low Molecular Weight Proteins, J. Clin. Invest. 51:2162-2174 (1972).
3. F. A. Carone and D. P. Peterson, Hydrolysis and Transport of Small Peptides by the Proximal Tubule, Am. J. Physiol. 238:F151-F158 (1980).
4. A. Röckel, S. Abdelhamid, P. Fliegel, and D. Walb, Elimination of Low Molecular Weight Proteins with High Flux Membranes, Contr. Nephrol. 46:69-74 (1985).
5. S. Jörstad, L. C. Smeby, T.-E. Widerøe, and K. J. Berg, Transport of Uremic Toxins through Conventional Hemodialysis Membranes, Clin. Nephrol. 12:168-173 (1979).
6. V. Wizemann, H. G. Velcovsky, H. Bleyl, S. Brüning, and G. Schütterle, Removal of Hormones by Hemofiltration and Hemodialysis with a Highly Permeable Polysulfone Membrane, Contr. Nephrol. 46:61-68 (1985).
7. O. H. Lowry, N. J. Rosebrough, A. L. Farr, and R. J. Randall, Protein Measurement with the Folin Phenol Reagent, J. Biol. Chem. 193: 265-275 (1952).
8. M. K. Laemmli, Cleavage of Structural Proteins during the Assembly of the Head of Bacteriophage T4, Nature 227:680-685 (1970).
9. B. R. Oakley, D. R. Kirsch, and N. R. Morris, A Simplified Ultrasensitive Silver Stain for Detecting Proteins in Polyacrylamide Gels, Anal. Biochem. 105:361-363 (1980).
10. J. Bergström and P. Fürst, Uremic Toxins, in: "Replacement of Renal Function by Dialysis", W. Drükker, F. M. Parsons, and J. F. Maher, eds., pp. 354-390, Martinus Nijhoff, (1983).

# HEMOFILTRATION-INDUCED SERUM MODIFICATIONS IN THE PATTERN OF LARGER MOLECULES.

S. David, R. Barani, C. Buzio, L. Arisi, and P. Quaretti

Clinical Medicine and Nephrology Dept. University of Parma, Via Gramsci 14-43100 Parma, Italy

In early clinical trials, hemofiltration (HF),unlike standard dialysis (HD), seemed able to improve some uremic abnormalities, such as neurophathy(1), lipid metabolism(2)and osteopathy(3). These results could have been the consequence of better removal of high molecular weight uremic toxins,according to the "Middle Molecules" hypothesis(4). However, in subsequent long-term studies, these observations were not univocally confirmed(5), and HF has not proved to be a more efficient treatment of the uremic syndrome than HD.

Recently, metabolic advantages were reported in some small patient group(6,7), but in these studies small solute clearances were similar to those of HD (high efficiency HF). Thus the toxicity of larger molecules retained in serum in HD, but easily removable by convection in HF, has once again become a topic of debate. Nevertheless the serum pattern of solutes weighing from 500 to 1500 D, the so-called "Middle Molecules" (MM), have not yet been investigated in long-term HF.

In order to elucidate the real HF-induced MM changes, these solutes were measured in a group of patients treated by HD and subsequently, by HF.


METHODS

Five clinically stable anuric patients (mean age = 59 yers ± 5 S.D.; mean body weight 59 Kg ± 3 S.D.), on HD for an average 54 months (±41 S.D.), were switched to HF for a 12-month period. HF was carried out with 32 l of on-line prepared sterile, pyrogen-free reinfusate, with the same composition of dialysate (Acetate= 40 mmol/l,Glucose = 1 g/l).

Mean hemofiltrate volume was 36(±3 S.D.)l/session. No drugs were allowed during the study, with the exception for $CaCO_3$ and phosphate binders. Serum, dialysate and hemofiltrate samples were taken during midweek treatment, performed after overnight fasting, on HD, as well as 3 and 12 months after the start of HF.

Solutes weighing from 500 to 1500 D were measured by gel-chromatography on a Sephadex G-15 column, and by high performance liquid chromatography (HPLC). As previously described in detail (8), serum samples were deproteinized by ultrafiltration through XM 50 Diaflo membranes (cut off=50,000 D),and lower molecular weight substances were removed by ultrafiltration through UM 50 Amicon membranes (cut off=500 D). Gel-chromatography analysis of residual material, containing 500 to 50,000 D solutes, was then performed on a Sephadex G-15 column. Chromatograms of collected eluates were obtained by spectrophotometry analysis at a wave length of 254 nm with continuous paper recording. The sum of the areas of recorded peaks was considerd as expression of total amount of MM in the sample. 350 ml samples of dialysate or hemofiltrate from a total volume of 120-130 and 33-39 l respectively,were dried; saline was then added in order to obtain the same dilution volume. Analysis was then carried out as described for serum samples. All but the first eluate fraction, corresponding to > 1500 D solutes, were processed by HPLC for both qualitative and quantitative analyses. Peaks with the same morphology and elution time were considered as formed by the same solute. Peaks eluted within 5 minutes were not considered, because they corresponded mainly to amino-acids and small solutes. HPLC features are described in Table I.

Predialysis MM body pool (MM-P) was calculated by multiplying serum concentration (MM-S) by distribution volume, which was considered as corresponding to total body water: MM-P = MM-S x ("Dry"body weight x 0.6)+ Weight gain. Errors in distribution volume determination were cancelled out by comparing the two different treatments in the same patient. MM removal was calculated form dialysate or hemofiltrate concentrations and their corresponding total volume. The extraction index (EI) was calculated as EI = Removal/MM-P % . EI for individual peaks, isolated by HPLC, was calculated in the same way. The Student's t paired test was carried out for statistical analyses.

RESULTS

HF did not changed neither clinical nor biochemical parameters, such as body weight, mean B.P., Hct, B.U.N.,Creatinine, Ca, P, Alcaline Phosphatase, PTH, acid-base status. Results of gel-chromatography (Table II) indicate that after 3 months, HF

## TABLE I - HPLC FEATURES

COLUMN: Ultrasphere - ODS 4.6x150 mm
STATIONARY PHASE: RP - 18.5 $\mu$
PRECOLUMN: 4.6 x 35 mm, RP- 18.5 $\mu$
PRESSURE: 1.20 $K_{psi}$
FLOW RATE: I.O ml/min.
TEMPERATURE: ambient
DETECTOR: UV 254 nm
SENSITIVITY : 0.050 A UFS
ELUENT: A - Water, 0.1 M $NaClO_4$+ 0.07% $H_3PO_4$

B - $CH_3CN$, 0.1 M $NaClO_4$+ 0.07% $H_3PO_4$

PROGRAM: Linear gradient from 0 to 15% B in 40 min.
INJECTION VOLUME : 40 $\mu$l
PAPER SPEED : 0.5 cm/min.

reduces the MM-P to 62.6% of HD values, considered as 100% . On the contrary, after 12 months of HF, an unexpected increase of up to 74% of HD values was detected. The extraction index changes in a specular way, showing an initial increase followed by a decrease after 12 months.

## TABLE II - RESULTS OF GEL-CHROMATOGRAPHY (see text)

|  | HD | HF1 (after 3 months) | HF2 (after 12 months) |
|---|---|---|---|
| MM-P % | 100 | 62.6± 15* | 74.4 ±22.2** |
| E I | 33.8±7.9 | 47.5± 4.6* | 43.0 +6.5**° |

* = HF1 vs. HD p $<$0.05;   **= HF2 vs. HF1 p $<$0.05;  °= HF2 vs. HD N.S.

Results of HPLC analyses are shown in Fig.1. In serum samples, 36 peaks, corresponding to 36 different solutes, were recorded, including both the HD and HF periods (upper part). The white bars indicate the solutes detected during HD; the black ones stand for HF. Twenty-nine peaks were detected in HD; six of them, reported as dotted bars, disappeared in HF. However, after 12 months of HF seven new peaks, previously absent, were found in the sera (gray bars). As indicated by the more numerous black bars, HF removes a larger amount of solutes than does HD.

The extraction index for each peak is shown in the lower part of the same figure; it is calculated on the basis of the amount removed by HD (white bars) or HF (black bars). HF removes 24 of 30 serum peaks, and is more efficient than HD which removes only 13 out of 29 peaks. However, as can be observed, some solutes, although having a molecular weight compatible with membrane cut off, are not ultrafiltered. On the other hand, the HF extraction index for some solutes is higher than 100%, suggesting that removal paradoxically surpasses the extimated predialysis pool.

In short, after 12 months of HF we observed the following: a) Some peaks completely disappeared (i.e. peaks 4 and 11); b) The concentration of some peaks was reduced. Their removal was lower (i.e.12 and 16) or nul (i.e. peaks 18 and 19) in HD. c) The concentration of some other peaks surprisingly increased. This occurred because of a lower removal (such as peaks 2 and 7) or in spite of a higher removal ( peaks 29 and 32); d) Seven new peaks appeared (gray bars), only 3 of which were ultrafiltered.


DISCUSSION

HF has been confirmed to be more efficient than HD in reducing the MM pool. This results from an increased removal by convection, evident after 3 months of treatment. Later on, the concentration of molecules unaffected by convection and probably selected on the basis of their molecular weight, increases. HPLC results furnish a clearer explanation of these phenomena: in fact the HF increased capacity for removal is in part counterbalanced not only by an increased production of some molecules, but also by the generation of new ones that are not ultrafiltered.

The extraction indexes above 100% suggest an unbalanced solute distribution between intra and extracellular compartments, causing an underestimation of the real predialysis body pool; however a treatment-induced solute generation is more probable. It should be emphasized that because the real extraction index was calculated by direct measurement of the extracted mass, our results were not affected by membrane resistance to solute diffusion.

In conclusion, increased MM removal by convection does not seem to be adequate theraphy for the control of MM production. This is probably related to both nutritional and biocompatibility factors. The final metabolic effects of the variation in concentration of individual solutes are not known,

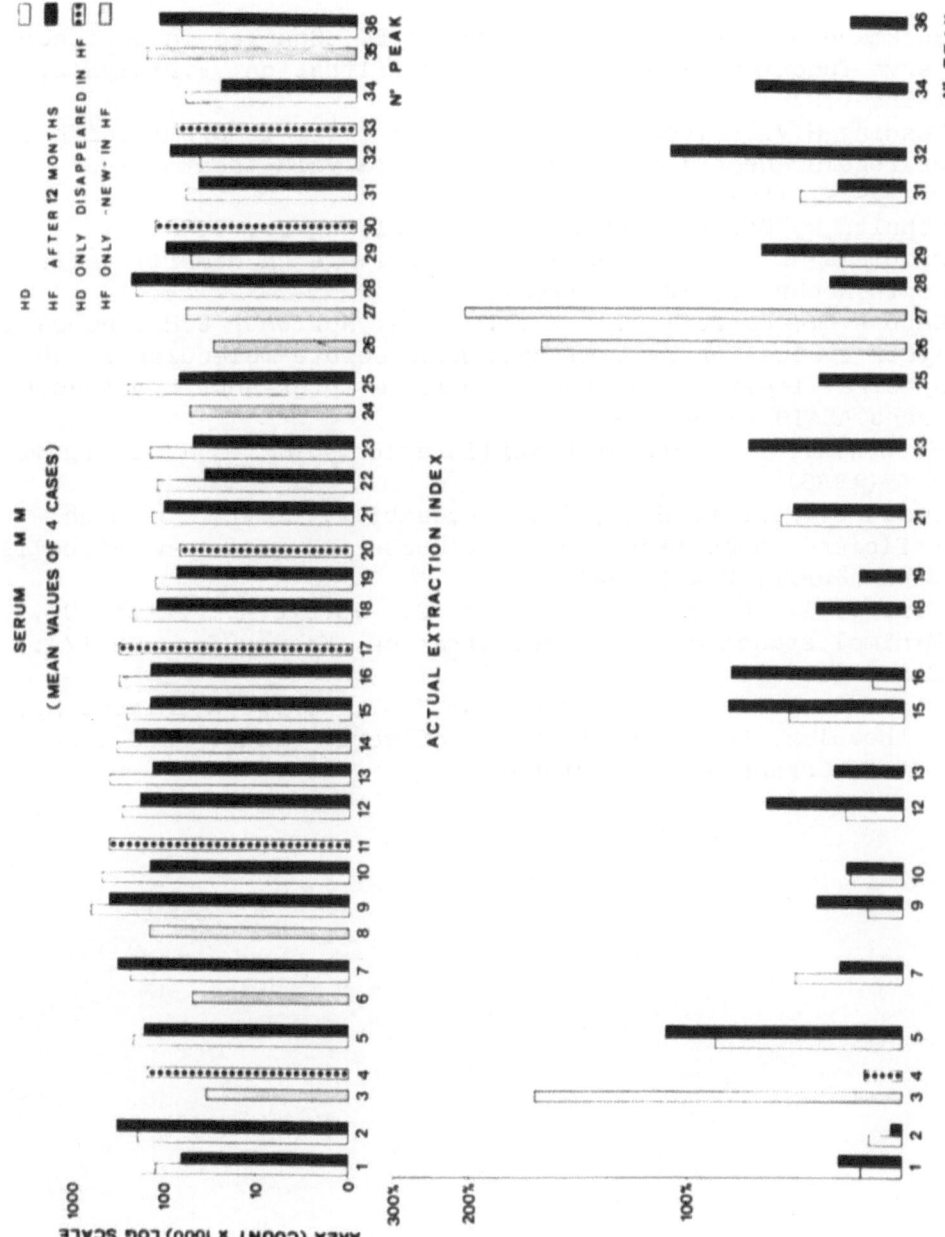

FIG.1   HPLC RESULTS (see text)

and therefore we cannot yet predict whether or not hemofiltration will provide metabolic advantages.

REFERENCES

1. Beckmann H.,Ossenkop C.,Quellhorst E., Changes in peripheral nerve function with long-term hemofiltration. J.Dialysis, 1:585 (1977)
2. Henning H.V.,Balusek E., Lipid metabolism in uremia:effect of regular hemofiltration and hemodialysis treatment.J. Dialysis, 1(6):595 (1977).
3. Schultz,W, Baier E.,Humpfner A., Delling G., Clinical problems of renal osteopathy in patients on hemofiltration. Contr.Nephrol., 32:86 (1982).
4. Babb A.L., Farrell P.C.,Uvelli D.,A.,Scribner B.H., Hemodialyzer evaluation by examination of solute molecular weight spectra: Implications for the square-meter-hour hypothesis. Trans.ASAIO 18:98 (1972).
5. Koch K.M.,Prospects of hemofiltration, Int. J.Artif.Organs 1:5 (1983)
6. David S.,Ferrari M.E., Arisi L,Cambi V.,Effects of high efficiency hemofiltration on glucose and nitrogen metabolism Blood Purif. 2:4 (1984)
7. Minetti l., Civati G., Guastoni C., Perego A.,Teatini U., Minimal standards for hemofiltration, Kidney Int. 28(17): S-116 (1985).
8. Buzio C.,Montagna G., Calderini M.C., Manari A.,Migone L., Methodology for identification of serum Middle Molecules. Artif. Organs 4(s):23 (1980)

# CARNITINE DEPLETION DURING CHRONIC HEMODIALYSIS :

# EFFECT OF SUBSTITUTION ON FREE CARNITINE PLASMA LEVELS

P.E. Broquet[*], C. Von Moos[+], J. Frei[+],
C. Gobelet[++] and J.P. Wauters[*]

Division of Nephrology[*], Laboratory of Clinical
Chemistry[+] and Service of Orthopedy[++]
University Hospital, Lausanne, Switzerland

## INTRODUCTION

Carnitine plays a role in the oxidative metabolism of free fatty-acids by transporting long chain fatty acids into the mitochondrial matrix (1). During chronic hemodialysis, a depletion of plasma carnitine levels has been observed and attributed to dialysis of this unbound low-molecular weight substance (M.W. 161) (2, 3). Due to the high efficiency of the presently used dialyzers, this may lead to long term carnitine depletion in the vascular and muscular compartments. Moreover, it appears that the synthesis of carnitine is altered in dialysis patients (4). Since some of the complications seen during chronic hemodialysis have been attributed to a depletion in carnitine, substitution by carnitine i.v. or in the dialysate has been logically applied. It appears however that a mixture of D, L-carnitine can lead to a myasthenia like syndrome due to the presence of D-carnitine (5). The aim of the present study was to evaluate the evolution of free carnitine plasma levels in a population of chronic hemodialysis patients before, during and after substitution with L-carnitine i.v. during 8 weeks.

## PATIENTS AND PROTOCOL

A group of 15 patients on chronic hemodialysis for at least 1 year (13 - 190 months) was involved in the study. The group consisted of 9 males and 6 females, aged between 46 to 73 years (mean age 61) and dialyzed 3 times/week. They were submitted to an open cross-over A/B/A study each period lasting 8 weeks. All the patients received L-carnitine 20 mg/kg intravenously at the end of each dialysis session during period B.

## METHODS

Free carnitine plasma levels were determined according to the enzymatic method of Pearson adapted by Gautschi and

Brogli for the centrifuge analyzer COBAS-Bio (5, 6). Normal values were established in a population of 115 blood donors. Patients were dialyzed on cuprophane or polysulfone capillary dialyzers, mostly with an glucose-free acetate dialysate, a blood flow of 300 ml/min and a dialysate flow of 500 ml/min for a duration of 3 hours.

RESULTS

In blood donors, the free carnitine plasma levels were at 54.1 ± 10.7 (mean ± SD) μmol/l in males and at 48.1 ± 11.9 in females. This corresponds to the usually reported values. In dialysis patients, the values observed pre-dialysis during the initial period were at 39.1 ± 15.4 μmol/l and diminished significantly (p 0.01) to 10.7 ± 3.5 after the dialysis session.

Four patients did not complete the study : 2 due to gastro-intestinal hemorrhage (1 patient with cirrhosis, 1 with a gastric ulcer), 1 due to non-compliance and 1 due to non specific side-effects (pruritus, headache). Therefore 11 patients only are considered in the cross-over evaluation.

The results after substitution with L-carnitine are illustrated in fig. 1. It is interesting to note that free plasma carnitine levels rapidly increased once substitution was initiated : at 2 weeks the mean pre-dialysis free carnitine plasma levels were at 139 μmol/l.

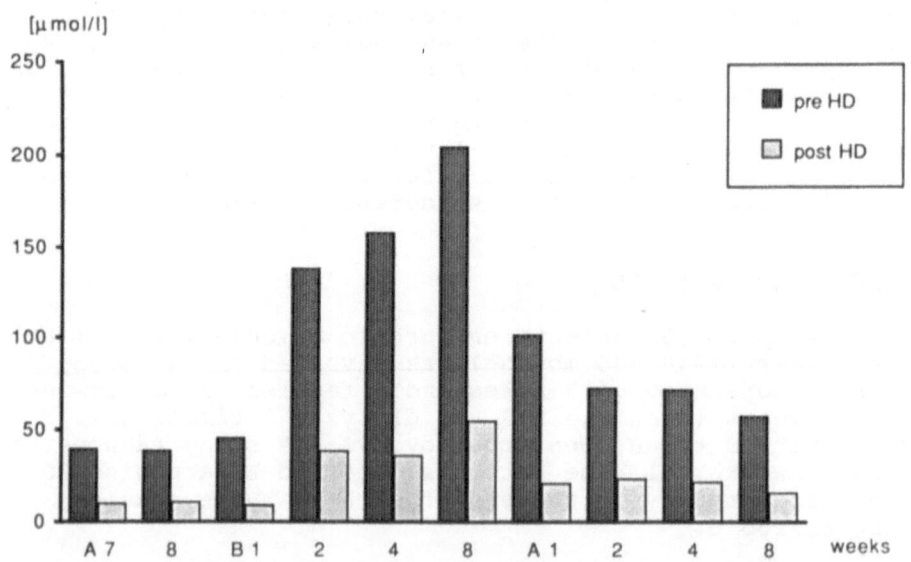

Figure 1 . Evolution of free carnitine plasma levels throughout study periods

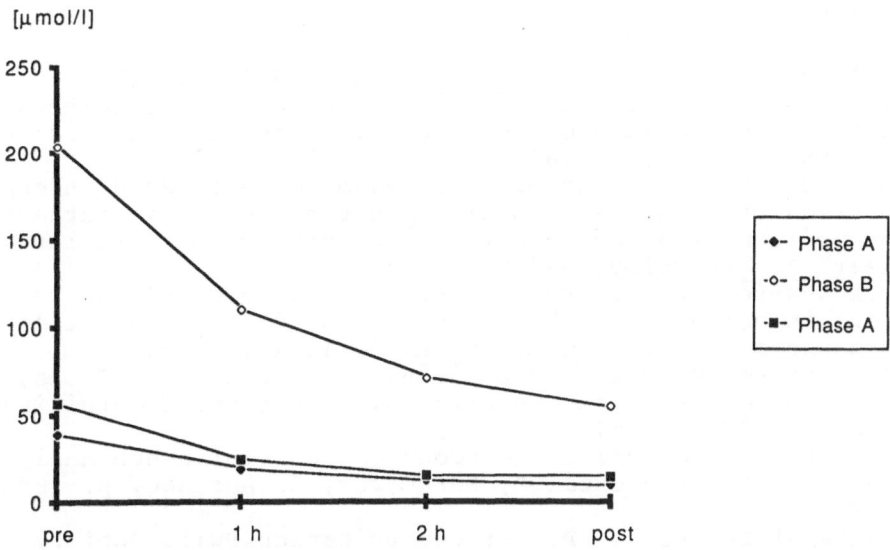

Figure 2 . Evolution of free carnitine plasma levels
           during dialysis.

Despite this increase the loss of carnitine during dialysis
remained important due to the increased concentration gradient. However, at the end of the dialysis session, the free
plasma carnitine levels remained in the normal limit of 39
μmol/l (fig. 2). At the end of the substitution period, free
plasma carnitine levels rapidly decreased and already 2
weeks after the end of the substitution, low carnitine values were again observed.


DISCUSSION

     The present study shows that the depletion in free car-
nitine plasma levels induced by a high efficiency hemodialysis schedule can be corrected by the intravenous administration of 20 mg/kg at the end of each dialysis session. This
correction appears within a few weeks and disappears 2 weeks
after stopping L-carnitine administration.

     A considerable increase of the amounts lost in the dia-
lysate was observed during the substitution period. Therefore, the administration of at least 20 mg/kg seems warranted since this dosage was just able to maintain post dialysis
values within the normal limits, at least during the 8 weeks
observation period. This substitution schedule did not lead
to severe side-effects.

# REFERENCES

1. P.R. Borum, Carnitine, Am. Rev. Nutr. 3:233 (1983).
2. Th. Bohmer, H. Bergrem, K. Eiklid, Carnitine deficiency induced during intermittent hemodialysis for renal failure, Lancet I:126 (1978).
3. A.V. Moorthy, M. Rosenbaum, R. Rajaram, A.L. Shug, A comparison of plasma and muscle carnitine levels in patients on peritoneal or hemodialysis for chronic renal failure, Amer. J. Nephrology 3:205 (1983).
4. L.L. Bartel et al., Perturbation of serum carnitine levels in human adults by chronic renal disease and dialysis therapy, Am. J. Clin. Nutr. 34:1314 (1981).
5. G. Bazzato, U. Coli, S. Landrini, C. Mezzina, M. Ciman, Myasthenia like syndrome after D, L- but not L-carnitine, Lancet I:1209 (1981).
6. D.Y. Pearson et coll., Methoden der enzymatischen Analyse, in : "Verlag Chemie", Bergmeyer, 3. Auflage, p. 1806: (1974).
7. Gautschi et Brogli, Poster communication Swiss Society of Clinical Chemistry (1984).